PROCEEDINGS

Scanning Imaging

Tony Wilson
Chair/Editor

ECO1
21-23 September 1988
Hamburg, Federal Republic of Germany

Sponsored by
EPS—European Physical Society
Europtica—The European Federation for Applied Optics
SPIE—The International Society for Optical Engineering

Cooperating Organizations
ANRT—Association Nationale de la Recherche Technique
Associazione Elettrotecnica ed Elettronica Italiana
Battelle Europe
British Pattern Recognition Association
CNES—Centre National d'Etudes Spatiales
Ecole Polytechnique Fédérale de Lausanne
EURASIP—European Association for Signal Processing
GIFO—Groupement Industriel Français pour l'Optique
International Solar Energy Society
Israel Laser and Electro-Optics Society
Office of Solar Heat Technologies, U.S. Department of Energy
Optical Engineering Society (Taiwan, China)
SEE—Société Française des Electriciens, des Electroniciens et des Radioeléctriciens

This publication sponsored and supported by UNESCO

Published by
SPIE—The International Society for Optical Engineering
P.O. Box 10, Bellingham, Washington 98227-0010 USA
Telephone 206/676-3290 (Pacific Time) • Telex 46-7053

Volume 1028

SPIE (The Society of Photo-Optical Instrumentation Engineers) is a nonprofit society dedicated to advancing engineering and scientific applications of optical, electro-optical, and optoelectronic instrumentation, systems, and technology.

The papers appearing in this book comprise the proceedings of the meeting mentioned on the cover and title page. They reflect the authors' opinions and are published as presented and without change, in the interests of timely dissemination. Their inclusion in this publication does not necessarily constitute endorsement by the editors or by SPIE.

Please use the following format to cite material from this book:
   Author(s), "Title of Paper," *Scanning Imaging*, Tony Wilson, Editor, Proc. SPIE 1028, page numbers (1989).

Library of Congress Catalog Card No. 88-63642
ISBN 0-8194-0063-7

Copyright © 1989, The Society of Photo-Optical Instrumentation Engineers.

Copying of material in this book for sale or for internal or personal use beyond the fair use provisions granted by the U.S. Copyright Law is subject to payment of copying fees. The Transactional Reporting Service base fee for this volume is $2.00 per article and should be paid directly to Copyright Clearance Center, 27 Congress Street, Salem, MA 01970. For those organizations that have been granted a photocopy license by CCC, a separate system of payment has been arranged. The fee code for users of the Transactional Reporting Service is 0-8194-0063-7/89/$2.00.

Individual readers of this book and nonprofit libraries acting for them are permitted to make fair use of the material in it, such as to copy an article for teaching or research, without payment of a fee. Republication or systematic or multiple reproduction of any material in this book (including abstracts) is prohibited except with the permission of SPIE and one of the authors.

Permission is granted to quote excerpts from articles in this book in other scientific or technical works with acknowledgment of the source, including the author's name, the title of the book, SPIE volume number, page number(s), and year. Reproduction of figures and tables is likewise permitted in other articles and books provided that the same acknowledgment of the source is printed with them, permission of one of the original authors is obtained, and notification is given to SPIE.

In the case of authors who are employees of the United States government, its contractors or grantees, SPIE recognizes the right of the United States government to retain a nonexclusive, royalty-free license to use the author's copyrighted article for United States government purposes.

Address inquiries and notices to Director of Publications, SPIE, P.O. Box 10, Bellingham, WA 98227-0010 USA.

ECO1
SCANNING IMAGING

Volume 1028

CONTENTS

Conference Committee vi
Technical Organizers ECO1 vii
Introduction viii

PLENARY SESSION
1028-PL05 **Overview of coherent optics applications in metrology**
J. Ebbeni, Univ. Libre de Bruxelle (Belgium). 2

SESSION 1 **CONFOCAL MICROSCOPY I**
1028-01 **Pupil filters in confocal imaging**
Z. S. Hegedus, CSIRO (Australia). 14
1028-02 **Axial resolution of a confocal scanning optical microscope**
M. J. Offside, C. W. See, M. G. Somekh, Univ. College London (UK). 18
1028-03 **Spherical aberration in confocal microscopy**
T. Hellmuth, P. Seidel, A. Siegel, Carl Zeiss (FRG). 28
1028-05 **Inverse problems in fluorescence confocal scanning microscopy**
M. Bertero, P. Boccacci, Univ. di Genova and Istituto Nazionale di Fisica Nucleare (Italy); E. R. Pike, King's College (UK). 33
1028-06 **Modeling of 3-D confocal imaging at high numerical aperture in fluorescence**
H. T. M. van der Voort, G. J. Brakenhoff, Univ. of Amsterdam (Netherlands). 39
1028-07 **Theoretical and experimental research on super-resolution of microscopes: I. partially coherent illumination and resolving power**
C.-M. Ma, R. W. Smith, Imperial College of Science, Technology, and Medicine (UK). 45

SESSION 2 **SCANNING MICROSCOPY I**
1028-08 **Scanning differential optical system for simultaneous phase and amplitude measurement**
R. K. Appel, C. W. See, M. G. Somekh, Univ. College London (UK). 54
1028-09 **Eccentricity errors combined with wavefront aberration in a coherent scanning microscope**
A. M. Hamed, Ain Shams Univ. (Egypt). 63
1028-10 **Optimization of recording conditions in laser scanning microscopy**
T. Damm, M. Kaschke, U. Stamm, Friedrich-Schiller Univ. of Jena (GDR). 69
1028-11 **Phase-shifting and Fourier transforming for submicron linewidth measurement**
Y. Xu, E. Hu, G. Wade, Univ. of California/Santa Barbara (USA). 76
1028-13 **Measurement of the degree of coherence in conventional microscopes**
A. Glindemann, Technical Univ. Berlin (FRG). 84
1028-40 **Confocal interference microscopy**
C. J. R. Sheppard, D. K. Hamilton, H. J. Matthews, Oxford Univ. (UK). 92

(continued)

ECO1
SCANNING IMAGING
Volume 1028

SESSION 3   **CONFOCAL MICROSCOPY II**

1028-14   **Confocal and conventional modes in tandem scanning reflected light microscopy**
A. Boyde, Univ. College London (UK). 98

1028-15   **Imaging theory for the scanning optical microscope**
G. S. Kino, C.-H. Chou, G. Q. Xiao, Stanford Univ. (USA). 104

1028-16   **Phase imaging in scanning optical microscopes**
T. R. Corle, G. S. Kino, Stanford Univ. (USA). 114

1028-17   **In vivo confocal imaging of the eye using tandem scanning confocal microscopy**
J. V. Jester, H. D. Cavanagh, M. A. Lemp, Georgetown Univ. (USA). 122

1028-18   **Confocal laser scanning microscopy for ophthalmology**
G. Zinser, R. W. Wijnaendts-van-Resandt, C. Ihrig, Heidelberg Instruments GmbH (FRG). ... 127

1028-19   **Scanning microscope for optically sectioning the living cornea**
B. R. Masters, Emory Univ. School of Medicine (USA). 133

SESSION 4   **BIOLOGICAL AND MATERIALS MICROSCOPY**

1028-20   **Confocal fluorescence microscopes for biological research**
E. H. K. Stelzer, R. Stricker, R. Pick, C. Storz, P. Hänninen, European Molecular Biology Lab. (FRG). 146

1028-21   **Apparatus for laser scanning microscopy and dynamic testing of muscle cells**
I. Hunter, S. Lafontaine, P. Nielsen, P. Hunter, McGill Univ. (Canada). 152

1028-22   **Laser scanning microscopy to study molecular transport in single cells**
M. Scholz, H. Sauer, H.-P. Rihs, R. Peters, Max-Planck-Institut für Biophysik (FRG). 160

1028-23   **Confocal fluorescence microscopy of epithelial cells**
E. H. K. Stelzer, R. Bacallao, European Molecular Biology Lab. (FRG). 167

1028-24   **Inverted confocal microscopy for biological and material applications**
G. J. Brakenhoff, Univ. of Amsterdam (Netherlands); R. W. Wijnaendts-van-Resandt, J. Engelhardt, W. Knebel, Heidelberg Instruments GmbH (FRG); H. T. M. van der Voort, Univ. of Amsterdam (Netherlands). 169

1028-25   **Applications of the microscope system LSM**
H.-G. Kapitza, V. Wilke, Carl Zeiss (FRG). 173

SESSION 5   **SEMICONDUCTOR MICROSCOPY**

1028-26   **Photoluminescence and optical beam-induced current imaging of defects**
P. D. Pester, T. Wilson, Oxford Univ. (UK). 182

1028-27   **Minority carrier lifetime mapping in gallium arsenide by time-resolved photoluminescence scanning microscopy**
T. A. Louis, Heriot-Watt Univ. (UK). 188

1028-28   **Topography of GaAs/AlGaAs heterostructures using the lateral photoeffect**
P. F. Fontein, P. Hendriks, J. Wolter, Eindhoven Univ. of Technology (Netherlands); A. Kucernak, R. Peat, D. E. Williams, U.K. Atomic Energy Authority. 197

1028-29   **Scanning laser photocurrent spectroscopy of electrochemically grown bismuth sulphide films**
A. R. Kucernak, R. Peat, D. E. Williams, U.K. Atomic Energy Authority (UK). 202

1028-30   **Circuit analysis in ICs using the scanning laser microscope**
J. Quincke, E. Plies, J. Otto, Siemens AG (FRG). 211

1028-31   **Automated latch-up measurement system using a laser scanning microscope**
J. Fritz, R. Lackmann, B. Rix, Fraunhofer Institute of Microelectronic Circuits and Systems (FRG). 217

ECO1
SCANNING IMAGING

Volume 1028

SESSION 6	**SCANNING MICROSCOPY II**	
1028-32	**Infrared laser scan microscope**	
	E. Ziegler, H. P. Feuerbaum, ICT GmbH (FRG).	226
1028-33	**Computerized surface plasmon microscopy**	
	E. M. Yeatman, E. A. Ash, Imperial College of Science and Technology (UK).	231
1028-34	**Computerized analysis of high resolution images by scanning acoustic microscopy**	
	D. D. Giusto, B. Bianco, A. Cambiaso, M. Grattarola, M. Tedesco, Univ. of Genoa (Italy).	237
1028-35	**Semiconductor laser digital scanner**	
	H. Sekii, A. Fujimoto, T. Takagi, K. Imanaka, M. Shimura, OMRON Tateisi Electronics Co. (Japan).	245
1028-36	**Measurements of optical waveguides by a near-field scanning technique**	
	J. Helsztyński, T. W. Kozek, Warsaw Univ. of Technology (Poland).	250
1028-37	**Reflectance and optical contrast of old manuscripts: wavelength dependence**	
	J. Bescós, Fundación Ramón Areces (Spain); F. Jaque, Univ. Autonoma de Madrid (Spain); L. Montoto, IBM Corp. (Spain).	258
1028-38	**Novel optoelectronic method of position measurement**	
	S. Gergely, A. J. Syson, Coventry Polytechnic (UK).	263
	Author Index	268

ECO1
SCANNING IMAGING

Volume 1028

CONFERENCE COMMITTEE

Chair
Tony Wilson, Oxford University (UK)

Cochairs
G. J. Brakenhoff, University of Amsterdam (Netherlands)
S. A. Siegel, Carl Zeiss GmbH (FRG)

Session Chairs
Session 1—Confocal Microscopy I
Tony Wilson, Oxford University (UK)

Session 2—Scanning Microscopy I
Zoltan S. Hegedus, CSIRO (Australia)

Session 3—Confocal Microscopy II
G. J. Brakenhoff, University of Amsterdam (Netherlands)

Session 4—Biological and Materials Microscopy
S. A. Siegel, Carl Zeiss GmbH (FRG)

Session 5—Semiconductor Microscopy
Tony Wilson, Oxford University (UK)

Session 6—Scanning Microscopy II
G. J. Brakenhoff, University of Amsterdam (Netherlands)

TECHNICAL ORGANIZERS ECO1

Technical Program Committee

EPS—European Physical Society
J.-P. Huignard, Thomson S.A. (France)
O. D. D. Soares, Universidade de Porto (Portugal)
H. Tiziani, Institut für Technische Optik/Universität Stuttgart (FRG)
H. A. Ferwerda, University of Groningen (Netherlands)
J. J. Stamnes, Central Institute for Industrial Research (Norway)

Europtica
J. Bulabois, Université de Franche-Comté, Besançon (France)
J. C. Dainty, Imperial College (UK)
R. Torge, Carl Zeiss (FRG)
H. Frankena, University of Technology (Netherlands)
A. Monfils, Université de Liège (Belgium)

SPIE—The International Society for Optical Engineering
L. H. J. F. Beckmann, Oldelft, TPC Chairman (Netherlands)
R. E. Fischer, Ernst Leitz Canada, Ltd., U.S. Operations (USA)
M. Kunt, EPFL (Switzerland)
H. Rottenkolber, Rottenkolber Holo-System GmbH (FRG)
K. Biedermann, The Royal Institute of Technology (Sweden)

Joint Policy Committee

EPS—European Physical Society
H. Tiziani, Institut für Technische Optik/Universität Stuttgart, JPC Vice Chairman (FRG)
H. A. Ferwerda, University of Groningen (Netherlands)
G. Thomas, Executive Secretary EPS (Switzerland)

Europtica
P. Bozec, ESSILOR (France)
H. Walter, Optische Werke G. Rodenstock (FRG)
P. Zaleski, ANRT (France)

SPIE—The International Society for Optical Engineering
W. L. Wolfe, Optical Sciences Center/University of Arizona (USA)
B. J. Thompson, University of Rochester (USA)
L. R. Baker, Sira Ltd., JPC Chairman (UK)

Technical Organizers

Jean M. Bulabois, Université de Franche-Comté Besançon and Technical Consultant, Europtica-Services I.C. (France)
Roy F. Potter, Western Washington University and Technical Consultant, SPIE (USA)

Technical Organizing Committee

G. A. Acket (Netherlands)
R. F. Agullo-Lopez (Spain)
Rolf-Jürgen Ahlers (FRG)
I. Appenzeller (FRG)
H. Arditty (France)
Ronald Baets (Belgium)
Lionel R. Baker (UK)
Dirk Basting (FRG)
Leo H. J. F. Beckmann (Netherlands)
Jean M. Bennett (USA)
P. Bozec (France)
K. Biedermann (Sweden)
Albrecht Brandenburg (FRG)
D. W. Braggins (UK)
G. J. Brakenhoff (Netherlands)
Gordon M. Brown (USA)
Ronald F. Burge (UK)
M. Cantello (Italy)
H. G. Cirkel (FRG)
Brian Culshaw (UK)
John P. Dakin (UK)
Hans Dammann (FRG)
Anthony Dandridge (USA)
F. Dausinger (FRG)
Alain Diard (France)
J. A. Dobrowolski (Canada)
H. Doetsch (FRG)
Victor J. Doherty (USA)
Jean Ebbeni (Belgium)
H. J. Eichler (FRG)
J. Encarnacao (FRG)
William P. Fagan (UK)
H. A. Ferwerda (Netherlands)
B. Fischer (Israel)
L. Garifo (Italy)
Claes Granqvist (Sweden)
P. Günter (Switzerland)

Karl H. Guenther (USA)
Rüdiger Haberland (FRG)
Alan L. Harmer (Switzerland)
Robert Hartel (FRG)
Klaus W. Hildebrand (Switzerland)
J.-P. Huignard (France)
Peter Hutzler (FRG)
R. Ifflander (FRG)
Kenichi Iga (Japan)
M. Jacobi (FRG)
J. Roland Jacobsson (Sweden)
J. Jaeger (Norway)
Werner P. O. Jüptner (FRG)
R. Th. Kersten (FRG)
R. Kist (FRG)
Ernst W. Kreutz (FRG)
H. Kunzmann (FRG)
Carl M. Lampert (USA)
P. Langenbeck (FRG)
L. D. Laude (Belgium)
J. C. Launay (France)
J. C. Laycok (UK)
Ole J. Løkberg (Norway)
H. Angus Macleod (USA)
André Masson (France)
Reginald S. Medlock (UK)
Carl Misiano (Italy)
Wolfgang Muckenheim (FRG)
G. Neukum (FRG)
R. J. F. Normandin (Canada)
G. Notenboom (Netherlands)
S. G. Oduluv (USSR)
André Oosterlinck (Belgium)
Daniel B. Ostrowsky (France)
J. L. Oudar (France)
M. Papuchon (France)
E. Pelletier (France)

R. Petit (France)
E. R. Pike (UK)
R. Propawe (FRG)
Ryszard J. Pryputniewicz (USA)
Hans Pulker (Liechtenstein)
A. Quenzer (France)
G. Rauscher (FRG)
Philip J. Rogers (UK)
Konrad A. Roider (Austria)
G. Roosen (France)
Hans Rottenkolber (FRG)
J. D. Rush (UK)
J. D. Russel (UK)
D. Rutovitz (UK)
J. Saedler (GDR)
W. D. Scharfe (FRG)
A. M. Scheggi (Italy)
J. Schulte in den Baeumen (FRG)
John Seeley (UK)
S. A. Siegel (FRG)
Alberto Sona (Italy)
I. J. Spalding (UK)
Kenneth H. Spring (UK)
H. Stadler (FRG)
S. Stauber (Switzerland)
Gunnar Svensson (Sweden)
A. Tanguay (USA)
Alfred J. Thelen (FRG)
H. Tietze (FRG)
H. G. Treusch (FRG)
Gordon J. Watt (USA)
M. Wautelet (Belgium)
H. Weber (FRG)
R. N. Wilson (FRG)
Tony Wilson (UK)
Hajime Yamashita (Japan)
Hanfried Zugge (FRG)

ECO1
SCANNING IMAGING

Volume 1028

INTRODUCTION

The last few years have seen a tremendous increase in interest in scanning optical microscopy in general and confocal microscopy in particular. In parallel with this, new application areas and new methods of optical implementation have been developed. This proceedings covers both aspects of this development.

The main property of confocal scanning microscopy is its ability to allow very accurate optical sectioning to be performed. This means that we can choose to image detail from a particular section within a thick specimen. The result of the technique is that we can obtain three-dimensional images. This form of imaging is ideally suited to many kinds of microscopy, but finds particular application in biology and medicine, where it may also be used in the fluorescence mode.

In the field of optical design and implementation, this proceedings includes new models of three-dimensional imaging in which various factors such as the size and shape of the confocal detector are taken into account. The question of the role of lens aberrations is also discussed. This is particularly important in confocal microscopy, as a very small amount of spherical aberration can have a quite dramatic effect on the axial imaging while leaving the lateral imaging essentially unchanged. New techniques to achieve high resolution imaging either by modification of the optical system or by computer post processing are also discussed.

A relatively new biomedical application of confocal microscopy is in the field of ophthalmology. This proceedings contains three papers on this exciting new development.

Light beams have been used traditionally to excite carriers in semiconductors. Scanning laser beam systems have now reached the stage of development where they can be used in applications as diverse as circuit analysis to detect areas of latch-up sensitivity and the mapping of material properties by scanning photoluminescence studies. New theories have also been developed to aid the understanding and interpretation of these results.

This proceedings contains papers presented in the Scanning Imaging conference held as part of the International Congress on Optical Science and Engineering held in Hamburg, 19–23 September, 1988. The papers cover, together with other novel techniques, the various topics mentioned above.

I would like to take this opportunity to thank the session chairs for help in ensuring the smooth running of the conference, as well as all the scientists who participated in it.

Tony Wilson
Oxford University (UK)

ECO1
SCANNING IMAGING

Volume 1028

PLENARY SESSION

Overview of coherent optics applications in metrology

Jean EBBENI

Université Libre de Bruxelles - Dpt Milieux Continus - CP 194/5
50, avenue Franklin Roosevelt, 1050 Bruxelles - Belgium.

A B S T R A C T

If optical methods are long time ago used in metrology, coming of laser sources has improved drastically the impact of optics in metrology. The progressive existence of more and more industrial optoelectronic components on the market is responsible of the actual introduction of optical technics in industrial processings like interferometric control, wide-ranging optical sensors, visual inspection..... Further partial coherence and guiding properties of the light field, non linear optical comportment of the medium offer also interesting metrological applications.

The aim of this paper is not to give a full description of all the optical methods used in metrology, but to draw some general specific properties and ideas illustrated by representative applications.

1.- I N T R O D U C T I O N

Metrology consists to CONTROL, to MEASURE or to DETECT QUANTITATIVELY or (and) QUALITATIVELY, in a WIDE-RANGING FIELDS, some TYPICAL PROPERTIES of a SCENIC GEOMETRY like

- distances - forms - displacements - rotations - accelerations - velocities - strains - buckling - state of surface - mapping contours - flaw detection - crack detection -
- level - densitometry - volume capacities - proximity detection - torque...
or
some EXTERNAL FIELD like

- temperature - pression - acoustic pression - gas detection - density - concentrations
- spectral absorption - humidity - moisture - magnetic field - electric currant...

in view of BETTER UNDERSTANDING OF PHYSICAL EXPERIENCES, IMPROVEMENT of PERFORMENCE, CONTROL OF QUALITY, SAFETY or ECONOMICAL CRITERIA...

The optical techniques are powerful in metrology because of their great precision, flexibility, non interaction (or small interaction) with the object to analyze, and the very wide-ranged possibilities based on the following light properties:

GEOMETRICAL POSITIONING: ALIGNEMENT (f.i. pipe-lines), TRACKING (f.i. for repetitive mechanism), TELEMETRY (classical and laser), POINTING, PHOTOGRAMETRY, DEFLECTION, SCANNERS...

INTENSITY: DETECTION (all photodetectors), ABSORPTIONS, IMAGING (f.i. image deference), INTENSITY MODULATORS, INTENSITY SENSORS, INSPECTION, CONTROL, DENSITOMETRY, SHORT PULSES PRODUCTION...

PHASE EFFECTS: INTERFEROMETERS (vibrations, deformations, index variations, coherent sensors, strains, non-destructive testing, mapping contours, high precision displacements set-ups...) FOURIER ANALYSIS (de) convolutions, correlations, filtering,...) SPECKLE METHODS, PHASE CONJUGATION EFFECTS...

POLARIZATION EFFECTS: PHOTOELASTICITY, STOKES PARAMETERS (f.i. study of roughness), SENSORS, MODULATORS (FARADAY, POCKELS CELLS,...)....

SPECTRAL ANALYSIS: ALL SPECTROSCOPICAL SYSTEMS, LIDARS (f.i. air pollution), CONCENTRATIONS MEASUREMENTS...

INCOHERENT SPATIAL FREQUENCY BEATING EFFECTS: all the MOIRE methods (rotations, strains, displacements, contours,...)

FREQUENCY MODULATION: all the HETERODYNE methods, DOPPLER effects.

PARTIAL COHERENCE STATE: some physical properties of the medium (f.i. turbulences, random diffusing, optical delays...) change the temporal, the spatial or the polarization state of coherence of the analyzing light beam and can be used in some applications like decorrelation estimation, size of light source measurement, noise estimation or reduction...

GUIDANCE EFFECTS: guidance effects are very important for concentration, parasite isolation, stability, compacity and flexibility performances. Further large optical distances in small volumes can be obtained, allowing particular sensitivity for sensor applications.

NON LINEAR EFFECTS: The more and more materials presenting a non-linear sensitized comportment for ELECTROPTIC applications (f.i. optical switching, double harmonique generation...) and PHOTOREFRACTIVE applications (light valves, autofocusing, multi-waves mixing (light amplification, phase conjugaison, associative memories...)

Those numerous properties are used in metrology and are selected following the general criteria

NATURE OF THE MEASUREMENT - ENVIRONMENT (COMPACITY, AGRESSIVITY, PERTURBATION WITH ENVIRONMENT) - PRECISION REQUESTED - DYNAMIC RANGE - DERIVES of the MEASURES - FREQUENCY BANDWIDTH - CREEPING EFFECTS - COST-MEASUREMENT TIME - DIFFICULTY OF INTERPRETATION - STORAGE POSSIBILITY OF THE DATA - TREATMENT OF THE DATA - RESISTANCE TO SHOCKS.

All the measurements can be classed roughly in 3 categories:

VERY HIGH RESOLUTION as in SPACE OPTICS, very small or very far TARGETS INDENTIFICATION, SUBMICRONIC GRIDS, SUBMARINE INSPECTION, SUBMICRONIC DEFECTS, LOW ACOUSTICS FIELDS, ASTRONOMY...

HIGH RESOLUTION as in AERONAUTICS, SOME METALLIC CONSTRUCTION (machines of precision, automobile...) AUTOMATIC INSPECTION OF ELEMENTS, NON-DESTRUCTIVE INSPECTION OF SOME MATERIALS (composite materials, brazing, cracks...), SPECTRAL ANALYSIS, CONCENTRATIONS and WEIGHTS in some chemical operations (biotechnology, biomedical, food...) ELECTRONICS, SURFACE ANALYSIS, GRANULOMETRY, VELOCIMETRY, PRESSION, TEMPERATURE...

COARSE RESOLUTION as in SOIL MOVEMENTS, SINKING OF CIVIL STRUCTURES, PLASTICITY, PHOTOGRAMETRY of LARGE OBJECTS, RHEOLOGY, CIVIL HYDRAULICS, ACOUSTICAL NOISE, FOOD PRODUCTS, TRIANGULATION...

This classification is not parallel to the degree of complexity of the technics, because of typical constraints and each case is always specific.

In this classification temporal and spatial coherence properties of LASER sources are directly responsible of all the INTERFEROMETRIC, GUIDANCE, FOURIER PROCESSING and NON LINEAR applications. Some typical examples are described in this paper.

2.- HOLOGRAPHIC INTERFEROMETRY.

In a general case the two object and reference beams incident on an hologram are expressed respectively by

$$\underline{E}_0 = A_0(r,t).EXP[i\phi_0(r,t)]\underline{n}_1 + B_0(r,t).EXP[i\psi_0(r,t)]\underline{n}_2$$

$$\underline{E}_R = A_R(r,t).EXP[i\phi_R(r,t)]\underline{n}_1 + B_R(r,t).EXP[i\psi_R(r,t)]\underline{n}_2$$

where $\underline{n}_1$, $\underline{n}_2$ are unit vectors along axis Ox and Oy in the hologram plane, and A_i, B_i, ϕ_i, ψ_i are real functions which fluctuate in time and space because of the states of coherence of the source, turbulences, vibrations...

The intensity light distribution is thus given by

$$I(x,y) = <|\underline{E}_0 + \underline{E}_R|^2>$$

$$= <A_0> + <A_R> + <B_0> + <B_R>$$

$$+ <A_0 A_R \text{EXP}[i(\emptyset_0 - \emptyset_R)]> + \text{C.C.}$$

$$+ <B_0 B_R \cdot \text{EXP}[i(\emptyset_0 - \emptyset_R)]> + \text{C.C.}$$

where < > is an average on time and space. It is clear that the interference pattern suffers of SPATIAL, TEMPORAL and POLARIZATION DEGRADATIONS.

All those perturbations decrease the interference contrast and thus the efficiency of the hologram. When recording conditions are good enough, the general reconstruction of a double exposed hologram generates the images of an object in two different mechanical sates, called neutral and deformed states.

Interference fringe pattern appears in the observation plane π whose the form was described by many authors ([1] to [4]).

With the help of Figure 1 it is possible to describe more precisely the amplitude distribution U(A) observed at a point A of the plane [5].

$$\underline{r}_1 = \underline{S}_0 Q \qquad \underline{R}_1 = \underline{S}_0 P$$

$$\underline{r}_2 = \underline{Q}A \qquad \underline{R}_2 = \underline{P}A$$

$$\underline{P}'_1 = \underline{S}_0 Q \qquad \underline{PP}' = \underline{u}(P)$$

$$\underline{r}'_2 = \underline{Q}'A \qquad \underline{QQ}' = \underline{u}(Q)$$

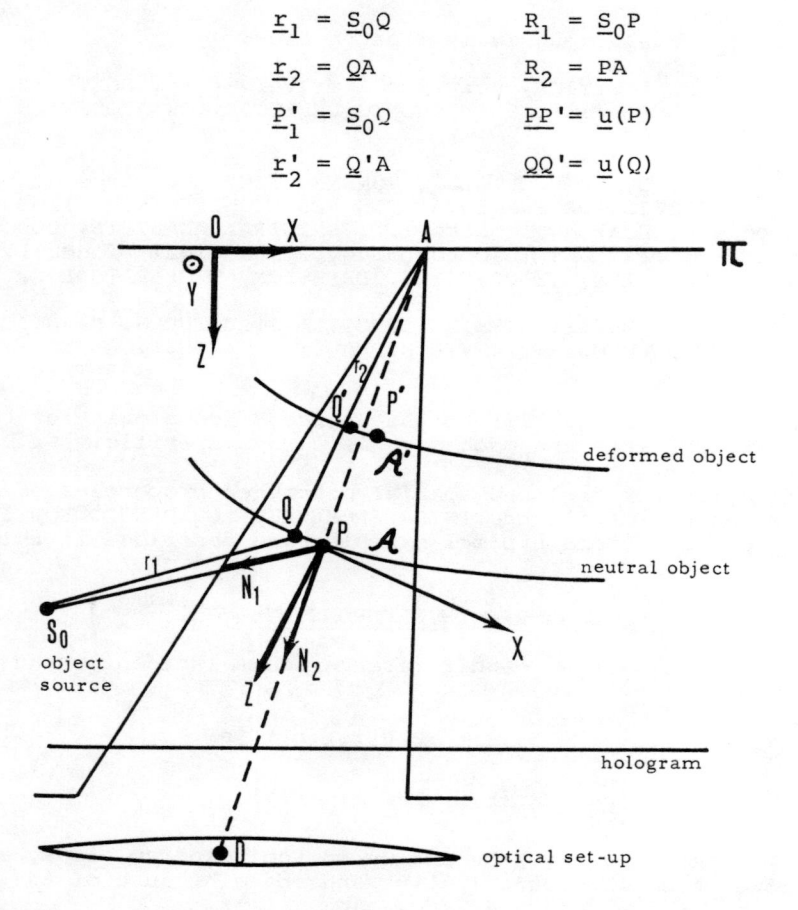

FIGURE 1.- Holographical reconstruction of the neutral and deformed objects. Interference fringes are looked in a plane.

$$U(A) \sim C_1 \int_A e^{ik(r_1 - r_2)} e^{i\psi(Q)} dQ$$
$$+ C_2 \int_{A'} e^{ik(r'_1 - r'_2)} e^{i\psi'(Q')} dQ'$$

where C_1, C_2 are constants.

A, A' are the projections of the optical pupil on the neutral and deformed objects with A taked as center of projection. They are assumed to be small enough to be plane.

ψ, ψ' are random phases introduced by the roughness of the same object at points Q and Q'.

r_1, r_2, r'_1, r'_2 are distances represented on the figure and k is the wave-vector.

For small elastic deformation we have $C_1 \sim C_2$, $A' \cap A = B = A$
and $\psi'(Q') = \psi(Q)$ if Q' is the new position of Q after the deformation, that is

$$x_i(Q') = x_i(Q) + u_i(Q)$$

where $\underline{u}(Q)$ is the displacement vector of Q. The optical center of the lens D and A gives the intersection point P on the neutral object surface. We define local axis at P with PZ normal at A an PX, PY in the plane A.
One can show that the intensity at A is ([5], [6]).

$$I(A) \simeq 2 A I_0 [1 + |\tilde{\beta}_n(B_I, B_{II})| \cos(\Delta_0 + \omega)]$$

where I_0 is a constant

$$B_J = \frac{u_J \left(\frac{1}{R_1} - \frac{1}{R_2}\right) - u_L \left(\frac{\alpha_L \alpha_J}{R_1} - \frac{\beta_L \beta_J}{R_2}\right) - u_{L,J}(\alpha_L + \beta_L)}{\lambda}$$

for $J = I, II$ and with summation on index L.

and α β are cosinus directors of unit vectors $\underline{n}_1$ $\underline{n}_2$
R_1, R_2 distances represented of figure 1.

$\Delta_0 = -k u_J (\alpha_J + \beta_J) = -k \underline{u} \cdot (\underline{n}_1 + \underline{n}_2)$ is the common part of A and A' (usually $A \simeq A'$)

$\tilde{\beta}_n$ = the Fourier normalized transform of B and $\varphi = A_r(\tilde{B})$.

$\varphi = 0$ if β has a symetrical form.

$|\beta_n(B_I, B_{II})| \leq 1$ presents a maximum for $B_I = B_{II} = 0$
which gives $|\beta_n| = \frac{B}{A}$

The fringe pattern of minimum intensity is given by

$\Delta_0 = (2N + 1)\pi$ with $N = 0, \pm 1, \pm 2 \ldots$

that is $-\underline{u} \cdot (\underline{n}_1 + \underline{n}_2) = (N + 1/2)\lambda$

It represents the classical fringe equation.

This fringe equation shows that from the analysis of their form, only the displacements fields can be deduced, the strain components being calculated by derivation. For full information 3 independent families are requested at the minimum and a lot of techniques where developed to access this information like scanning views from one hologram [7], use of different holograms, moiré techniques [8], [9], phase shifting [10], with different rigid displacements compensation and automatical processing [11], [12], [13], [14]...

If the deformation is too important, it is possible that the state of surface is changed (plastic deformation), Ψ' becomes independent of Ψ, there is no more optical correlation and thus no fringes. It is thesame if local displacement is too important because $A \cap A' = 0$.

In real time the contrast of the fringes is worse because the roughness reconstructed holographically is smoothed and thus not exactly the same: this smoothing is due to finite dimensions of lighted hologram, inhomogeneites, parasite vibrations...

The fringes have the best contrast in the plane where $|\beta_n|$ is maximum that is when $B_I = B_{II} = 0$. Usually this plane is near the surface of the object because rotation about an axis parallel to the surface of the object, presents a great sensitivity in the fringe formation and gives fringes localized near the object surface [6].

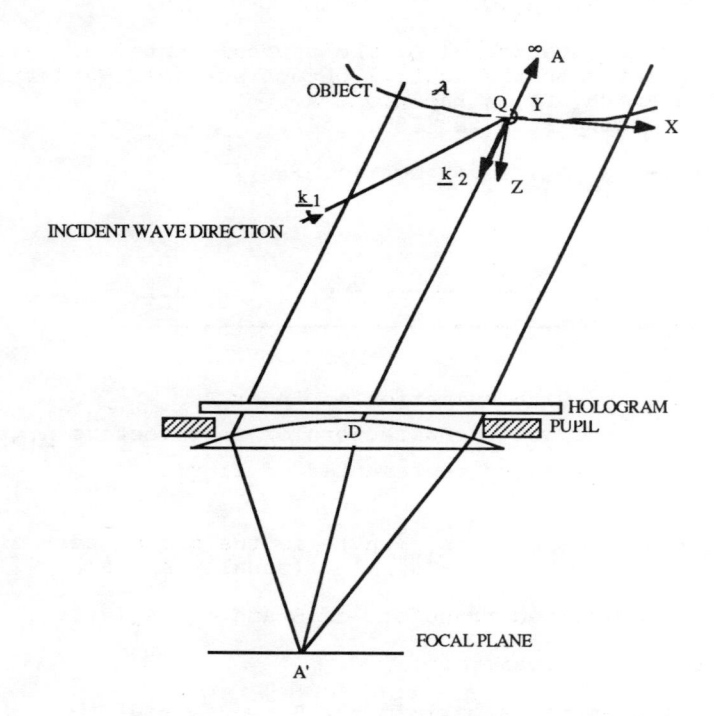

FIGURE 2.- The CONTRAST METHOD GEOMETRY.

The fringe contrast allow a direct determination of a strain component, remembering that the contrast function is a normalized Fourier transform of the pupil projection. If a variable slit mask is disposed in front of the lens, used for fringe observation, this contrast function is a sinc function

$$\text{sinc}[2\pi a' B_I]$$

where 2a' is the projection of the slit width 2a on the object, B_I the variable defined previously in the perpendicular direction of the slit. If further the object illumination and observation are established to infinity (i.e. $R_1 = R_2 = \infty$), with the requested cautions for depth field and speckle elimination [6], this relation becomes (cfr. Fig.2)

$$\text{sinc } [2\pi a'/\lambda \{U_{I,I} \cdot (\alpha_I + \beta_I) + U_{III,I} \cdot (\alpha_{III} + \beta_{III})\}]$$

By changing continuously the slit width, the sinc function becomes null for [ ] = π, allowing a measure of the strain component $\varepsilon = U_{I,I}$ if the rotation $U_{III,I}$ is small enough or measured separately.

This technique is full operated when the fringes present a adequate density and are oriented parallely to the slit. This can be obtained by use of a fringe control (sandwich, two reference beams, real time methods).

Strain accuracy is around 15 μS and the strain field can be determined on the whole surface of interest. Of course for the total strain distribution it is necessary to proceed to the "rosette" method and thus to use different angular geometries.

3.- OPTICAL PROCESSING USED FOR PATTERN RECOGNITION.

If classical holographic pattern recognition is a powerfull technique because it allows direct analogical Fourier correlation access [15], its practical application suffers from some drastic drawbacks as

- decorrelation due to variations of sizes, roughness, scale, orientation, position contour profile...

- partial obstruction

- resolution in time and space

- automatical processing of the data.

If some of those limitations are overcome optically by use of light-valves and various systems for the different requested invariances, detection and storage, all those techniques suffer of lack of precision and do not allow to satisfy simultaneously all the preceeding properties. This is namely difficult because of the drastic dependance of the Fourier analysis on any small variation of the object to recognize.

Those distorsions external to the correlator need special matched filters which maximize the SNR. The correlator must be stable enough to avoid defect of position of the matched filters [16], roughness and phase variations and that with the adequate bandwidth.

To overcome to much light concentration in the Fourier central peak, some phase adapted filters [17] [18] can be used. If different objects must be recognized together with various positions, each object will be represented by a class $\{f_{n_i}(x,y)\}$ of n_i constitutive images.

A non redundant synthetic discriminator function will be established with different algorythms [19].

MULTICHANNEL NON REDUNDANT OPTICAL CORRELATORS can compare one input image to multiplexed filters, by use of multiplexed holograms and different sources [20].

The binar simple approach [21] is usefull for a great number of applications, N filters allowing 2^N codable images.

As an example, let us consider a rectangular object from 8 images O_k are selected for 8 different regularly spaced orientations.

3 filters f_i i = 1,2,3 are constructed in so a manner to satisfy the values specified on Figure 3.

	O_1	O_2	O_3	O_4	O_5	O_6	O_7	O_8
f_1	0	1	1	0	0	1	1	0
f_2	0	0	1	1	1	1	0	0
f_3	0	0	0	0	1	1	1	1

FIGURE 3.- Filters in parallel

Figures 4, 5, 6 and 7 show respectively the coded filters, the computer simulation actions, the holographic applications on photoplate and BGO crystal.

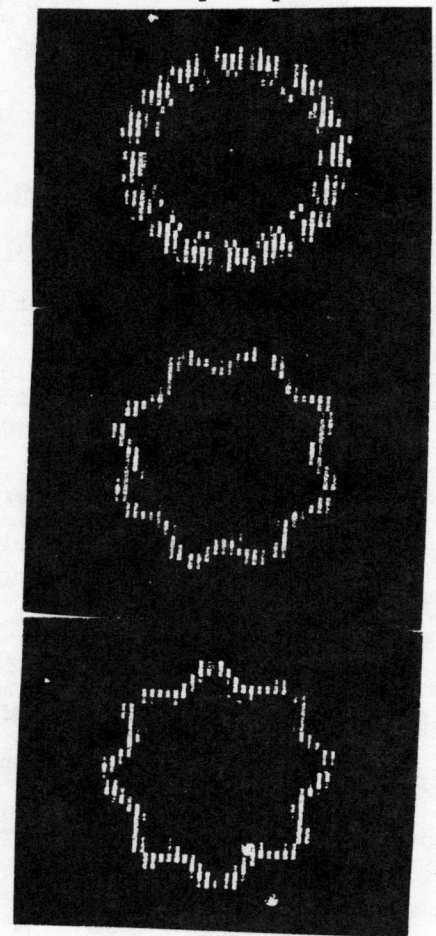

FIGURE 4.- Coded filters.

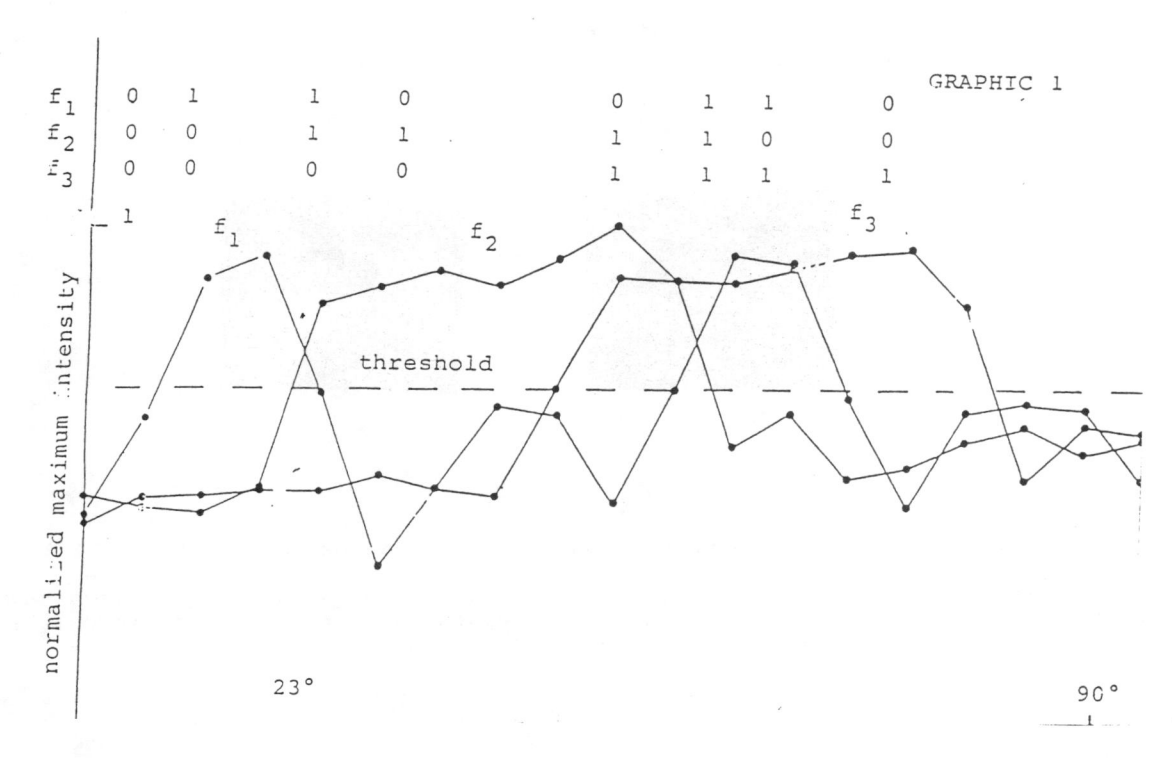

FIGURE 5.- Computer simulation.

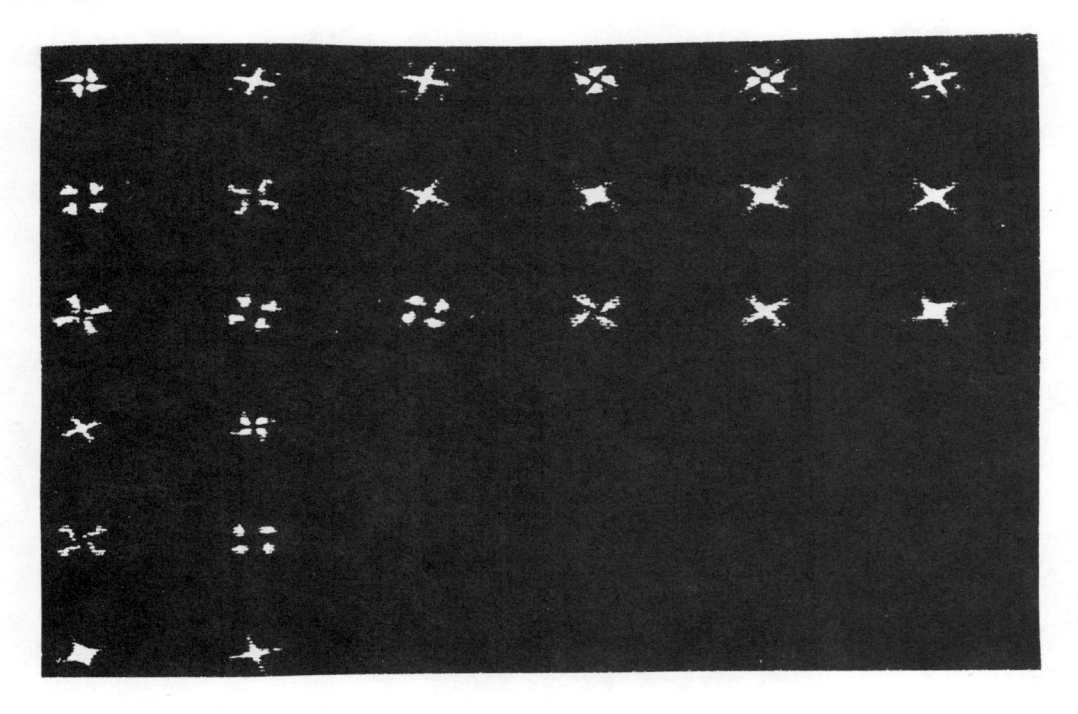

FIGURE 6.- Holographic results on photoplate.

FIGURE 7.- Two typical bright and black central peaks obtaines on BGO.

Optical implentation is possible by introducing a half-tone screen technique with appropriate shifting to obtain adequate dephazing in the Fourier plane [22] and by use of phase conjugation holograms to compensate aberrations.

In practice, it is evident that some differences between the original reference object and the test object are greatly accentuated in the correlation product. It is the reason why an ITERATIVE proceesing, using CONJUGATE MIRRORS and adequate AMPLIFICATION can be applied as shown on Figure 8 [23].

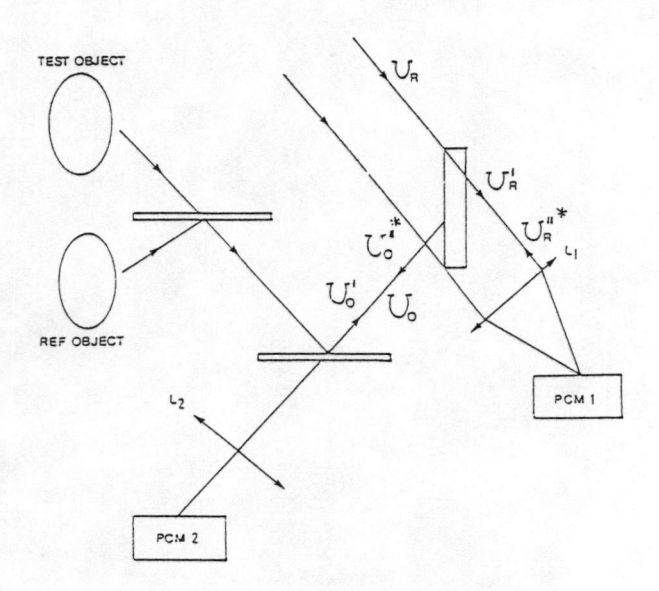

FIGURE 8.- Associative memory system.

The original object beam U_O is holographically recorded with waveplane reference U_R. If this hologram is lighted by a test object beam U'_O, slightly different from U_O, the reconstructed reference beam U'_R, which is no more a perfect waveplane beam, is focused by a lens L_1 on a first PCM1.

It is evident that the light distribution presents a maximum centered on the focus point of the perfect reference beam U_R. The amplification gain depending of the intensity with a tresholding processing, the reflected conjugate beam U'''^*_R is thus more closed to U^*_R than U'^*_R and in consequence the resulting diffracted beam U'''^*_R is also improved. This beam is imaged by a lens L_R on a second PCM2 which generates a beam U'''_O and the processus is applied again with more improvement of the signal.

Suppose now that a class of objects are multiplexed on the same Fourier hologram, each Fourier spectrum interfering with a reference waveplane in such a way that the total number of recorded filters is $N = N_1 \cdot N_2$ where N_1 is the number of spatial zones and N_2 the number of local multiplexing. An object input is lighted simultaneously by N_1 incident waveplanes with the adequate orientations to create the lightening of the different zones. The former technics can be used to select the more matched storaged image. This final image is projected to the output channel by help of another beamsplitter as shown on Figure 9.

The advantage of this configuration is that a large number of object classes can be analysed simultaneously in function of the resolution of the components. It is further possible to compare quantitatively the values of the different peaks of the correlation function and to calculate some intermediate situations.

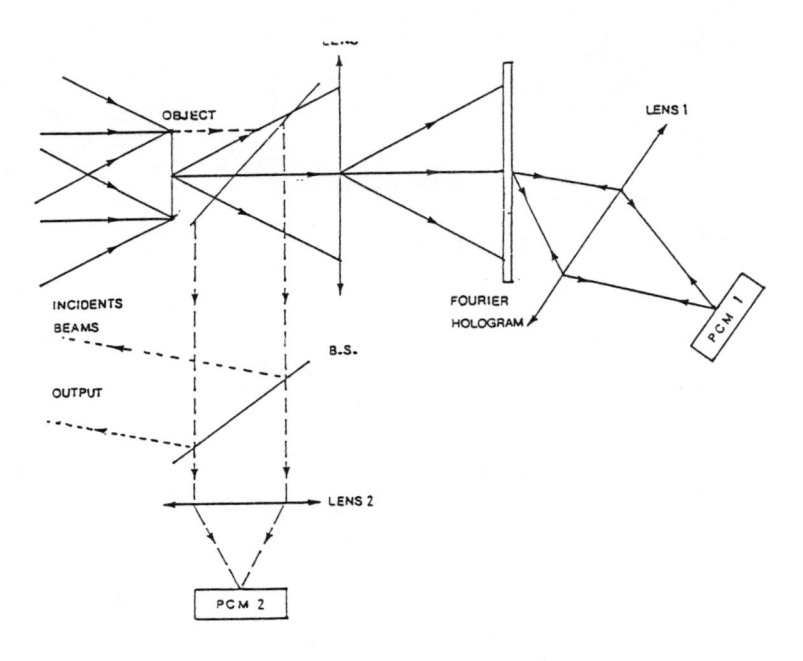

FIGURE 9.- Proposed pattern Recognition set-up.

One of the mist interesting application of the pattern recognition is probably FRINGE ANALYSIS and namely for holographic non destructive testing, where the decision of acceptation or reject is particularly important. Those applications present typical problems because the DENSITY, FORM and PROFILE of fringes are influenced by rigid displacement fields and deformation intensity. It is important to try to compensate those eventual variations by use of moiré techniques and electronic processings.

4.- CONCLUSIONS

It appears that coherent optical processing is very usefull in metrology, but that some limitations remain as availability of low cost, stable compact and low dissipation power sources, compact systems and limited optical processing.

The flexibility of hybrid opto-electronic systems, namely in parallel neuronic processing, will constitute the best way for next future applications.

5.- REFERENCES

[1] R.A. Stetson, R.L. Powell , Josa, vol. 55, n° 12, 1965.
[2] K.A. Haines, B.P. Hildebrand, App. Opt. Vol.5, n° 5, 1969.
[3] J.E. Sollid, App. Opt. Vol. 8, n° 8, 1969.
[4] A.E. Ennos, Jour. Phys. E, Sc. Inst. Ser.2, Vol. 1, 1968.
[5] J. Ebbeni, Porc. IUTAM, 1979, Sijthoff & Noordhoff, U.S.A.
[6] J. Ebbeni, J.C. Charmet, App. Opt., vol. 16, n° 9, 1977.
[7] E.B. Aleksandrov, A.M. Bonch Bruevich, Sov. Phys. Tech. Vol. 12, n° 258, 1967.
[8] C. Aleksoff, App. Opt. Vol. 10, 1971.
[9] J.N. Butters, The eng. uses of holography, Cambridge Univ. Press, 1978.
[10] R. Dandliker & Al, Josa, vol. 66, 1976.
[11] E. Champagne, Proc. SPIE, 1972.
[12] P.M. Boone, The eng. uses of holography, Cambridge Univ. Press, 1978.
[13] N. Abramson, App. Opt. Vol. 16, n° 9, 1977 [6]
[14] A. Stimpfling, Proc. SPIE 672, 1987.
[15] D. CASASENT, A. FORMAN, Appl. Opt. Vol. 16, n° 6, 1977.
[16] J.W. Goodman, Introduction to Fourier Optics,Perg. Press, 1964.
[17] J.L.Horner, P.D. Gianino, Appl. Opt., vol. 23, n° 6, 1984.
[18] J.L.Horner, H.O. Bartelt, Appl. Opt. Vol. 24, n° 18, 1985.
[19] D. Casasent, D. Saltis, Progress in Optics, Vol. XVI, North Holland Publ. 1978.
[20] Z.H. Gu, S.H. Lee, Opt. Eng. Vol. 23, n°6,1984.
[21] B. Braunecker, R. Hauck, A.W. Lohman, Appl. Opt., Vol. 11, n° 2, 1974.
[22] F. Dubois, Private communication Dpt. Opt. ULB, 194/5 Brussels.
[23] B.H. Soffer & Al, Opt. Lett. 11, n° 2, 1986.

SESSION 1

Confocal Microscopy I

Chair
Tony Wilson
Oxford University (UK)

Pupil filters in confocal imaging

Zoltan S. Hegedus

CSIRO Division of Applied Physics, Lindfield, Australia, 2070

ABSTRACT

The confocal scanning microscope is a coherent system with significantly improved imaging characteristics. The effective pupil function of such an instrument is a composite one, formed by the convolution of the pupil functions associated with the illuminating and imaging optical systems. Because of this additional degree of freedom, the transfer function of the system can be modified with comparative ease.

Confocal imaging will be analysed within a very simple but quantitative physical model. Also, the effect of various pupil combinations on the imaging properties of such systems will be presented with special emphasis on the use of superresolving pupils.

1. INTRODUCTION

To satisfy true confocality, the scanning imaging system must satisfy two conditions. These conditions set the size of the detector and the source with respect to the diffraction spot produced by the imaging and illuminating optical systems respectively.

The image must be collected through a pinhole, which, is significantly smaller than the central spot of the diffraction disk produced by the imaging objective.[1] The source illuminating the object must also be restricted in size, however it must satisfy only the normal spatial coherence requirements for the illuminating system. If these conditions are satisfied, the scanning system behaves as a true confocal system and its imaging performance shows an improvement over a conventional microscope both in the lateral and longitudinal sense.

2. MODEL DESCRIPTION

The imaging improvement in the confocal microscope manifests itself in improving the lateral and longitudinal discrimination of the amplitude transmitted by the object. Fig. 1 shows a simple semi-geometrical argument with the aid of which the improvement can be understood qualitatively.

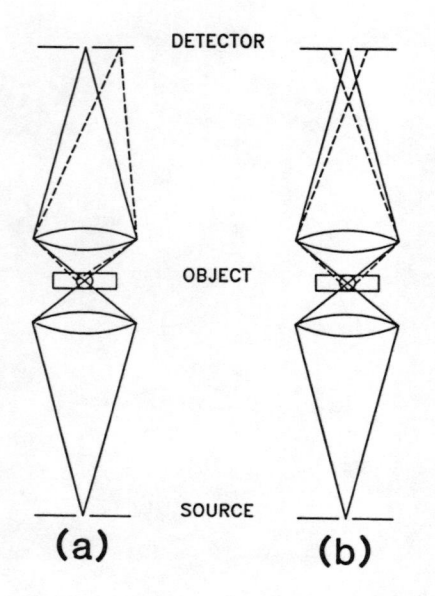

Figure 1. Lateral (a) and longitudinal (b) discrimination of a confocal microscope. The circle within the object indicates the diffraction spread.

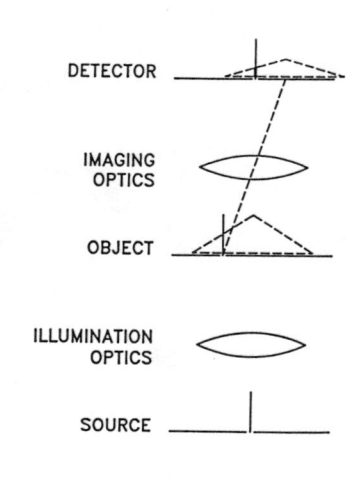

Figure 2. Simplified model of light propagation through the scanning imaging system

From the figure it is clear, that if there is no obstruction at the detector plane, all the light emerging from the illuminated region of the object will be registered. On the other hand if a pinhole is placed on-axis, the off-axis contributions of the light scattered by the object will be prevented from reaching the detector. The same process will occur with light scattered on-axis but not in the plane of best focus.

Based on this geometrical model a more quantitative description can be developed if we introduce the appropriate point spread functions associated with the optical components of the system. For simplicity, we analyse a one-dimensional system, whose components have identical and triangular spread functions. Following Fig. 2 we can trace the propagation of light through the system.

The point source produces through the illumination optics the first triangular spread (at the object plane). This spread is intercepted by the scanned object, in this case a single pinhole. The light amplitude transmitted by this pinhole at this given scan position will generate another spread at the detector plane. The position and amplitude of this second spread is determined by the object position and the amplitude transmitted by the object. As the object traverses the first spread function, the second spread function traverses the detector and is appropriately modulated. A detector, which is smaller than the second spread will limit the recorded signal.

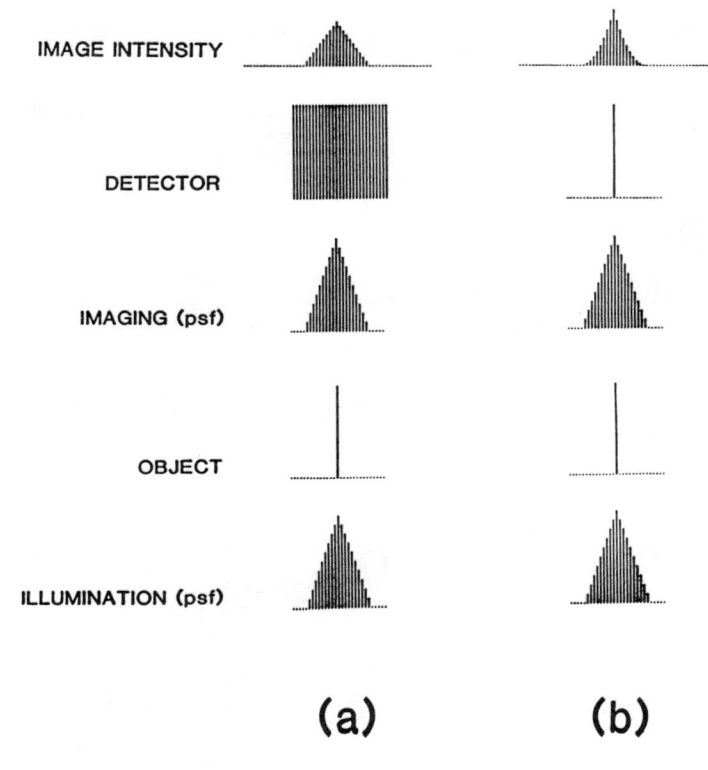

Figure 3 shows the final result of the implementation of this algorithm. Here, as in Fig. 1, the light propagates through the system from the bottom. The corresponding functions are shown for the on-axis position of the object. The resulting image intensity is shown at the top and it is clear, that the confocal instrument sharpens the point spread function of the total system. Detailed analysis shows, that the shape of the curve at the top of Fig. 3 (b) is a parabola, that is the confocal mode multiplies the point spread functions (in our case linear functions) of the two optical systems.

Figure 3. Calculated image intensity for an object which approximates a delta function. The image therefore corresponds to the point spread function of the whole instrument: (a) type-1 imaging and (b) confocal imaging.

3. PUPIL FUNCTIONS

The confocal microscope is a coherent imaging system whose effective amplitude point spread function is the product of the point spread functions of the illuminating and imaging optics of the system.[2] Conversely, the effective pupil of such an instrument may be described as the convolution of the two pupil functions associated with these two optical components. The result of a convolution operation leads to a function which is defined by the domains of both functions involved, that is to say, for two functions defined in equal circular domains, their convolution will be another circle of twice the radius.

Since for coherent imaging the pupil function is equivalent to the transfer function, it is evident that the confocal microscope doubles the effective pupil size of a conventional coherent image forming instrument. It must be noted that this is not equivalent to doubling the clear pupil radius. Due to the convolution operation, the transmittance of the effective pupil will gradually drop from a maximum to zero.

Apodising and superresolving pupils can modify the effective pupil transmittance and the imaging characteristics of the confocal microscope. Fig. 4 shows some examples of the effect of modifying pupil functions on the effective pupil (or transfer) function of the confocal microscope.

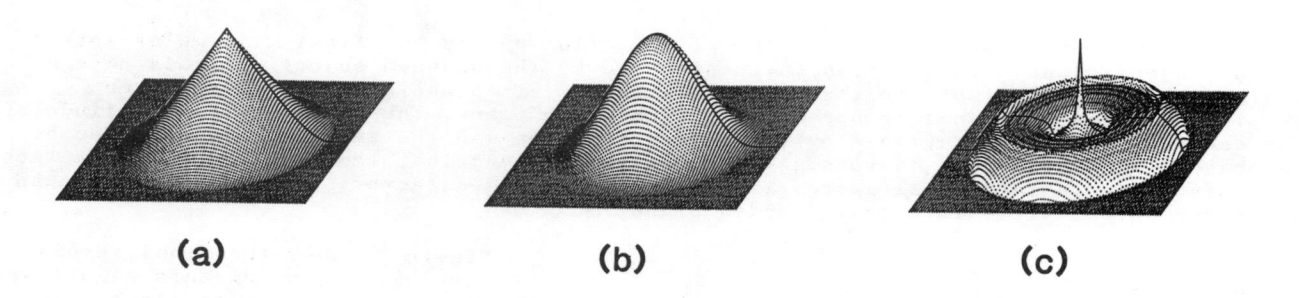

Figure 4. Pupil functions of a confocal microscope: (a) both pupils are clear, (b) one of the pupils is a quadratic apodiser and (c) one of the pupils is a complex superresolving filter.

4. IMAGING

The model outlined in Part 2 can predict the general behaviour of the system. Figure 5 shows the calculated results for a periodic object whose period was chosen to be smaller than the extent of the central diffraction disc associated with the amplitude spread functions of the two optical systems.

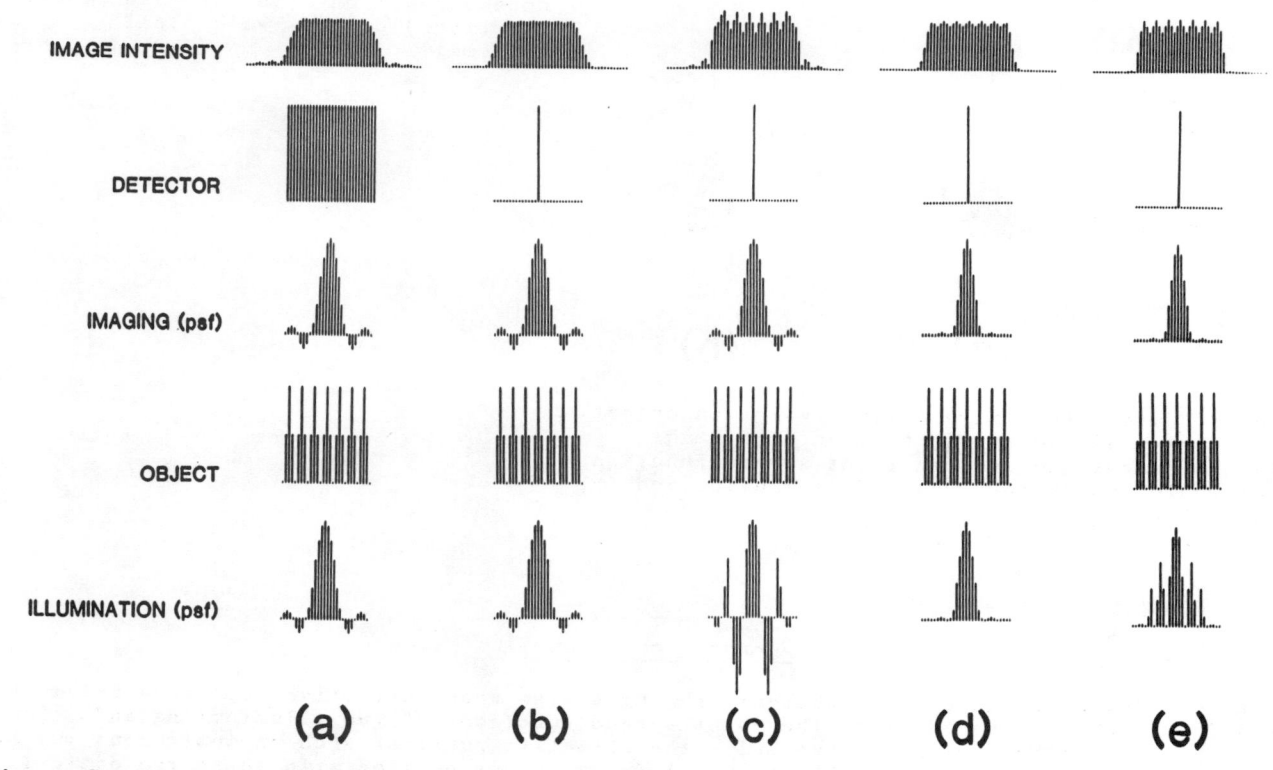

Figure 5. Imaging at the diffraction limit for various imaging modes: (a) type-1 imaging, (b) confocal imaging with clear pupils, (c) one pupil is a complex superresolving filter, (d) and (e) as (b) and (c) in the fluorescent mode.

As expected, the object is not resolved with either the type-1 or with the confocal microscope. If the illuminating pupil is replaced with the superresolving pupil, shown in Fig. 4 (c), strong signal modulation occurs, however the signal is not true to the object, the intensity is reversed! The amplitude point spread function of the illuminating system in Figure 5 (c) follows the distribution associated with a two-ring complex filter.[3] A significant improvement is observed, when the system operates in the fluorescent mode. The amplitude spread functions are now replaced with their intensity counterparts and the images show very clearly the true intensity distribution of the original object. We can see that the filter has improved the contrast of the signal significantly. This is an important feature when imaging near the diffraction limit, where the noise may be comparable with the signal in an instrument with clear pupils.

5. CONCLUSIONS

Modification of the pupil transmittance in confocal systems leads to changes in the imaging characteristics of a scanning microscope. Application of superresolving filters leads to improvements to the coherent transfer function towards the cutoff frequency, hence offering imaging capabilities not achievable with post-processing techniques.

6. REFERENCES

1. T. Wilson and A.R. Carlini, "Size of the detector in confocal imaging systems," Opt. Lett. 12(4), 227-229 (1987).
2. C.J.R. Sheppard and A. Choudhury, "Image formation in the scanning microscope," Opt. Acta 24(10), 1051-1073 (1977).
3. Z.S. Hegedus and V. Sarafis, "Superresolving filters in confocally scanned imaging systems," J. Opt. Soc. Am. A3, 1892-1896 (1986).

Axial resolution of a confocal scanning optical microscope

Marlo J. Offside, Chung Wah See and Michael G. Somekh

University College London, Department of Electrical and Electronic Engineering, Torrington Place,
London WC1E 7JE United Kingdom

ABSTRACT

It is well known that the type II scanning optical microscope has important advantages over conventional optical microscopes. These advantages lie primarily in their ability to eliminate out of focus signals so that the image contains information from a thin axial slice. In this paper we quantify this effect for several typical microscope configurations.

We have calculated the axial response of the microscope as a function of the numerical aperture and the focal length of each lens in the system. The effect of misalignment of the point detector has been investigated. In addition the effect of feature size has been analysed; we have demonstrated that the depth discrimination differs substantially for a point and large area object. These analytical results have been extended numerically to account for arbitrary size objects.

The computations have been performed using both Fresnel diffraction theory and a hybrid approach which combines diffraction theory and ray optics. The latter method greatly reduces computational complexity and the results agree with those obtained using diffraction theory in cases where comparison is applicable.

1. INTRODUCTION

It is well known that type II scanning optical microscope has important advantages over conventional optical microscopes.[1] One such advantage is the improvement in lateral resolution. The other is their ability to eliminate out of focus signals so that the image contains information from a thin axial slice. The effect of the finite size of the point detector has been analysed.[2] In this paper, we address ourselves to the latter property of the system. We have investigated the axial response for various feature sizes using a simple microscope configuration. Two methods are employed in this study. The first involves the application of Fresnel diffraction theory through each stage in the system. In the second approach, a hybrid model combining geometrical optics and Fresnel diffraction theory is used. This greatly reduces the effort required for the computation. Where comparison is applicable, the results obtained using the hybrid model agree very well with those obtained with Fresnel diffraction theory. Having established the validity of the hybrid approach, it was used to investigate the axial response of a more complex microscope configuration. For this system, successive application of Fresnel diffraction theory would lead to serious computational difficulties.

Analytical expressions, describing the axial response as a function of different system parameters, will be given. In addition, we have considered the effect on the system output when there is a second, partially transparent object, situated in close proximity to the in focus sample. Computer simulations obtained with this analysis will be presented.

2. CONFIGURATION 1 : POINT SOURCE SITUATED AT FINITE DISTANCE FROM LENS

Fig. 1 shows the arrangement of the system. A point source is situated at a distance u from the lens. The distance to the object plane at 0 is $(v + d)$, where v is the image distance of the point source, and d is the object defocus distance. Light returning from the sample passes back through the lens and is directed towards the point detector by the beamsplitter. For clarity, the system is represented by the equivalent configuration illustrated in fig. 2. This is effectively a transmission system with two identical lenses. However, it is important to note that a defocus d results in an increase in the lens separation of $2d$. In a true transmission system, defocusing does not alter the separation of the lenses. We will first consider the system response by using Fresnel diffraction theory.

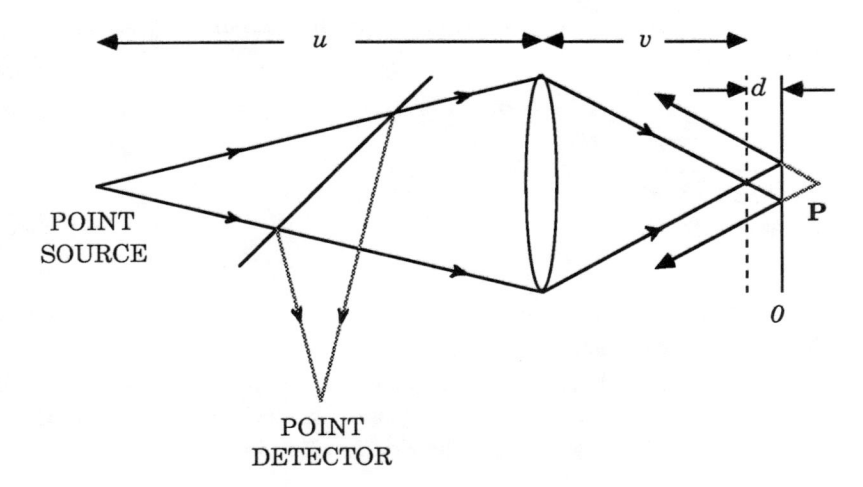

Figure 1. System configuration for a one lens microscope

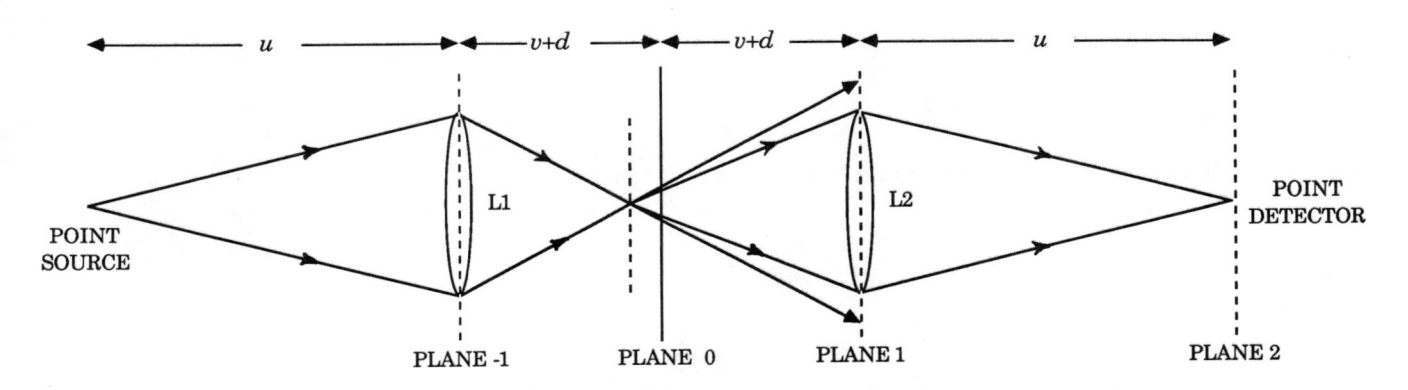

Figure 2. Equivalent configuration for the one lens system

2.1. Analysis using Fresnel diffraction theory

Since the system is circularly symmetrical about the optical axis, the Fresnel diffraction formula can be written as [3]

$$U_m(r_m) = \frac{2\pi \exp(jkz)}{j\lambda z} \exp\left\{\frac{jkr_m^2}{2z}\right\} \int_0^\infty U_n(r_n) J_0\left\{\frac{kr_n r_m}{z}\right\} \exp\left\{\frac{jkr_n^2}{2z}\right\} r_n dr_n \qquad (1)$$

where $U_n(r_n)$ and $U_m(r_m)$ are the amplitude and phase distributions of the fields at plane n and m respectively, r_n and r_m are the radial coordinates, z is the distance between the two planes, and J_0 is the zero order Bessel function of the first kind. We will assume that the lens function takes the form

$$T_l = C \exp\left\{\frac{-jkr^2}{2f}\right\} \qquad (2)$$

where f is the focal length of the lens. We can apply the Fresnel diffraction formula successively to propagate the field from the point source to the detector plane. By doing this, we will arrive at an expression, $U_2(r_2)$ which describes the field distribution at the detector plane. The photodetector, situated at $r_2 = 0$, detects the intensity of the light. Its output is therefore proportional to $|U_2(0)|^2$. If we assume a point object, one whose transmissivity or reflectivity is described by a Delta function, the photodetector output I_{pt} takes the form

$$I_{Pt} = \xi_1 \frac{1}{(v+d)^4} \left\{ \frac{\sin\left(\frac{kdR^2}{4v(v+d)}\right)}{\frac{kdR^2}{4v(v+d)}} \right\}^4 \qquad (3)$$

where ξ_1 is a constant, and R is the radius of the lens aperture. By using the following parameters : $R =$ 2.5mm, $f = 1$cm, $u = 10f$, the variation of I_{pt} as a function of the object defocus d is computed. The result is shown in fig. 3. The sidelobes are not visible because the output is proportional to the 4th power of the Sinc function. The distance between the central peak and the first zero, which is a measure of the axial resolution of the system, can be obtained by equating the argument of the sine function in eq. (3) to π. Thus we have

$$d_{Pt} = \frac{2\lambda v^2}{(R^2 - 2\lambda v)} \approx \frac{2\lambda v^2}{R^2}$$

$$= \frac{2\lambda}{NA_e^2} \qquad (4)$$

where NA_e is defined as the effective numerical aperture of the lens.

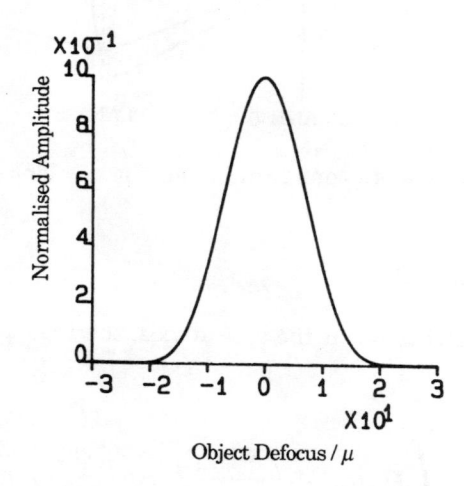

Figure 3. System axial response for a point object

If the object has finite size (assumed to be circular), the expression $U_2(0)$ is more complicated

$$U_2(0) = \xi_2 \frac{1}{(v+d)} \int_0^R \int_0^{R_0} U_0(r_0) \exp\left\{\frac{-jkdr_1^2}{2v(v+d)}\right\} \exp\left\{\frac{jkr_0^2}{2(v+d)}\right\} J_0\left(\frac{kr_0 r_1}{(v+d)}\right) r_0 r_1 \, dr_0 \, dr_1 \qquad (5)$$

where

$$U_0(r_0) = \xi_3 \frac{1}{(v+d)} \exp\left\{ \frac{jkr_0^2}{2(v+d)} \right\} \int_0^R \exp\left\{ \frac{-jkdr_{-1}^2}{2v(v+d)} \right\} J_0\left(\frac{kr_{-1}r_0}{(v+d)} \right) r_{-1} dr_{-1} \qquad (6)$$

and ξ_2 again is a constant. Figure 4 shows the detector output as a function of the object defocus for different object sizes. The system parameters listed previously are retained. The object radius ranges from 0.01 μ to 10 μ. When the object size is small (fig. 4a), the system response is similar to that of a point object. Figure 4b shows the response when the object radius is 2.0 μ. This is slightly larger than the spot size of the light (the spot radius of the light at the object plane is about 1.7μ). The response begins to depart significantly from that of the point object. The main lobe becomes narrower and sidelobes begin to appear. As the object size increases, the depth discrimination improves continuously. This improvement approaches saturation for an object size of about twice that of the spot size. Figure 4c shows the response for an object size of 10 μ. This is a very interesting result. The axial resolution can thus be seen to improve when distinguishing features much greater than the lateral resolution. The practical consequences of this is that the elimination of out of focus information may vary throughout a sample depending on the local structure. For instance, on an integrated circuit the depth discrimination will be better on largely homogeneous regions such as bond pads compared to regions containing narrow tracks whose width is comparable to the spot size.

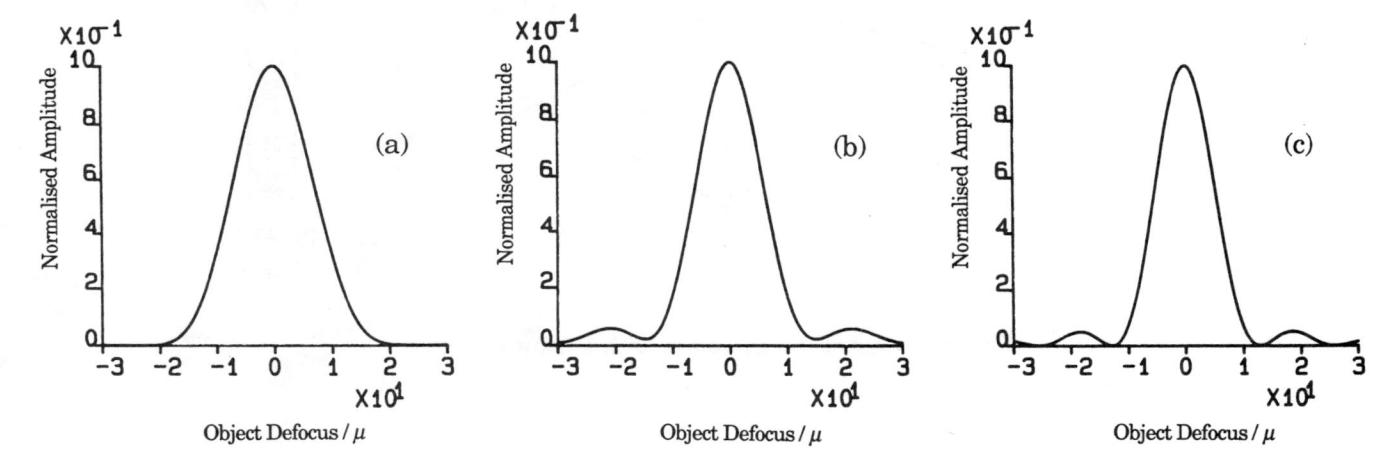

Figure 4. System axial response for features with different sizes.
Feature sizes: (a) = 0.01μ radius, (b) = 2.0μ, (c) = 10μ.

We will examine this effect more closely. In fig. 1, the light reflecting from the object surface can be considered to be emitted from a point source situated at **P**. If the object is defocused by a distance d, the effective point source will be displaced by a distance $2d$ from the plane **O**. We therefore have a moving point source and a stationary point detector. This is very similar to the situation depicted in fig. 5 where we have a stationary point source, a moving point object and a large area detector. (This is of course a Type I microscope). The movement or defocus of this point object, D, has to be equal to $2d$ in order for the two systems to be equivalent. If we now apply Fresnel diffraction analysis to this model, we obtain, for the photodetector output

$$I_{Lr} = \xi_4 \frac{1}{(v+2d)^2} \left\{ \frac{\sin\left(\frac{kdR^2}{2v(v+2d)} \right)}{\frac{kdR^2}{2v(v+2d)}} \right\}^2 \qquad (7)$$

Using the same parameters as before, we plot I_{Lr} as a function of d which is shown in fig. 6. As one can see, it is practically indistinguishable from fig. 4c. We can therefore use this simple model to describe the system response for a large area object. The distance between the peak and the first zero is now

$$d_{Lr} = \frac{\lambda v^2}{R^2 - 2\lambda v} \approx \frac{\lambda v^2}{R^2}$$

$$= \frac{\lambda}{NA_e^2} \tag{8}$$

We now have two expressions (eq. 3 & 7) which describe the relationship between the system output and the object defocus for small and large area objects. In addition, eq. 4 & 8 can be used to predict the occurrence of the first zero. However, as eq. 5 has shown, Fresnel diffraction theory may lead to a complicated expression for the field distribution. This is especially true for systems containing many optical components. For such systems, an alternative approach is required. In the next section, a novel approach to this problem will be demonstrated.

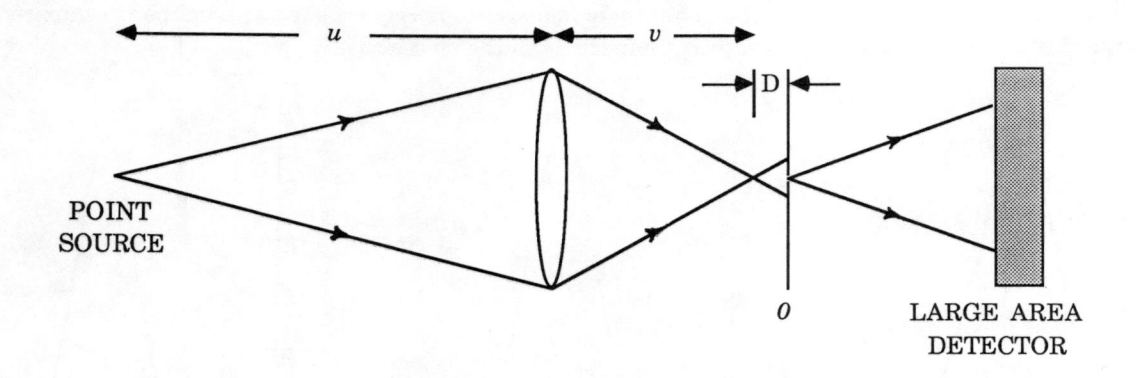

Figure 5. Type I configuration with a point object.

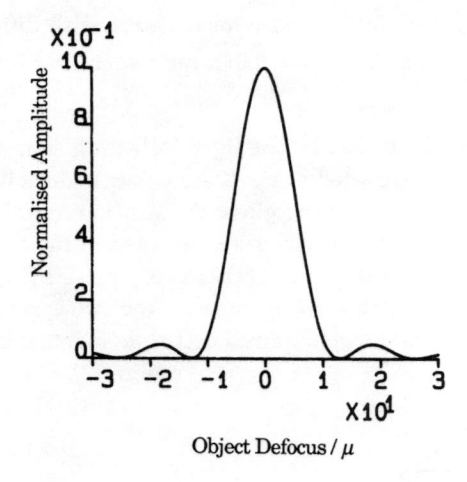

Figure 6. System axial response for the Type I microscope.

2.2. Analysis using hybrid approach

Referring to fig. 2, a geometrical formulation was determined for the phase and amplitude distribution over plane 1 for a large area object. Neglecting a constant amplitude factor, the field distribution is given by

$$U_1(r_1) = \frac{1}{\sqrt{p^2+r_1^2}} \exp\left\{ jk\left(\sqrt{p^2+r_1^2} - p\right)\right\} \tag{9}$$

where p is defined by

$$p = v + 2d \tag{10}$$

It has been assumed that the image point of lens L1 acts as a secondary point source. The exponential term in eq. 9 represents the phase distribution over plane 1 resulting from a spherical wavefront emanating from the secondary point source. Similarly the preceding term in eq. 9 represents the amplitude distribution over plane 1, accounting for the decay in amplitude as the spherical wave expands.

This distribution is modified by the lens function and is then propagated a distance u to the detector using Fresnel diffraction theory. The resulting field distribution is

$$U_2(r_2) = \frac{1}{u}\int_0^R \frac{J_0\left(\frac{kr_1 r_2}{u}\right)}{\sqrt{p^2+r_1^2}} \exp\left\{jk\left(\sqrt{p^2+r_1^2}-p\right)\right\} \exp\left\{\frac{jkr_1^2}{2}(1/u - 1/f)\right\} r_1 dr_1 \tag{11}$$

At $r_2 = 0$, this reduces to

$$U_2(0) = \frac{1}{u}\int_0^R \frac{1}{\sqrt{p^2+r_1^2}} \exp\left\{jk\left[\sqrt{p^2+r_1^2}-p+\frac{r_1^2}{2}(1/u-1/f)\right]\right\} r_1 dr_1 \tag{12}$$

This has the following solution when $p \gg R$

$$|U_2(0)| = \frac{R^2}{2u} \operatorname{Sinc} \theta \qquad \text{where} \quad \theta = \frac{kdR^2}{2v(v+2d)} \tag{13}$$

The axial response is again obtained by taking the square of eq. 13. The result, shown in fig. 7, agrees very well with that obtained using Fresnel diffraction theory throughout the system. Compared with eq.5, eq. 12 is much more simple. The hybrid approach therefore allows us to analyse more complex and practical configurations.

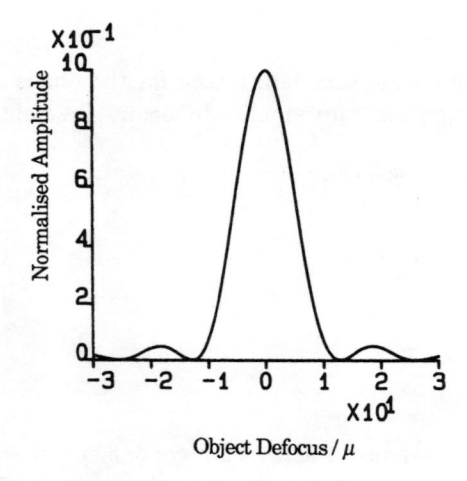

Figure 7. System axial response for a large area object by using hybrid approach.

3. CONFIGURATION 2 : TWO LENS SYSTEM

The system considered is shown in fig. 8. It consists of two lenses separated by a distance s. The collimated light beam is focused onto the object plane by lens L3. The returned beam is focused onto the detector by L4. For this configuration, application of Fresnel diffraction theory throughout the system would result in an expression containing many integrals. Thus the hybrid model proves to be a far more convenient method of tackling the problem.

After passing through lens L3 twice, the light beam can be regarded as emitted from an effective point source (or converging to a point). By considering the spherical wavefront emanating from this effective point source, the geometrical formulation for the phase and amplitude distribution over plane 3, just before lens L4, is

$$U_3(r_3) = \left[1 - \frac{2r_3^2 d^2}{\{f^2 + 2d(f-s)\}^2} \right] \exp\left\{ jk \left[\frac{r_3^2 d}{\{f^2 + 2d(f-s)\}} + s + 2d \right] \right\} \tag{14}$$

This distribution is then propagated to the detector plane using Fresnel diffraction theory as described previously. We have conducted a detailed study of the effects of changing the parameters of the two lenses on depth discrimination. These parameters include the aperture and the focal length of each lens, as well as the lens separation. Some of the results obtained are presented below.

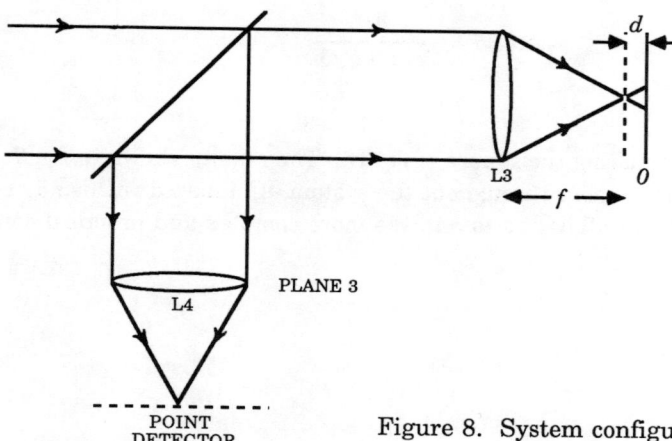

Figure 8. System configuration for a two lens microscope.

3.1. Axial response as a function of lens focal length

Figure 9 is the system axial response when the two lenses are identical and the separation between the two lenses is 40 times the lens focal length. The response is symmetrical about the in focus position. This is always the case as long as the apertures of the lenses are matched, irrespective of their individual focal lengths. Furthermore, the depth discrimination is independent of the focal length of the detector lens. This conclusion, derived from the numerical results, can be explained analytically as follows:

A given object defocus will give rise to a cone of converging or diverging rays emerging from the object lens. If this effective source is located at a distance u from the detector lens. This effective point source is imaged at a distance δz from the focal plane of the detector lens. Simple application of the lens formula yields

$$\delta z = \left(\frac{f^2}{u}\right) \tag{15}$$

For modest values of δz the Lommel function distribution is centered about the image plane at $f + \delta z$. The axial distribution about this plane is given by [4]

$$I(q) = I(0) \left[\frac{\sin(q/4)}{(q/4)}\right]^2 \tag{16}$$

where the optical parameter q is defined as

$$q = \frac{2\pi}{\lambda}\left(\frac{a}{f}\right)^2 \delta z \tag{17}$$

where a is the aperture radius of the detector lens. Using eq. 15, the optical parameter q becomes

$$q = \frac{2\pi a^2}{\lambda u} \tag{18}$$

Clearly, q, which is a measure of the axial resolution of the system, is independent of the focal length of the detector lens. Although depth discrimination is not affected by the focal length of the detector lens, it is strongly dependent on the focal length of the object lens. It has been shown numerically that the radius of the first zero of the axial response varies directly as the square of the object lens focal length.

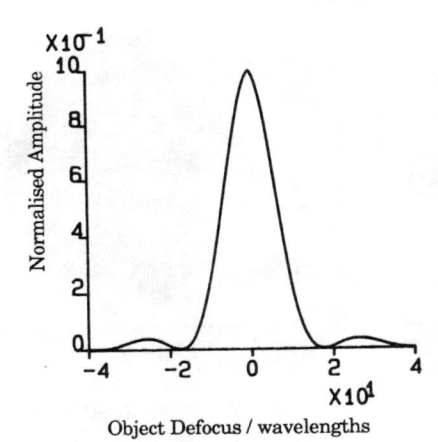

Figure 9. System axial response with identical lenses.

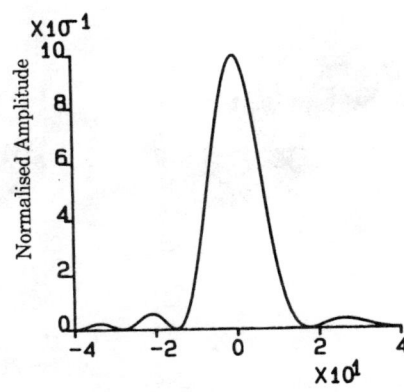

Figure 10. System axial response with detector lens aperture twice that of the object lens.

3.2 Axial response as a function of lens aperture

Figure 10 shows the system response when the aperture of the detector lens is twice that of the object lens. The graph is no longer symmetrical about the zero defocus position. At negative defocus (ie, object distance less than the focal length), the response has a steeper gradient. In addition, the periodicity of the sidelobes increases. This phenomenon can be explained by considering the optical parameter q discussed in section 3.1. It can be shown

$$q = \frac{2kdb^2}{f_o^4}(f_o^2 - 2ds)$$

where b is the aperture radius of the object lens and f_o is the object lens focal length. It is clear that the asymmetry is noticeable only if the lens separation s is sufficiently large compared to f_o.

It is clear that the asymmetry is noticeable only if the lens separation s is sufficiently large compared to the focal length f.

3.3 Lens separation

We have also investigated the effect of changing the lens separation s on the system response. Reducing the value of s leads to a small but significant improvement in depth discrimination. It is interesting to consider the case when the s is reduced to zero. This situation is equivalent to the one lens configuration discussed in section 2. The result obtained here shows no visible difference to that predicted by eq. 7. We have therefore demonstrated that our methods and results are entirely consistent with each other.

4. EXTENSION TO HETERODYNE CONFIGURATION

By adding a reference arm to the two lens configuration shown in fig. 8, the hybrid approach was used to investigate the performance of a heterodyne interference system. We have considered the effect on the system output when there is a second, partially transparent object, situated in close proximity to the in focus sample. The results obtained for three detector sizes are shown in figures 11 and 12. Figure 11a is the amplitude of the heterodyne signal as a function of the position of the second object. In this case, a point detector is used. The position of the first null in the envelope is an indication the system's ability to reject the signal from the out of focus feature. As the detector size increases, the depth discrimination of the system deteriorates as shown in fig. 11b. Figure 11c is obtained with a large area detector (six times the focal spot diameter). It is clear that the system exhibits confocal behaviour. Figure 12 shows the corresponding results for the phase of the heterodyne signal. From these two sets of results, we can conclude that, in an interference system, confocality can be achieved by either using a point detector or by collecting the entire interfering wavefronts.

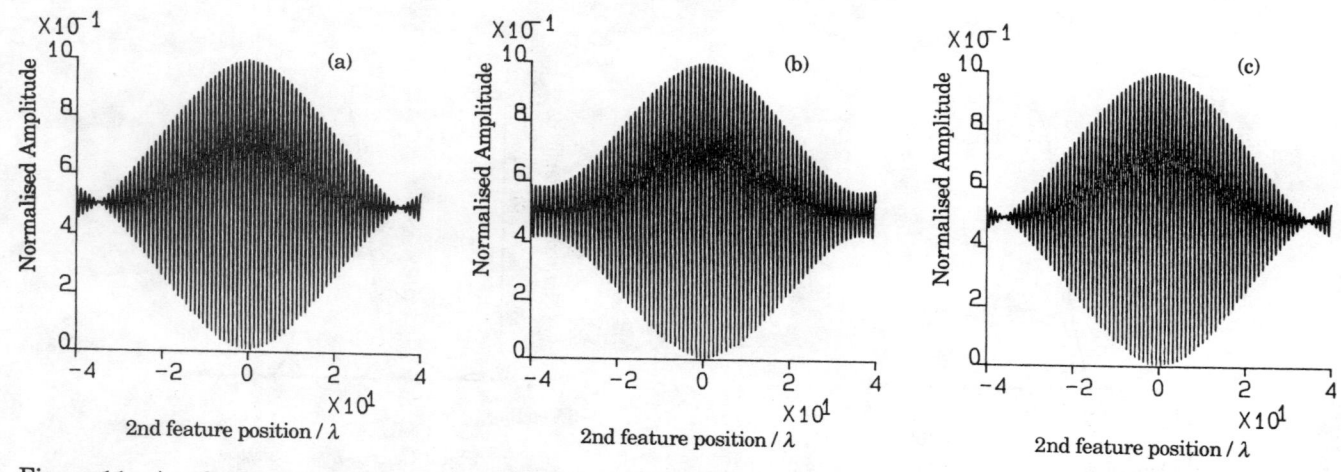

Figure 11. Amplitude of the heterodyne signal as a function of the position of the out of focus feature measured in wavelengths. Detector size: (a) = point detector, (b) = 1 focal spot diameter, (c) = 6 focal spot diameters.

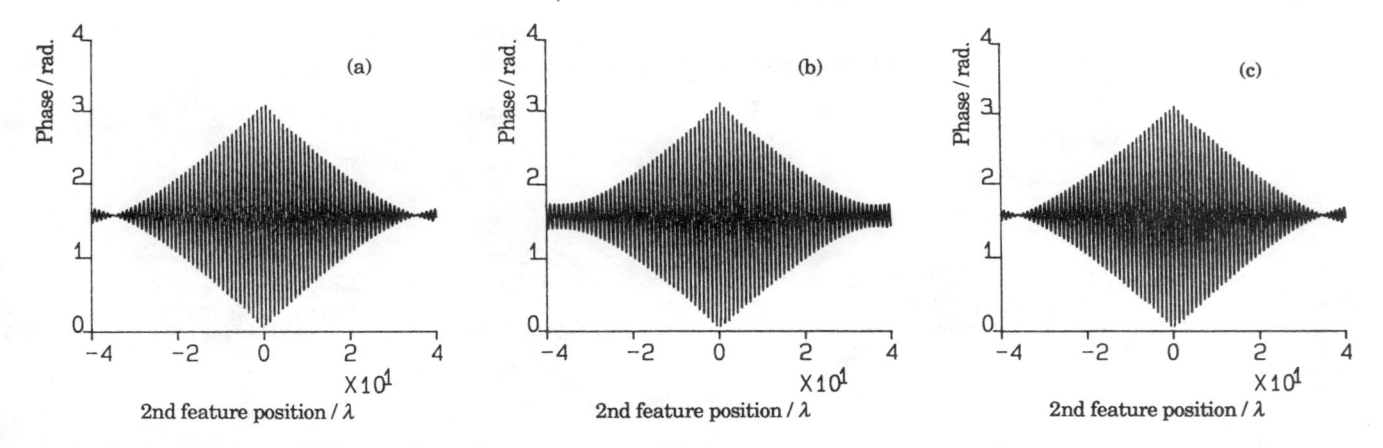

Figure 12. Phase of the heterodyne signal as a function of the position of the out of focus feature measured in wavelengths Detector size: (a) = point detector, (b) = 1 focal spot diameter, (c) = 6 focal spot diameters.

5. CONCLUSIONS

We have demonstrated that geometrical optics provides a very satisfactory means of modelling the response of a confocal scanning optical microscope. Diffraction theory need only be considered in the regions where the sample or the detector interacts with a tightly focused light beam. Elsewhere, geometrical optics provides an adequate solution.

We have used both Fresnel diffraction theory and the hybrid approach to analyse the performance of a simple microscope configuration. We have shown that the system has a superior axial resolution if the object feature is large compared with the lateral resolution of the microscope. There is a factor of two difference in the radius of the first zero of the response between a large area feature and one whose size is comparable or less than a focal spot size. We have also used the hybrid approach to investigate the performance of a more complicated microscope configuration. We have demonstrated that the depth discrimination of the system is independent of the focal length of the detector lens. Furthermore, if the separation of the two lenses is large compared with the focal length of the object lens, the system response is asymmetrical when the lens apertures do not match. The hybrid model has been applied to an interference microscope. It is concluded that confocal behaviour is achieved by using either a point detector or a large area detector. The system performance deteriorates if the detector size is not large enough to collect all the light.

6. ACKNOWLEDGEMENTS

The authors would like to thank the Science and Engineering Research Council (SERC) and Unilever Research for a CASE studentship for M.J.O., and SERC - Alvey for financial assistance for C.W.S.

7. REFERENCES

1. T. Wilson and C. J. R. Sheppard, Theory and Practice of Scanning Optical Microscopy, Academic Press, London (1984).
2. T. Wilson and A. R. Carlini, "Size of the Detector in Confocal Imaging Systems", Opt. Letts., 12 (4), 227-229, (1987).
3. J. W. Goodman, Introduction to Fourier Optics, McGraw Hill, New York (1968).
4. M. Born and E. Wolf, Principles of Optics, fifth Edition, Pergamon Press, Oxford (1975).
5. A detailed proof will be presented in a future publication.

Spherical Aberration in Confocal Microscopy

T. Hellmuth, P. Seidel, A. Siegel

Carl Zeiss, 7082 Oberkochen, Federal Republic of Germany

ABSTRACT

Scanning confocal microscopy has become an important tool in the field of optical sectioning and 3D-reconstruction of biological objects. Usually these samples are embedded in dispersive media, introducing depth dependent spherical aberration. Depth and lateral resolution are therefore impaired. We discuss this problem theoretically and show experimental results.

1. INTRODUCTION

Optical sectioning has become an important field of confocal microscopy. With this method it is possible to generate images of extended objects with a high depth of focus. By using image processing equipment three-dimensional pictures of cells and other biological objects can be generated. Other applications are line-width measurements of microlithographic structures and stereo imaging. As in conventional microscopy the picture quality depends on the correct application of the microscope objective. In confocal microscopy not only the lateral resolution but also the depth resolution of the objective is important, because it determines the extension of the picture cell in the z-direction. Here we discuss especially the influence of spherical aberrations.

2. SPHERICAL ABERRATION IN MICROSCOPY

Microscope objectives are designed with a minimum of abberations in the image of the object plane and in a certain range around this plane. In the case of defocussing beyond this range or in the case of incorrect use aberrations like spherical aberration become noticable:

2.1 <u>Spherical abberation and defocussing</u>. In conventional microscopy spherical aberration for the out-of-focus case is usually masked by the defocussing error. So it does not play a significant role in conventional microscopy. As we will see later, this effect has to be considered in the context of confocal microscopy.

2.2 <u>Spherical aberration and cover glasses</u>. It is well known from conventional microscopy that spherical aberration is caused by a glass plate between objective and object, i.e. when a cover plate is used for example. Figure 1 shows the geometry of such an arrangement.

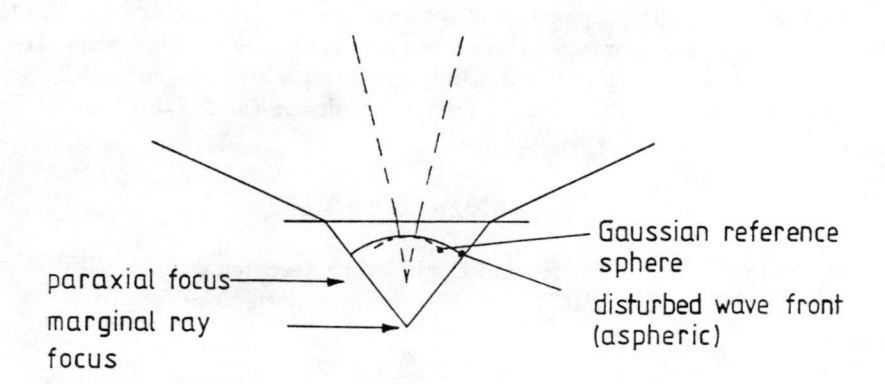

Figure 1. Spherical aberration caused by a glass plate.

The paraxial rays are focussed to the Gaussian focus by the objective. This ray pencil can be characterized by the Gaussian reference sphere, which is defined as their normal surface. With the glass plate the marginal rays are focussed to a different point on the optical axis than the paraxial rays. The focus is smeared out in the z- as well as in the x-y-direction. The disturbed ray pencil can also be characterized by the wave front which is normal to the rays of the disturbed pencil. It can be interpreted as a disturbed Gaussian reference sphere, where marginal zones are delayed relative to the paraxial zones.

Objectives, which are designed to be used together with cover plates, are already undercorrected, that is, the marginal zones of the wavefront leaving the objective are advanced relatively to the paraxial zones, so that the delay of the glass plate is just compensated by this undercorrection.

However, the standard thickness of cover plates (170 µm) varies by about ± 10 µm, which introduces spherical aberration, that can be considerable with high aperture objectives. One can get rid of that problem with immersion objectives, if the refractive index of the immersion liquid (oil) is equal to the refractive index of glass. If one uses however microscope objectives which are designed to be used with various immersion liquids, as for example water or glycerine, the spherical aberration depends on the coverglass thickness and on the refractive index of the immersion medium.

In this case it is recommended to use correction objectives. The design principle is shown in Fig. 2. The objective consists of two subsystems, one with low (I) and one with a high refractivity (II). By varying their distance the marginal rays pass the system II either in an outer zone (a) or in an inner zone (b), so that the contribution of system II to the undercorrection of the whole system can be varied.

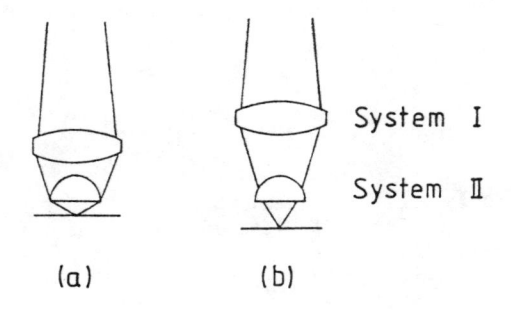

Figure 2. Principle of a correction objective.

3. DEPTH RESOLUTION OF THE CONFOCAL MICROSCOPE WITHOUT ABERRATION

There are several ways to define the depth resolution of a confocal microscope.[1] The common problem of all methods is, that depth resolution depends on the object. Figure 3 shows a simple setup to measure the depth resolution with a mirror object.

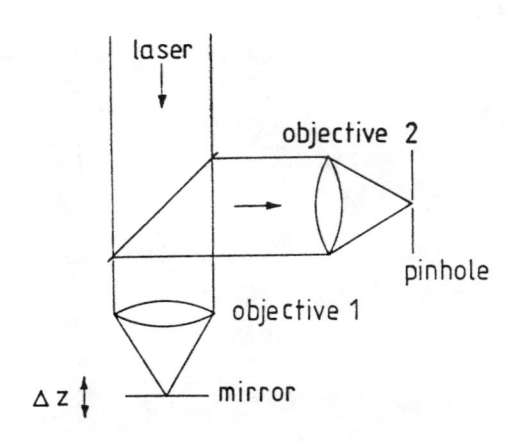

Figure 3. Setup for measuring the depth resolution.

Objective 2 is focussed onto the pinhole. Objective 1 can be defocussed by Δz. (The following discussion is valid for beam-scanning-type as well as for stage-scanning-type microscopes). A laser beam is focussed onto a mirror. A beamsplitter reflects the rays leaving the objective in the backward direction to a further objective, which focusses the beam onto a pinhole. Behind that a detector registers the transmitted light. The near focus intensity distribution of the laser beam is transformed via the mirror, the objectives O_1 and O_2 to the pinhole, which together with the photomultiplier can be regarded as a pointlike detector. Therefore by moving the mirror in the z-direction the near focus intensity distribution along the optical axis of objective O_2 is measured with the pinhole detector as a function of the mirror displacement Δz. One expects a function of the type[2]:

$$I(\Delta z) = \left[\frac{\sin \frac{\pi}{2\lambda}(NA)^2 \cdot \Delta z}{\frac{\pi}{2\lambda}(NA)^2 \cdot \Delta z} \right]^2 \tag{1}$$

The half width (FWHM) of this function is given by[3]

$$\Delta z_{1/2} = 0.89 \cdot n \cdot \frac{\lambda}{(NA)^2} \tag{2}$$

where NA is the numerical aperture of O_1, λ is the wavelength and n is the index of refraction of the immersion medium of O_1. In a similar way one derives an expression for pointlike objects[3].

$$\Delta z_{1/2} = 1.27 \cdot n \frac{\lambda}{(NA)^2} \tag{3}$$

Relations (1) - (3) however are only valid for ideal objectives and pinholes with infinitesimal small size. The influence of a finite pinhole size on the half width of a measured intensity distribution has been described elsewhere.[4] Another influence is, that real objectives introduce spherical aberration when the object is defocussed. This is done when we move the mirror object in the z-direction. We can not assume, that the near focus intensity distribution of objective 1 is transformed into the focal space of objective 2 without aberrations. Figure 4 shows the results of a measurement for the ZEISS Epiplan 40/0.85. The assymmetry of the curve is due to the spherical aberration of the defocussed objective.

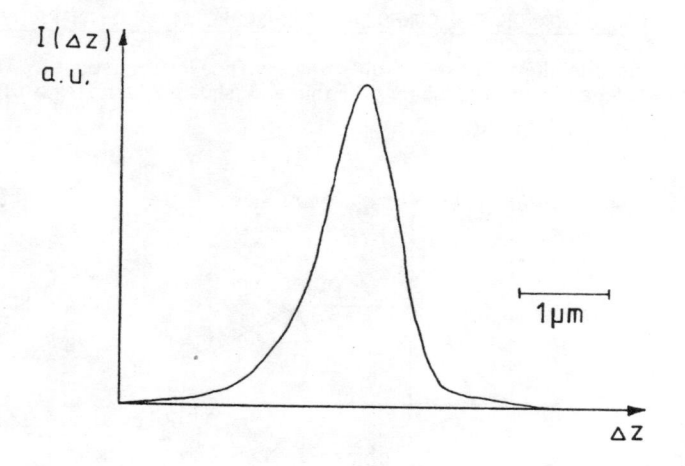

Figure 4.

The intensity distribution measured with the pinhole detector as a function of the defocussing parameter Δz. Increasing Δz means that the mirror is moved towards the objective (objective ZEISS Epiplan 40/0.85).

4. DEPTH RESOLUTION IN CONFOCAL MICROSCOPY WITH SPHERICAL ABERRATION

Figure 5 shows the geometry of a confocal setup when a glass plate is put between objective and the mirror object. It is evident, that the glass plate causes a deformation of the wave on its way to as well as on the way back from the object. The wave leaving the objective in the backward direction is no longer a plane one. As a consequence one gets a disturbed intensity distribution in the focal region of objective 2.

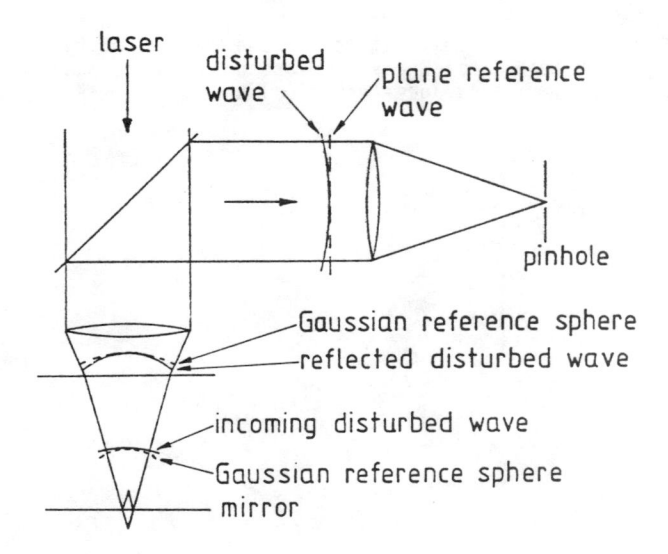

Figure 5. Geometry of a confocal system with spherical aberration caused by a glass plate.

As in section 3 we are interested in the intensity distribution along the optical axis in the focal region of objective 2. Particularly we want to know how the half width of this distribution varies as a function of the plate thickness.

5. MEASUREMENTS

The experimental setup is schematically the same as shown in Fig. 5. The beam of a He-Ne-Laser is expanded by a 7x telescope, so that the objective pupil is filled. The beam is focussed onto a mirror, which can be moved in the z-direction by a piezoelectric translator. The reflected light is focussed onto a pinhole in front of a photomultiplier. Because the microscope objectives, which are used in the measurement, are corrected for a tubelength of 160 mm, correction lenses are necessary, to guarantee a parallel beam geometry when the objective is focussed.

In order to measure the influence of spherical aberration on the depth resolution the mirror is moved along the z-axis by Δz and the power transmitted through the pinhole is registered with the photomultiplier as a function of Δz. In this setup the pinhole diameter is 20 µm. The focussing objective is a ZEISS Planapo 2.5/0.08.

We use objectives with a correction mounting, which usually compensate the spherical aberration of a refracting plate between object and objective. Instead of varying the glass plate thickness it is easier to vary the correction of the objective by turning a set screw with a scale ranging for glass plate thicknesses between 120 and 220 µm. Figures 6 a,b,c show the results for the water immersion objective Planapo 63/1.2 W.

It can be seen that due to spherical aberration the width of the intensity distribution increases significantly. Calculations show that the assymmetry of the distribution is due to spherical aberration of higher than third order.

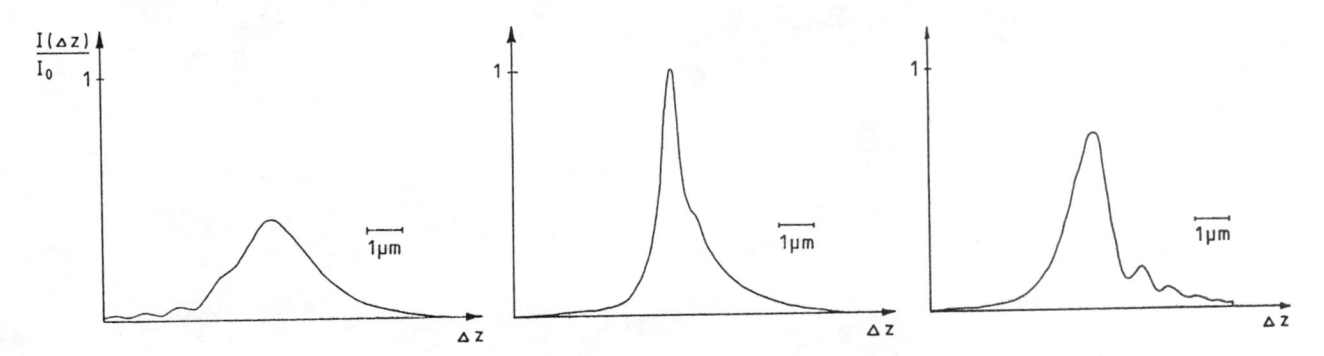

Figure 6.

The intensity distribution is measured with the pinholedetector for three corrections, (a) 120 µm, (b) 170 µm, (c) 220 µm. Increasing Δz means that the mirror is moved towards the objective. The intensity is normalized to the maximum intensity of (b). Figure 7 shows the halfwidth plotted against the effective glass plate thickness. The tolerance of the thickness of glass plates of the highest quality standard is ± 10 µm. From Fig. 7 we see, that with these tolerances spherical aberration might deteriorate depth resolution by a factor greater than two.

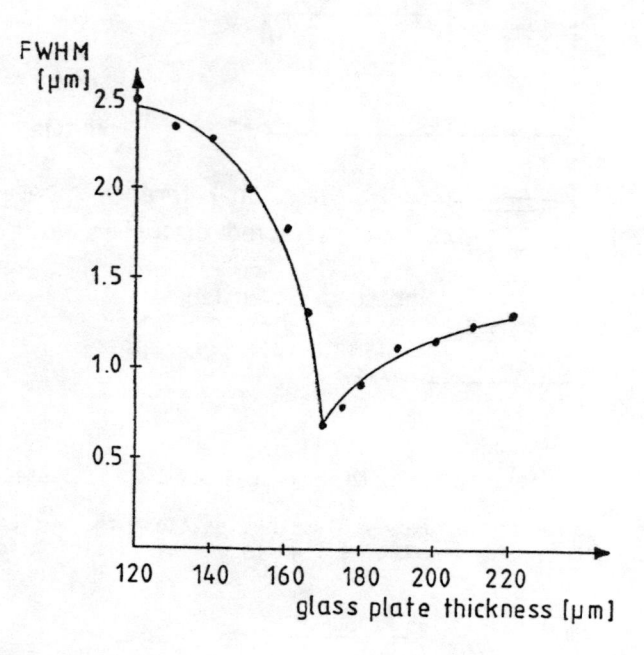

Figure 7. The halfwidth (FWHM) as a function of the correction.

6. CONCLUSION

The results show, that in certain cases spherical aberration deteriorates not only the lateral but also the depth resolution of confocal microscopes. In those cases it is useful to choose either oil immersion objectives or when this is not possible, to use objectives with a correction mounting.

7. REFERENCES

1. T. Wilson and A. R. Carlini, "Depth Discrimination Criteria in Confocal Optical Systems", Optik 76(4), 164-166 (1987).
2. T. Wilson and C. J. R. Sheppard, Theory and Practice of Scanning Optical Microscopy, Academic Press (London) (1984).
3. T. Hellmuth, A. Siegel, P. Seidel, "Generation of 3-D Images via Laser Scanning Microscopy", Int. Workshop on 3-D-Image Reconstruction 3/88 (Giessen).
4. T. Wilson and A. R. Carlini, "Size of the Detector in Confocal Imaging Systems", Opt. Lett. 12(4), 227-229 (1987).

Inverse problems in fluorescence confocal scanning microscopy

M. Bertero[+], P. Boccacci[+], E. R. Pike[o]

[+] Dipartimento di Fisica, Universita' di Genova, and Istituto Nazionale di Fisica Nucleare, 16146 Genova, Italy
[o] Dept. of Physics, King's College, London WC2R 2LS, and RSRE, Great Malvern WR14 3PS, England

ABSTRACT

It has been recently demonstrated that one of the most important applications of Confocal Scanning Laser Microscopy (CSLM) is fluorescence microscopy. In this paper we discuss the application to this case of a general method suggested by two of the authors (M.B. and E.R.P.) for improving resolution. The method requires that the full image is recorded by means of a suitable array of detectors and that a linear integral equation is solved for determining the object at each step of the scanning procedure. We investigate several integral equations related to 1D, 2D and 3D imaging.

1. INTRODUCTION

It has been recently demonstrated that a very high resolution can be obtained by means of Confocal Scanning Laser Microscopy (CSLM) in the case of fluorescent objects. In fact, if one assumes complete incoherence of the fluorescence light, it is possible to prove that a confocal incoherent system has a bandwidth four times that of a conventional coherent microscope[1,2]. This result, however, does not imply an improvement of a factor 4 in resolution, because the transfer function is very small in a large region near the boundary of the band.

Another interesting feature of CSLM is the possibility of obtaining a microtomoscopy of three-dimensional fluorescent objects[3,4]. In such a case the main problem is the fact that the axial resolution distance is approximately four times the lateral one.

In recent years it has been suggested by two of the authors[5] that a broader transfer function, and therefore a better resolution, can be obtained if, at each step of the scanning procedure, the full image of the confocal system is recorded and suitable mathematical techniques are used for recovering the object. In the case of coherent objects it has been demonstrated that, by means of this method, one can effectively obtain an improvement of a factor 2 with respect to the resolution of the conventional microscope[6,7]. Practical applications, however, are difficult in that case because of the lack of phase information.

Let us denote by $S_1(\underline{\rho})$ and $S_2(\underline{\rho})$ ($\underline{\rho} = \{x,y\}$) the point spread function respectively of the illuminating and imaging lens. In the case of the reflection mode of operation[4], the two point-spread functions are different because the primary and the fluorescent wavelength are different. For simplicity we neglect this effect and we assume that $S_1(\underline{\rho}) = S_2(\underline{\rho}) = S(\underline{\rho})$. Then, if $f(\underline{\rho})$ is the effective density of the fluorescent centres in the focal plane, the intensity distribution $g(\underline{\rho})$ in the image plane, for a fixed scanning position, is given by

$$g(\underline{\rho}) = \int |S(\underline{\rho} - \underline{\rho}')|^2 |S(\underline{\rho}')|^2 f(\underline{\rho}') d\underline{\rho}'. \qquad (1)$$

We need only to recover $f(\underline{0})$ from given values of $g(\underline{\rho})$. Then the complete function $f(\underline{\rho})$ can be obtained by solving this problem for each scanning position.

In the case of a thick, weakly fluorescent object, eq.(1) can be replaced by the following one

$$g(\underline{\rho}) = \int |S(\underline{\rho} - \underline{\rho}', z')|^2 |S(\underline{\rho}', z')|^2 f(\underline{\rho}', z') d\underline{\rho}' dz' \qquad (2)$$

where z' is the axial cohordinate and $S(\underline{\rho}, z)$ is the amplitude of the field distribution near the focus. Notice that we have a three-dimensional object and a two-dimensional image. Again the problem is the recovery of $f(\underline{0},0)$ and $f(\underline{\rho},z)$ can be obtained by three-dimensional scanning.

2. THE ONE-DIMENSIONAL CASE

We consider firstly the one-dimensional problem, assuming that the lenses are uniformly filled and have the same aperture. Then eq.(1) takes the following form

$$g = Af \tag{3}$$

where A is the integral operator

$$(Af)(x) = \int_{-\infty}^{+\infty} \text{sinc}^2(x-x') \, \text{sinc}^2(x') \, f(x') \, dx' \tag{4}$$

and $\text{sinc}(x) = \sin(\pi x)/(\pi x)$.

It is easy to prove that all the noise-free images, i.e. the images g given by eq.(3), are band-limited with bandwidth 2π. As a consequence they can be represented by means of the sampling expansion

$$g(x) = \sum_{n=-\infty}^{+\infty} g(\tfrac{1}{2}n) \, \text{sinc}[2(x - \tfrac{1}{2}n)] \, . \tag{5}$$

Moreover, from general results of the theory of first kind integral equations, it follows[6,7] that one can only recover the projection of f(x) (which will also be denoted by f(x) in the following) on the closure of the range of the adjoint operator A*. The latter is given by

$$(A^*g)(x') = \text{sinc}^2(x') \int_{-\infty}^{+\infty} \text{sinc}^2(x' - x) \, g(x) \, dx \tag{6}$$

and therefore the recoverable component is band-limited, with bandwidth 4π, and is zero at the integers samplint points. As a consequence it can be represented by means of the following sampling expansion

$$f(x') = \sum_{m=-\infty}^{+\infty} f(x'_m) \, \text{sinc}[4(x' - x'_m)] \tag{7}$$

where $x'_0 = 0$ and $x'_m = \pm(k+j/4)$ if $m = \pm(3k+j)$ ($k = 0,1,2,...$; $j = 1,2,3$).

One can prove that the operator (4) is compact. Thus one can introduce its singular system which is the set of the solutions of the coupled equations

$$Au_k = \alpha_k v_k \, , \quad A^* v_k = \alpha_k u_k \, . \tag{8}$$

In terms of the singular system of the operator A the recovered value of f(x') at the origin is given by

$$f(0) = \sum_{k=0}^{+\infty} \frac{1}{\alpha_k} (g, v_k) \, u_k(0) \tag{9}$$

where (g, v_k) is the usual scalar product of $L^2(-\infty, +\infty)$. Since α_k tends to zero when k tends to infinity, one must truncate the series in eq.(9) in order to limit the amplification of the noise affecting the image g(x). This is the method of <u>truncated singular function expansion</u> and the truncation is essentially determined by the signal-to-noise ratio[6].

In the coherent case it has been proved[6] that very good approximations of the singular values and singular functions of the operator A can be obtained using the sampling expansions of the image and of the object for discretizing the integral equation. This suggests that the same result might also be true in the incoherent case. If we introduce the quantities

$$b_n = \frac{1}{\sqrt{2}} \, g(\tfrac{1}{2}n) \, , \quad a_m = \tfrac{1}{2} f(x'_m) \tag{10}$$

(in this way the L^2-norm of $g(x)$ is just the sum of the b_n^2 while the L^2-norm of $f(x')$ is the sum of the a_m^2) and if we substitute the sampling expansion (5) and (6) into the integral equation (3), this equation is transformed into the infinite-dimensional linear system

$$b_n = \sum_{m=-\infty}^{+\infty} A_{nm} a_m \quad ; \quad n = 0, \pm 1, \pm 2, \ldots \qquad (11)$$

where

$$A_{nm} = \frac{\sqrt{2}}{4} \operatorname{sinc}^2(x_m' - \frac{1}{2} n) \operatorname{sinc}^2(x_m') \qquad (12)$$

This expression of the matrix elements A_{nm} can be easily obtained using the projection properties of the function $\operatorname{sinc}(4x)$.

In order to compute the singular values one can apply standard routines to a finite section of the system (11). For example, using 25 sampling points for the image and 49 sampling points for the object we obtain the following values (we also give, for comparison, the singular values of the coherent case[6], obtained using the same number of sampling points):

k	COHERENT	INCOHERENT
0	0.82190	0.60336
1	0.44286	0.25675
2	0.20642	0.10761
3	0.14761	0.40294×10^{-1}
4	0.10984	0.23774×10^{-1}
5	0.88560×10^{-1}	0.19872×10^{-1}
6	0.74199×10^{-1}	0.10131×10^{-1}
7	0.63249×10^{-1}	0.80676×10^{-2}
8	0.55914×10^{-1}	0.76978×10^{-2}
9	0.49184×10^{-1}	0.43231×10^{-2}

We notice that the singular values of the incoherent case tend to zero more rapidly than the singular values of the coherent one and this means that, in the incoherent case we must use a smaller number of terms in eq.(9). For example, if we use 5 singular values, the condition number α_0/α_4 is 7.5 in the coherent case and 25.4 in the incoherent one.

The previous result implies that it can be rather difficult to fill the available band $[-4\pi, 4\pi]$. A more precise estimate can be obtained by computing the impulse response function describing the total effect of imaging, scanning and subsequent recovery procedure. If we use K singular functions in the recovery procedure, this impulse response function is given by[6,7]

$$A(x) = \sum_{k=0}^{K-1} u_k(x) u_k(0) . \qquad (13)$$

Since the singular functions are alternatively even and odd, only singular functions with an even value of the index contribute to this sum. For example, if we use the singular functions with k=0,2,4, the transfer function, i.e. the Fourier transform of the impulse response function (13), is significantly different from zero on the interval $[3.7\pi, 3.7\pi]$ and this means an improvement of a factor 3.7 with respect to the resolution of the conventional coherent microscope.

3. THE TWO-DIMENSIONAL CASE

The solution of the one-dimensional problem discussed in the previous Section provides also the solution of the two-dimensional problem in the case of square pupils. In fact in such a case we have

$$S(\underline{\rho}) = \text{sinc}(x)\,\text{sinc}(y) \tag{14}$$

and the singular system of the integral operator obtained by replacing (14) in (1) can be obtained from the singular system of the integral operator (4) as follows: the singular values are all the possible products of the singular values of (4)

$$\alpha_{i,k} = \alpha_i \alpha_k \; ; \; i,k = 0,1,2... \tag{15}$$

and the corresponding singular functions are given by the tensor products of the singular functions of (4)

$$u_{i,k}(\underline{\rho}) = u_i(x)\,u_k(y) \quad , v_{i,k}(\underline{\rho}) = v_i(x)\,v_k(y) \tag{16}$$

It is obvious that the singular values with $i \neq k$ have multiplicity 2 while the singular values with $i=k$ have multiplicity 1.

Only the singular functions corresponding to even values of both indices contribute to the reconstruction of the object at the origin. The two-dimensional problem is however more ill-conditioned than the one-dimensional problem. As we have seen, if we use the singular functions with $k=0,2,4$, the condition number is 25.4 in the 1D case. In the 2D case, in order to obtain the same resolution, we must use nine singular functions corresponding to all the possible pairs of the values 0,2,4. But the condition number is now $(25.4)^2 \cong 645$ and this value can be exceedingly high.

The case of circular pupils is more difficult from the computational point of view even if it exhibits some interesting theoretical properties. We assume again that the lenses are uniformly filled and that they have the same aperture. Then in eq.(1) we have

$$S(\rho) = 2\,\frac{J_1(\pi\rho)}{\pi\rho}\;, \quad \rho = |\underline{\rho}| \tag{17}$$

(the normalization has been chosen in such a way that $0 \leq S(\rho) \leq 1$).

The two-dimensional integral equation (1) can be reduced to an infinite se of uncoupled one-dimensional equations by means of the Fourier series expansions of the image, of the object and of the kernel[5]. If we remember, however, that our problem is the restoration of the object at the origin, we recognize that this problem has circular symmetry and therefore only the circularly symmetric singular functions contribute to its solution. The same conclusion is reached by noticing that all the singular functions which depend also on the angle have nodal lines passing through the focus.

Thus, if we put

$$g_0(\rho) = \frac{1}{2\pi}\int_0^{2\pi} g(\rho,\phi)\,d\phi \;, \quad f_0(\rho') = \frac{1}{2\pi}\int_0^{2\pi} f(\rho',\phi')\,d\phi' \tag{18}$$

(notice that $f_0(0) = f(\underline{0})$) and also

$$H_0(\rho,\rho') = \int_0^{2\pi} S^2\left(\sqrt{\rho^2 + \rho'^2 - 2\rho\rho'\cos\vartheta}\right) d\vartheta \;, \tag{19}$$

where $\vartheta = \phi - \phi'$, from eq.(1) we obtain

$$g_0(\rho) = \int_0^{+\infty} \rho'\,H_0(\rho,\rho')\,S^2(\rho')\,f_0(\rho')\,d\rho' \;. \tag{20}$$

Recalling the relationship between the Fourier transform and the Hankel transform[8], it is easy to recognize that the Hankel transform of $g_0(\rho)$ is non-zero only on the interval $[0,2\pi]$ while the Hankel transform of the recoverable component of $f_0(\rho')$ is non-zero only on the interval $[0,4\pi]$. Therefore both the image and the object can be represented by means of a sampling expansion of the following kind[8]

$$h(\rho) = \sum_{n=1}^{\infty} \frac{\sqrt{2}}{c J_1(x_n)} h\left(\frac{x_n}{c}\right) S_n(\rho) \qquad (21)$$

where the x_n are the zeros of $J_0(x)$ and

$$S_n(\rho) = \sqrt{2}\, c\, x_n \frac{J_0(c\rho)}{x_n^2 - (c\rho)^2} \qquad (22)$$

The quantity c is the bandwidth of $h(\rho)$ and therefore $c=2\pi$ in the case of $g_0(\rho)$ and $c=4\pi$ in the case of $f_0(\rho)$. The sampling functions $S_n(\rho)$ form an orthonormal system in the following sense

$$\int_0^{+\infty} \rho\, S_n(\rho) S_m(\rho)\, d\rho = \delta_{nm} \qquad (23)$$

and they have the following projection property

$$\int_0^{+\infty} h(\rho) S_n(\rho)\, d\rho = \frac{\sqrt{2}}{c J_1(x_n)} h\left(\frac{x_n}{c}\right) \qquad (24)$$

By replacing the sampling expansions for $g_0(\rho)$ and $f_0(\rho)$ in eq.(20) and using eqs.(23),(24), the integral equation (20) can be transformed into an infinite dimensional linear system. The latter can be truncated in order to get approximations for the singular values and singular functions, as in the case of the problem of Section 2.

Preliminary computations in the case of the coherent problem indicate that, using 4 singular values (the corresponding condition number is 19), it is possible to obtain a transfer function which is significantly different from zero over the full band (in this case the circle of radius 2π).

4. THE THREE-DIMENSIONAL CASE

As we noticed in the Introduction, in the case of a thick, weakly fluorescent object, the image of a confocal system, for a fixed scanning position, is given by eq.(2). If we consider again the case of uniformly filled, circular lenses, then the function $S(\underline{\rho}, z)$ is given by[9]

$$S(\rho, z) = 2 \int_0^1 t\, J_0(\pi \rho t)\, e^{-i\frac{\pi}{2} z t^2}\, dt, \quad \rho = |\underline{\rho}| \qquad (25)$$

Here the normalization has been chosen in such a way that $S(\rho, 0)$ coincides with the point spread function (17).

Since, for fixed z, the function (25) has circular symmetry, all the remarks of the previous Section apply also to this case and therefore one can reduce the dimension of the problem by integrating over the angular variable in the transverse planes.

Preliminary computations have been performed in the case of a two-dimensional model with one lateral and one axial variable. In such a case the appropriate integral operator is the following one

$$(Af)(x) = \int_{-\infty}^{+\infty} |S(x-x', z')|^2 |S(x', z')|^2 f(x', z')\, dx'\, dz' \qquad (26)$$

where

$$S(x,z) = \frac{1}{2} \int_{-1}^{1} e^{-i\pi xt - i\frac{\pi}{2}zt^2} dt \quad . \tag{27}$$

Also in this case the normalization has been chosen in such a way that $S(x,0) = \text{sinc}(x)$, the point spread function related to the integral operator (4). Preliminary computations show that the number of significant singular values of the operator (26,27) is larger than that of the integral operator (4). For instance, in the case of a condition number of the order of 25, we can use 11 singular values instead of 5. Also in this case, however, only the singular functions with even index contribute so that we have 6 terms (instead of 3) in the truncated singular function expansion.

5. ACKNOWLEDGMENTS

This work has been partly supported by NATO Grant No 463/84, by EEC Contract No. BAP-0293-NL (GDF) and by Ministero Pubblica Istruzione, Italy.

6. REFERENCES

1. C.W. McCutchen, "Superresolution in microscopy and the Abbe resolution limit", J.Opt.Soc.Am.56, 1190-1192 (1967)
2. I.J.Cox, C.J.R.Sheppard and T.Wilson, "Super-resolution by confocal fluorescent microscopy", Optik 60, 391-396 (1982)
3. R.W.Wijnaendts van Resandt, H.J.B.Marsman, R.Kaplan, J.Davoust, E.H.K.Stelzer and R.Stricker, "Optical fluorescence microscopy in three dimensions: microtomoscopy", J.Microx. 138, 29-34 (1985)
4. G.J.Brakenhoff, H.T.M. van der Voort, E.A. van Spronsen and N.Nanninga, "Three-dimensional imaging by confocal scanning fluorescence microscopy", Annals N.Y.Acad. of Sci.483, 405-415 (1986)
5. M.Bertero and E.R.Pike, "Resolution in diffraction-limited imaging, a singular value analysis", Opt.Acta 29, 727-746 (1982)
6. M.Bertero, P.Brianzi and E.R.Pike, "Super-resolution in confocal scanning microscopy", Inverse Problems 3, 195-212 (1987)
7. M.Bertero, P.Boccacci, P.Brianzi and E.R.Pike, "Inverse problems in confocal scanning microscopy", in <u>Inverse Problems: and Interdisciplinary Study</u>, P.C.Sabatier, ed., Adv.Electr. and Electron Phys., Supplement 19, 225-239, Academic Press, London (1987)
8. A.Papoulis, <u>Systems and Transforms with Applications to Optics</u>, pp.140-175, McGraw-Hill, New York (1968)
9. M.Born and E.Wolf, <u>Principles of Optics</u>, pp.434-448, Pergamon Press, London (1959)

Modelling of 3-D confocal imaging at high numerical aperture in fluorescence

H.T.M. van der Voort and G.J Brakenhoff.

Department of Molecular Cell Biology, University of Amsterdam.
Plantage Muidergracht 14, 1018 TV Amsterdam, The Netherlands.

Abstract

The imaging properties of a model high aperture fluorescence confocal microscope are studied by way of numerical methods. The model consists of an aberration-free high aperture objective illuminated by a monochromatic linear or circular polarized light source, and a detection system equipped with a finite sized pinhole. The computation of the intensity distribution near the confocal plane is based on electromagnetic diffraction theory.

We study the 3-D imaging properties of the model by examining the images of three different objects; points, lines and planes. We calculate the resolving power for these objects as a function of their orientation and the size of the detector pinhole. The results indicate the necessity for image analysis schemes to take these effects into account.

1. Introduction

Several important applications of Confocal microscopy require good knowledge of the three-dimensional imaging properties of the instrument. Examples of these are resolution improvement by regularized inversion (1) or singular value decomposition (2), and the segmentation and measurement phases in image processing.

When an image is to be segmented into a set of objects, it is necessary to take the orientation dependency of confocal imaging into account. A typical example of this problem is the segmentation of an image containing fluorescent labeled cellular or nuclear membranes into single objects. The orientation dependency causes in this case the brightness of the membrane in the image to be dependent on its local orientation. The segmentation algorithm must in this case compensate for this effect in order to localize the membrane correctly. Fortunately, a number of biological objects which are currently being studied in our lab fall into simple categories: small labeled sites such as centromeres on chromosomes (point-like objects), cytoskeletons (line-like objects) and membranes (plane-like objects).

We have developed a software package which is able to compute the instrumental properties on the basis of a model microscope and use these properties to determine the dependency of the imaging on the object type and its orientation. We have incorporated a number of the most important experimental aspects of a practical instrument into our model: high aperture lenses, a finite detector pinhole, the polarization state of the excitation source and incoherent, random polarized fluorescence light. As scalar paraxial diffraction theory is not well suited for high aperture applications, we have used electromagnetic diffraction theory to calculate the intensity distribution near the confocal plane. At the same time, this allows us to study polarization effects.

In fluorescence CSLM, finite sized detector pinholes are generally used to achieve a sufficient signal-to-noise ratio in the resulting image. The size of the pinhole should be adjustable in order to obtain optimal images from specimen with different bleaching and quantum efficiency properties. The performance of the microscope is dependent on the size of the pinhole, and has therefore recently attracted considerable interest (3, 4).

As the lateral resolution in CSLM is three to four times better than the axial resolution (5), confocal imaging is orientation dependent. We will study this dependency by calculating the resolving power for point, line and plane objects as a function of their orientation. In addition we will discuss the trade-off problem between pinhole size and resolution by computing the maximum signal strength of these objects as a function of the pinhole size. Our calculations of the various images will be based on the assumption that the imaging process is linear. This is true in the case of a weakly absorbing object which emits fully incoherent, random polarized fluorescence light.

2. Theoretical Considerations

As stated before (6,7), the optical sectioning capability of a confocal microscope is strongly dependent on the aperture of the objectives used. In paraxial approximation theory, full width half maximum (*FWHM*) of the point spread function (*PSF*) is inversely proportional to the square of the aperture angle of the objective. This relationship is reflected in the definition of the so-called optical coordinates (see below, equation (3)). Our model confocal system (Fig. 1) especially takes high aperture objectives into account as almost all practical instruments are equipped with high aperture objectives. At these

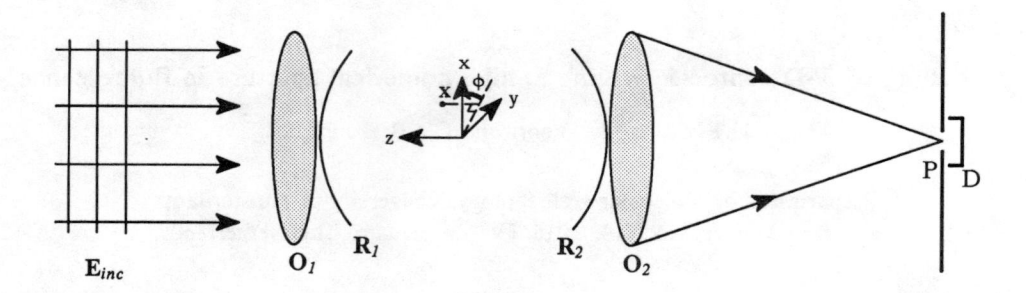

Fig. 1 The model confocal system. A circular or plane polarized plane wave E_{inc} originating from a laser is focused by an aberration free high aperture objective O1 onto the confocal plane. Incoherent random polarized fluorescent light is focused by a second objective O2 on the finite detector pinhole P. R1, R2: reference spheres, D: detector (photomultiplier).

high numerical apertures, there is a considerable deviation in the shape and size of the point spread function (PSF) as predicted by paraxial theory and electromagnetic diffraction theory (8). To investigate confocal imaging, we therefore based our calculations on the full vector wave theory of the electromagnetic field behind a refracting surface as developed and described by Richards and Wolf and others (8, 9, 11, 12, 13).

To compute the overall PSF of our model we must first determine the time-averaged electrical energy distribution near focus of the excitation and detection systems. For an aberration-free lens illuminated by a plane-polarized wave, the spatial component of the electrical field near focus $\mathbf{E}_\parallel(\mathbf{x})$ can be expressed in terms of a surface integral over the field $\mathbf{E}_s$ at the reference sphere S (8,9):

$$\mathbf{E}_\parallel(\mathbf{x}) = z_0 \int_S \mathbf{E}_s \, e^{-ik\,\mathbf{n}\cdot\mathbf{x}} \, d\sigma , \qquad (1)$$

with z_0 a complex constant. The field at the lens reference sphere can be obtained by geometric reasoning (9). The time-averaged energy distribution is (10):

$$\langle W_\parallel(u, v, \phi)\rangle = \frac{\epsilon}{16\pi} (\mathbf{E}\cdot\mathbf{E}^*) \qquad (2)$$

For reasons of convention we have written this distribution as a function of optical coordinates (10):

$$u = nk\sin^2\alpha \, z, \quad v = nk\sin\alpha \, r \quad \text{with} \quad r = \sqrt{x^2+y^2} \qquad (3)$$

As distribution (2) is not axially symmetric, the overall PSF of the microscope will also show non axial symmetry. This will result in considerable complications when the resulting images are to be processed or analyzed. These problems can be avoided by the use of circular polarized excitation light (14). We will therefore continue our considerations assuming the excitation light to be circular polarized.

Equations (1) and (2) can be also used when lenses with small aberrations (high quality microscope objectives) are to be studied. For these lenses the surface S is slightly non-spherical and therefore the field $\mathbf{E}_s$ and eventually $\mathbf{E}(\mathbf{x})$ will be altered. For the present purpose of studying the imaging of point, line and plane objects we have equipped our model microscope with aberration-free objectives.

In the case of circular polarized light the electrical field near the confocal plane may be written as

$$\mathbf{E}_{circ} = 1/2(\mathbf{E}_\parallel + i\mathbf{E}_\perp) \qquad (4)$$

with $\mathbf{E}_\parallel$ the field as calculated above and

$$\mathbf{E}_\perp = (-e_{\parallel y}(u, v, \phi-\tfrac{\pi}{2}), e_{\parallel x}(u, v, \phi-\tfrac{\pi}{2}), e_{\parallel z}(\phi-\tfrac{\pi}{2})) \qquad (5)$$

The time-averaged electrical energy distribution is now

$$\langle W_{circ}(u, v)\rangle = \frac{\epsilon}{32\pi}(\mathbf{E}_\parallel\mathbf{E}_\parallel^* + \mathbf{E}_\perp\mathbf{E}_\perp^* + 2\Re(i\mathbf{E}_\perp\mathbf{E}_\parallel^*)) = \frac{\epsilon}{32\pi}(\mathbf{E}_\parallel\mathbf{E}_\parallel^* + \mathbf{E}_\perp\mathbf{E}_\perp^*) \qquad (6)$$

as $E_\perp E_\parallel^*$ can be shown to be real in the case of an aberration-free objective (to be published). This distribution represents the effective excitation field near the confocal plane. It is axially symmetric around the u-axis, and mirror symmetric with respect to the the u=0 plane.

To calculate the detector efficiency distribution $D(u,v)$ we consider the emitted fluorescence light to be incoherent and random polarized. To determine the detector distribution equation (2) must be integrated first over all polarization directions:

$$\langle W_{rand}(u,v)\rangle = \frac{1}{2\pi}\int_0^{2\pi}\langle W_\parallel(u,v,\phi)\rangle d\phi \tag{7}$$

The main part of the emitted fluorescent light is generally contained in a relatively small band. Therefore the error in the overall PSF introduced by replacement of this band by an effective wave number k_{em} is only small. In the worst case of an infinitely small detector pinhole we have calculated this error to be less than 1%. For these reasons we may now write the distribution of a detector system with an infinitely small pinhole as

$$\langle W_{em}(u,v;k_r)\rangle = \langle W_{rand}(\frac{u}{k_r},\frac{v}{k_r})\rangle \text{ with } k_r = \frac{k_{exc}}{k_{em}} \text{ and } k_r > 1 \tag{8}$$

with k_{exc} the wave number of the excitation light. For a finite pinhole we convolve (8) with the pinhole function P:

$$D(u,v) = W_{em}(u,v;k_r) \otimes P \tag{9}$$

where the $\otimes$ symbol denotes the convolution operation and P is the geometric image of the pinhole in the confocal plane. The overall PSF of the model microscope is the product of the excitation distribution and the detector efficiency distribution:

$$P_{tot}(u,v) = \langle W_{circ}(u,v)\rangle . D(u,v) \tag{10}$$

The images of the line and point objects were obtained by direct convolution of the object function with (10). All programs were written in the Pascal programming language and run on an Apollo DN 4000 workstation.

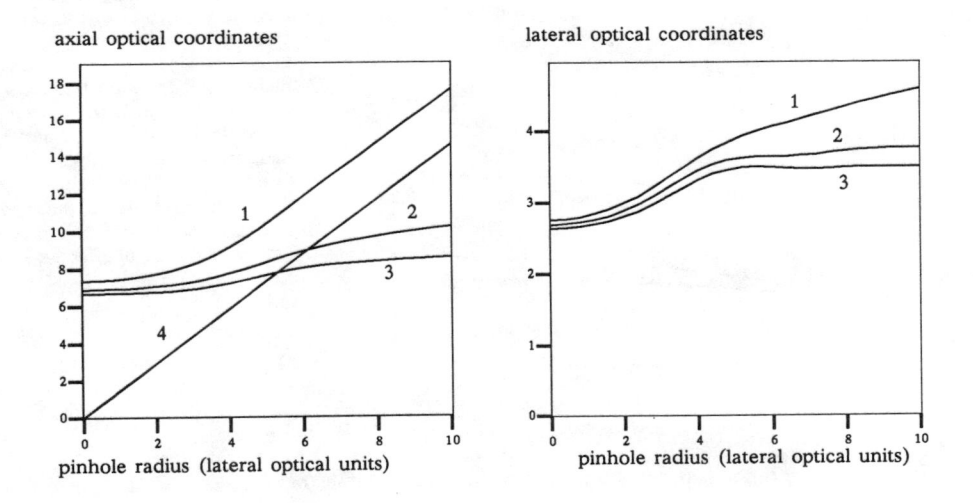

Fig. 2 Resolution as a function of the pinhole radius in the case of plane (curve 1), line (curve 2) and line (curve 3) objects. Fig. 2a: axial resolution (horizontally oriented objects). Curve 4: axial resolution for horizontal planes as calculated from geometrical optics. Fig. 2b: lateral resolution (vertical oriented objects).

3. Results and Discussion

We have calculated the resolving power for the three different objects by measuring the *FWHM* of the image. In the case of line and plane objects the *FWHM* was measured in a direction perpendicular to the object. Fig. 2 shows the dependency of the resolving power in both axial and lateral directions as a function of the pinhole. In the case of line and point objects it is clear from this figure that the resolving power does not suffer greatly from enlargement of the pinhole. The resolving power for horizontal planes can be associated with the power to suppress contributions from neighboring planes to the signal from the confocal plane (the sectioning capability). Fig. 3 shows the dependency of an objects maximum signal strength on the pinhole radius. Only the curve representing the signal from horizontal planes shows a continuing increase in signal strength above 6 optical units. In the other cases, opening of the pinhole above 6 optical units does not contribute to significantly to the signal. This means that in cases were the objects of interest are composed of small point and line-like objects, a pinhole radius of 4 optical units is already nearly optimal from the signal to noise ratio point of view. At this radius, the resolution for both horizontal and vertical planes is still nearly optimal giving rise to efficient suppression of contributions from adjacent areas in the specimen. Fig. 2a shows a strong dependency of this resolving power to the pinhole radius. The same holds for vertical oriented planes (Fig. 2b), although the slope of the curve above 6 optical units is less. The results in Fig. 2 for horizontal planes and lateral point-resolution are in general agreement with the results obtained in paraxial approximation by Wilson and Carlini (4).

To summarize the findings from Fig. 2 and Fig. 3, 1) there is a direct tradeoff between the signal strength and therefore the signal to noise ration from plane-like objects and the resolving power (equivalent to the sectioning capability) for these objects, 2) when the objects of interest are point or line-like, the pinhole radius should not be larger than 4 optical units, 3) true confocal operation can be achieved with detector pinholes up to a radius of 2 optical units (see Fig. 2b). In a typical CSLM, equipped with NA=1.3 oil immersion objectives and operated at the 488nm. line of an argon-ion laser, this corresponds with a 120nm radius.

To compare the results of our electromagnetic diffraction theory based model with the results from geometrical optics, we

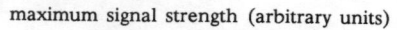

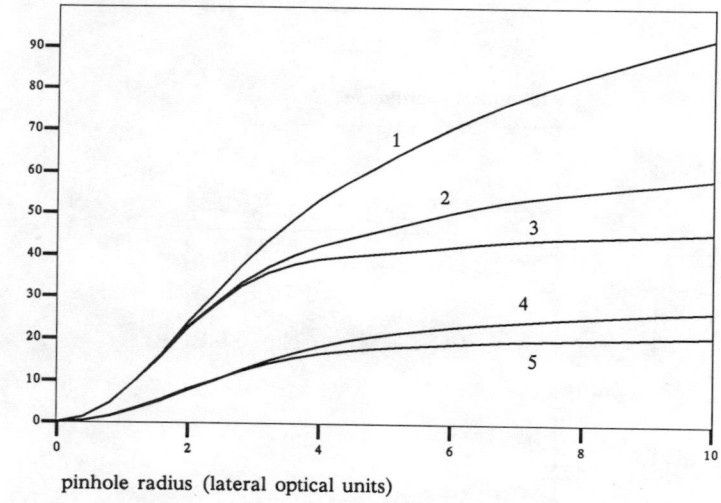

Fig. 3 Maximum signal strength of various objects as a function of the detector pinhole size and their orientation. Curve 1: vertical planes, 2: vertical lines, 3: points objects, 4: horizontal planes, 5: horizontal lines. As signal strengths of point, line and plane objects cannot be compared directly, curves 1,2 and 3 have been normalized to fit each other in the pinhole radius interval (0,2). The relative signal strength between curves 1 and 2 (horizontal and vertical planes), and curves 2 and 5 (horizontal and vertical lines) is unaffected by this normalization.

have plotted in Fig. 2a also the geometrical resolving power for planes (*FWHM*) (5).

$$FWMH_{geom} = 2\sqrt{2}\ p\cos a \qquad (11)$$

with $FWHM_{geom}$ in axial optical units and the pinhole radius p in lateral optical units. The slopes of both curves (curves 1 and 4 in Fig. 2a) for a pinhole radius above 6 optical units appears to be identical with a constant offset of around 3 optical units of curve 1 with respect to curve 4.

A prerequisite for reliable measurement of small objects with CSLM is knowledge of the resolving power for the objects as a function of their orientation. We computed the resolution for point, line and plane objects assuming a CSLM in typical

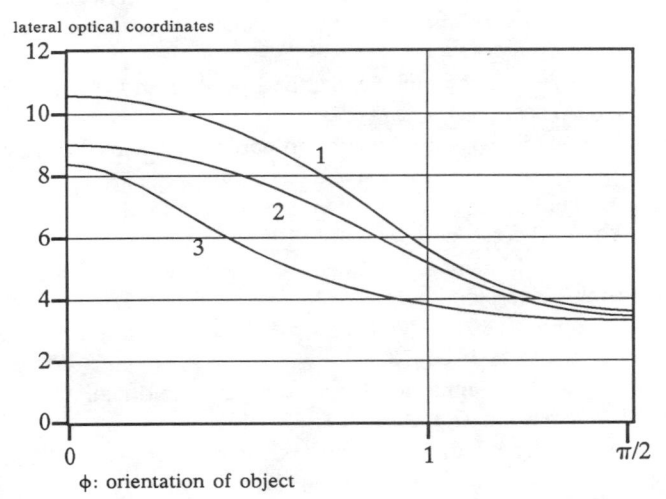

Fig. 4 Resolution of a CSLM in typical working conditions (pinhole radius 4 opt. units, $k_{em}= 0.9\ k_{exc}$) as a function of the orientation for three different objects: 1:plane-like objects, 2: line-like objects, 3: point-like objects. The horizontal axis corresponds with the angle ϕ between the planes or lines and the $z=0$ plane. Resolution for line and plane-like objects is measured in the $y=0$ plane and perpendicular to the object. Curve 3 corresponds with the point resolution in the same direction.

working conditions (Fig. 4). The resolving power in this figure is measured in the x-z plane and perpendicular to the objects. In the case of planes and lines this corresponds with the distance of two just-resolved objects. For comparison, we have also given the two-point resolution in the same direction. The results clearly show the object and orientation dependent character of confocal imaging.

4. Conclusions and outlook

In this paper we have studied confocal imaging of three types of objects using a theoretical model of a confocal microscope comprising a number of important features of a practical instrument: high aperture lenses, fluorescence imaging, a finite detector pinhole and the polarization state of the excitation light. We have based our numerical computations of the instrumental functions on electromagnetic diffraction theory. The use of this theory combined with numerical techniques allows us to study aspects of a high aperture confocal system like object type and object orientation dependent imaging. The results can be use directly in image processing schemes designed to segment and measure biological specimen which can be classified as point, line or plane-like objects.

We have structured the software in such a way that further elaboration of the model by for instance incorporation of spherical aberration can be carried out by adding a single procedure to the program. In a similar way the effects of annular filters, non-uniform collection efficiency in the detector system or axial mis-alignment of the microscope through chromatic aberration can be simulated.

To conclude we may say that we have developed a powerful tool which will allow us to study many aspects of confocal imaging in future.

Acknowledgments.

We would like to thank M.W. Baarslag, J. Krol, N. Vischer and P. Weiss for their assistance in the software development, and dr. H.G.P. van der Voort for reading the manuscript. This work has been supported by the Stichting Technische Wetenschappen (STW), the Foundation for Fundamental Biological Research (BION/NWO) and E.C contract BAP 0293-NL.

5. References

1 Fay, F.S. K.E.F. Fogarty and J.M. Coggins. Analysis of Molecular Distributions in Single cells using a digital imaging microscope. in *Optical Methods in Cell Physiology*. P. de Weer and B. Salzburg (eds). John Wiley (1985) 51-63.

2 Bertero, M.,P. Boccacci, P. Brianzi and E.R. Pike. (1987) Inverse Problems in Confocal Scanning Microscopy. *Suppl. Inverse Problems, an interdisciplinary study.* Ed. P.C. Sabatier. *in Advances in Electronics and Electrom Phys.* **19** pp 225-239.

3 Carlini, A.R. and T. Wilson. The role of the pinhole size and position in Confocal Scanning Imaging system (1987) *SPIE* **809** Scanning Image Techn.

4 Wilson, T. and A.R. Carlini. Three-dimensional imaging in confocal systems with finite size detectors. (1988) *J. Microsc.* **149**, pp 51-66.

5 van der Voort, H.T.M., G.J. Brakenhoff & G.C.A.M. Janssen. Determination of the 3-dimensional optical properties of a Confocal Scanning Laser Microscope. *Optik.* **78** (1988) 48-53.

6 Wijnaends van Resandt, R.W. H.J.B. Marsman, R. Kaplan, J. Davoust, E.H.K. Stelzer, R. Sticker: Optical fluorescence microscopy in 3 dimensions: microtomoscopy. *J. Microsc.* **138** (1985) 29-34.

7 van der Voort, H.T.M, G.J. Brakenhoff, J.A.C. Valkenburg & N.Nanninga. 1985. Design and use of a computer-controlled confocal microscope. *Scanning* **7**: 66-78.

8 Sheppard, C.J.R., A. Choudhury and J. Gannaway. Electromagnetic field near the focus of wide-angular lens and mirror systems. (1977) *IEE J. Microwaves, Opt. Acoust.* **1**, 129-132.

9 Richards, B. and E. Wolf. Electromagnetic diffraction in optical systems. Pt. II- Structure of the image field in an aplanatic system.(1959) *Proc. Roy. Soc.* (A), **253**, pp. 358-379.

10 Born, M. & E. Wolf. Principles of Optics. 5th edn. Pergamon Press, Oxford. (1975) 435-448.

11 Wolf, E. Electromagnetic diffraction in optical systems Pt. I-An integral representation of the image field. (1959) *Proc. Roy. Soc.*, (A), **253**, pp. 349-357

12 Jackson, J.D. Classical Electrodynamics. Second ed. J. Wiley and Sons, New York. (1962).

13 Stamnes, J.J. Waves in focal regions. Adam Hilger, Bristol-Boston. (1986).

14 Brakenhoff, G.J. P. Blom & P. Barends. Confocal scanning light microscopy with high aperture immersion lenses. *J. of Microsc.* **117** (1979) 219-232.

Theoretical and Experimental Research on Super-resolution of Microscopes
I. Partially Coherent Illumination and Resolving Power

Chang-ming Ma * Robin W. Smith

Optics Section, Blackett Laboratory
Imperial College of Science, Technology and Medicine
London SW7 2BZ U.K.

ABSTRACT

This paper presents theoretical and experimental research on super-resolution in optical microscopy. The theoretical analysis is based on partially coherent imaging theory. A source of annular form is used to achieved super-resolving imaging and a significant improvement in the resolving power has been observed experimentally. The imaging of a two-point object and two-bar objects of different spaces and widths are discussed. The displacement of the Gaussian image point and problems associated with this technique are also considered.

1. INTRODUCTION

The limitation imposed by diffraction on the resolving power of optical systems has been a central problem in optics since the fundamental studies by Abbé and Rayleigh. The maximum resolution obtainable with a given imaging system is usually taken to begiven by the Rayleigh criterion. Although it may be shown mathematically that if measurements were noise free then an arbitrarily fine object could be reconstructed exactly from the diffraction-limited image[1], it has been found that the Rayleigh criterion represents a resolution which may hardly, in practice, be surpassed optically even for extremely low noise levels. A mathematical analysis in which it is assumed that both the object and image domains are finite has been developed in terms of the eigen-functions and eigen-values of the imaging equation. This analysis has been extended to deal with a finite object and an image sampled over the entired image plane, in terms of singular functions and singular values.[2] At low Shannon numbers, the singular value analysis predicts that reconstruction of an object from a diffraction-limited image can be achieved with a resolution better than the Rayleigh criterion (super-resolution). Applications of this analysis to real imaging systems have been considered and object details finer than the diffraction limit have been observed numerically and optically by making use of the frequency information contained in the image and a priori information about the object.

In this paper we report our research on super-resolution in optical microscopes. A new system using partially coherent tilted or oblique illumination is proposed and its imaging performance is discussed. The relation between the source shape and resolving power is deduced. Numerical simulation of partially coherent imaging of different objects has been carried out. Theoretical and numerical computing results indicate that super-resolution can be achieved not only at low Shannon numbers but at high Shannon numbers provided the object details are non-periodic. The resolving power for the case of a two-point object in the proposed system can be 1.5 times as high as that in traditional microscopes. The resolution achieved in this system varies with the number of the object details to be resolved in the object and their distributions. Comparision of different methods used in super-resolving imaging research and the problems of the displacement of the Gaussian image point will be discussed in a later contribution.

2. PARTIALLY COHERENT IMAGING IN OPTICAL MICROSCOPES

The typical setup is shown in Fig.1.

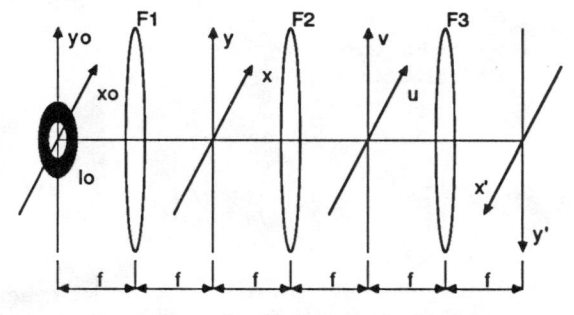

Fig.1 Geometry of a simple optical microscope

---

* Permanent Address: College of Instrumentation, Shanghai Institute of Mechanical Engineering, Shanghai, China

In the two-dimensional case and in the absence of aberrations and noise, according to the Van Cittert-Zernike theoram[3], a quasi-monochromatic extended source will give rise to a correlation between vibrations at any pair of points in the object plane. With the usual approximations the mutual intentisity is given by

$$J(x_1 - x_2, y_1 - y_2) = c \int \int I(x_0, y_0) \exp[-\frac{2\pi i}{\lambda f}(x_0 x_1 - x_0 x_2 + y_0 y_1 - y_0 y_2)] dx_0 dy_0, \tag{1}$$

Assuming that the object transmission function is $O(x, y)$, we get the mutual intensity in the image plane

$$J'(x_1', x_2', y_1', y_2') = J(x_1 - x_2, y_1 - y_2) O(x_1, y_1) O^*(x_2, y_2) \otimes \tau(x_1, y_1) \tau^*(x_2, y_2) \tag{2}$$

where the asterisk denotes the complex conjugate, $\otimes$ denotes the convolution operation, and $\tau(x, y)$ is the Fourier transform of the optical pupil function. If $x' = x_1' = x_2'$ and $y' = y_1' = y_2'$, the intensity distribution in the image plane is

$$I(x', y') = c' \int \int \int \int J(x_1 - x_2, y_1 - y_2) O(x_1, y_1) O^*(x_2, y_2) \tau(x_1 - x', y_1 - y') \tau^*(x_2 - x', y_2 - y') dx_1 dx_2 dy_1 dy_2 \tag{3}$$

where c' is a complex constant which for simplicity, we will omit.

For a two-point object, that is, $O(x, y) = \delta(x - \frac{L}{2}) + \delta(x + \frac{L}{2})$, we get

$$I(x', y') = \left| \tau\left(x' - \frac{L}{2}, y'\right) \right|^2 + \left| \tau\left(x' + \frac{L}{2}, y'\right) \right|^2$$
$$+ \left[\tau\left(x' - \frac{L}{2}, y'\right) \tau^*\left(x' + \frac{L}{2}, y'\right) + \tau^*\left(x' - \frac{L}{2}, y'\right) \tau\left(x' + \frac{L}{2}, y'\right) \right] J(L, 0) \tag{4}$$

Equation (4) shows that the intensity distribution in the image plane depends on both $\tau(x, y)$ and $J(L, 0)$, the mutual intensity between the pair of points P_1 and P_2. This means that for a certain optical system the way in which partially coherent illumination is formed has a strong effect on imaging performance, especially on the resolution of two-point objects. For incoherent illumination, $J(L, 0) = 0$, and equation (4) becomes

$$I(x', y') = \left| \tau\left(x' - \frac{L}{2}, y'\right) \right|^2 + \left| \tau\left(x' + \frac{L}{2}, y'\right) \right|^2 \tag{5}$$

According to the Rayleigh criterion two separated points may be distinguished when the maximum of the diffraction pattern of one point falls exactly on the first dark ring of the other. In practice, however, we can never have an infinite source to ensure that $J(L, 0) = 0$ for all L. The Rayleigh criterion is just the theoretical resolution which can be achieved only on the assumption that an infinite source is used in the system. For a real imaging system, the resolution obtainable will depend on the illumination.

For ideal coherent illumination, we have $J(L, 0) = 1$, then equation (4) becomes

$$I(x', y') = \left\{ \left| \tau\left(x' - \frac{L}{2}, y'\right) \right| + \left| \tau\left(x' + \frac{L}{2}, y'\right) \right| \right\}^2 \tag{6}$$

According to information theory, the spatial frequency transfer function for coherent imaging differs from that for incoherent imaging. The incoherent cutoff frequency f_i is twice as high as the coherent cutoff frequency f_c when the coherent point source is at the origin. However if we shift the coherent point source from the origin, the coherent frequency transfer function changes and the cutoff frequency is increased. It is found that the total integrated area under the frequency transfer function curve is conserved..

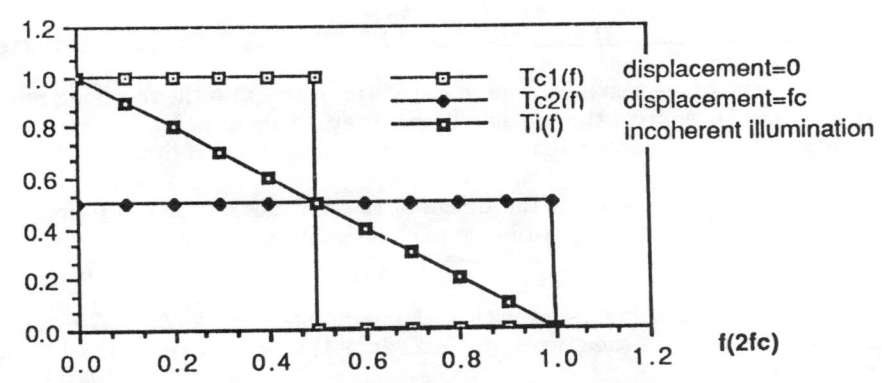

Fig.2 Relation between frequency transfer function and the displacement of the source

Figure 2 shows that the coherent spatial cutoff frequency will double if the displacement of the image of the coherent source in the spatial frequency plane is equal to f_c. This causes a reduction by a factor of two in the value of the frequency transfer function. This explains why the resolving power can be increased by using tilted illumination. Unfortunately, because of serious coherent noise coherent illumination has not been widely used in optical microscopy.

Under partially coherent illumination, the image intensity distribution depends on the mutual intensity $J(L,0)$ between P_1 and P_2. Therefore we can design an extended source with a special shape so that we can get a mutual intensity with negative value at a pair of points separated by a distance which is smaller than the conventional diffraction limit. In this case, the third term in equation (4) is subtracted from the other two and this results in a super-resolved image of the two-point object, that is

$$I(x',y') = \left|\tau\left(x'-\frac{L}{2},y'\right)\right|^2 + \left|\tau\left(x'+\frac{L}{2},y'\right)\right|^2 \\ - \left[\tau\left(x'-\frac{L}{2},y'\right)\tau^*\left(x'+\frac{L}{2},y'\right) + \tau^*\left(x'-\frac{L}{2},y'\right)\tau\left(x'+\frac{L}{2},y'\right)\right]J(L,0) \qquad (7)$$

Figure 3 shows the distributions of intensity under different illumination conditions.

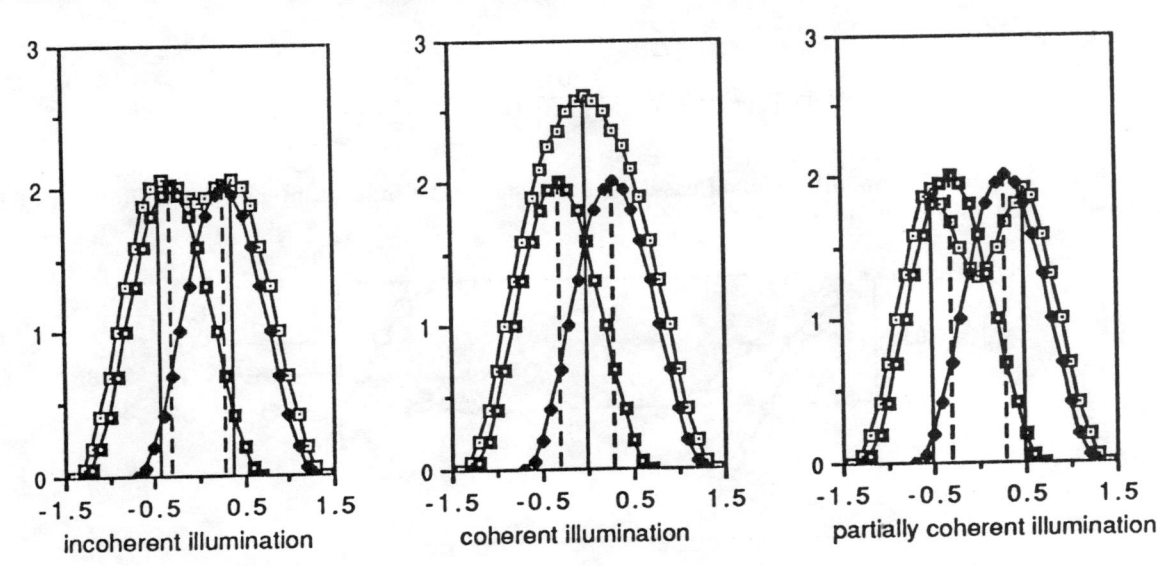

Fig.3 Intensity distribution under different illumination conditions.

3. EFFECT OF THE SOURCE SHAPE ON THE RESOLVING POWER

As discussed above, the shape of the extended source has a strong effect on the resolving power of a partially coherent imaging system. For a circular source, the mutual intensity at a pair of points P_1 and P_2 can never be negative if the distance between P_1 and P_2 is smaller than the conventional diffraction limit L_d (Fig. 6).

In the special case when the diameter D_i of the image of the circular source in the frequency plane equals that of the optical pupil and if $L = nL_d$ where n is an integer the intensity distribution of a two-point object take the form given by equation (5).

But if in this case $L \neq nLd$, the imaging varies with L. For example when $2nL_d < L < (2n+1)L_d$, the third term in equation (4) is subtracted from the other two and when $(2n+1)L_d < L < 2nL_d$, it adds. A similar analysis applies to a rectangular source.

For super-resolution of the two-point object we have considered the two slit shaped source shown in Fig. 4(a) which is defined by the values of e, β and b.

Fig.4 Sources of special shape for superresolution; (a) A two slit shaped source, (b) An annular source.

In this case, the mutual intensity in the object plane is given by

$$J(x_1 - x_2, y_1 - y_2) = [\mathrm{sinc}\, \omega_b(x_1 - x_2) - \beta \mathrm{sinc}\, \omega_a(x_1 - x_2)] \mathrm{sinc}\, \omega_e(y_1 - y_2) \qquad (8)$$

or

$$J(x_1 - x_2, y_1 - y_2) = \mathrm{sinc}\, \omega_c(x_1 - x_2) \cos \omega_d(x_1 - x_2) \mathrm{sinc}\, \omega_e(y_1 - y_2) \qquad (9)$$

where

$$\omega_a = \frac{\beta b}{\lambda f}, \quad \omega_b = \frac{b}{\lambda f}, \quad \omega_c = \frac{(1-\beta)b}{2\lambda f}, \quad \omega_d = \frac{\pi(1+\beta)b}{2\lambda f}, \quad \omega_e = \frac{e}{\lambda f} \qquad (10)$$

For simplicity we consider the one dimensional case. Figure 5 shows the mutual intensity as a function of L for the two slit shaped source with different values of β and b.

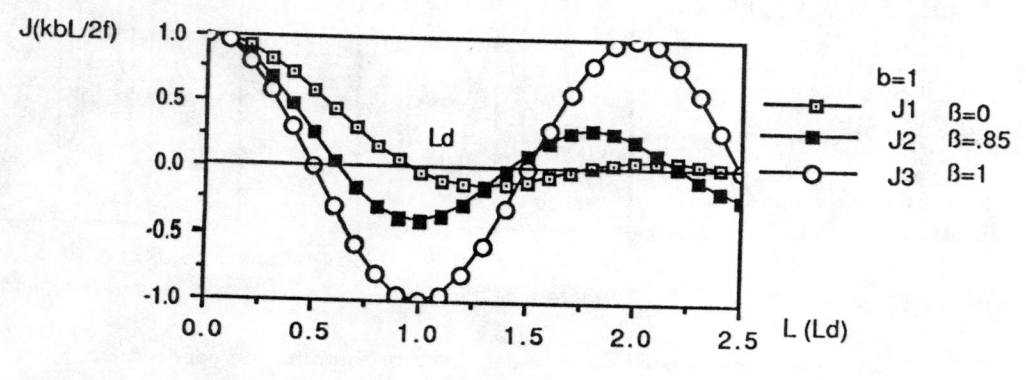

Fig.5 The Mutual Intensity as a function of L for the two slit shaped source.

If we let $b = f_c$ and $\beta > 0$, the smallest resolvable distance of the two points will be smaller than L_d. For the extreme case that $\beta \to 1$ equation (9) becomes

$$J(x_1 - x_2, y_1 - y_2) = \cos\frac{\pi b}{\lambda f}(x_1 - x_2) \tag{11}$$

The maximum resolution obtainable in this case will be about 1.5 times as high as that in the case of a rectangular source. Although the resolving power has been improved by making use of a two-slit source, two problems remain. One is the low level of illumination at the object and the other is that only a small region of the object receives the special illumination needed for super-resolution to occur. This will be discussed elsewhere.

Experimentally, as might be expected, it is found that such a system of illumination only improves the resolution in one dimension, the resolution in the orthogonal dimension being unchanged. To avoid this an annular shaped source shown in Fig. 4(b) is proposed.

The mutual intensity in the object plane is now given by

$$J(x_1 - x_2, y_1 - y_2) = \left\{\frac{J_1(\omega_b r)}{\omega_b r} - \beta^2 \frac{J_1(\omega_a r)}{\omega_a r}\right\} \exp\left[\frac{i\pi}{\lambda f}(x_1^2 - x_2^2 + y_1^2 - y_2^2)\right] \tag{12}$$

where $J_1(\omega r)$ is Bessel function of first kind order 1 and

$$\omega_a = \frac{\pi 2\beta R}{\lambda f}, \qquad \omega_b = \frac{\pi 2R}{\lambda f}, \qquad r = \sqrt{(x_1 - x_2)^2 + (y_1 - y_2)^2} \tag{13}$$

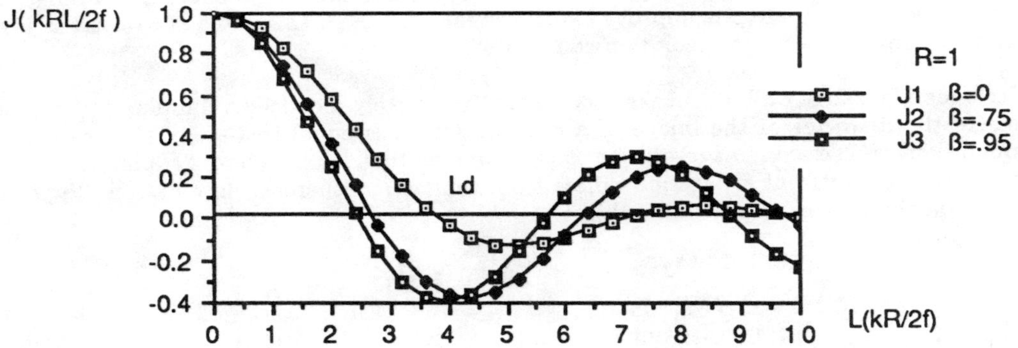

Fig.6 The mutual intensity as a function of L for the annular source.

The phase factor can be neglected when the difference between the distances of P_1 and P_2 from the origin is small. Figure 6 shows the mutual intensity as a function of L for the annular source with different inner and outer diameters. Let $R = L_d$, the first null point will get nearer to the origin with the increase of the inner diameter $2\beta R$. In the extreme case that $\beta \to 1$ equation (12) becomes

$$J(x_1 - x_2, y_1 - y_2) = J_0(\frac{\pi 2R}{\lambda f}r) \tag{14}$$

The maximum resolution obtainable now will be about 1.3 times as high as that in the case of a circular source. The problems of low illumination and the limitation on effective super-resolving area still remain, but now two-dimensional super-resolution can be achieved in this proposed system. It is possible to design an annular source with suitable values for β and R so that adequate illuminantion and reasonable effective super-resolving area occur with significant improvement in resolution.

4. NUMERICAL RESULTS.

In this section, we discuss the numerical computing results of super-resolving imaging obtained in the proposed system. All image intensity distributions were computed assuming a two-slit source and a rectangular optical pupil. We first consider the effect of the source shape on the resolving power.

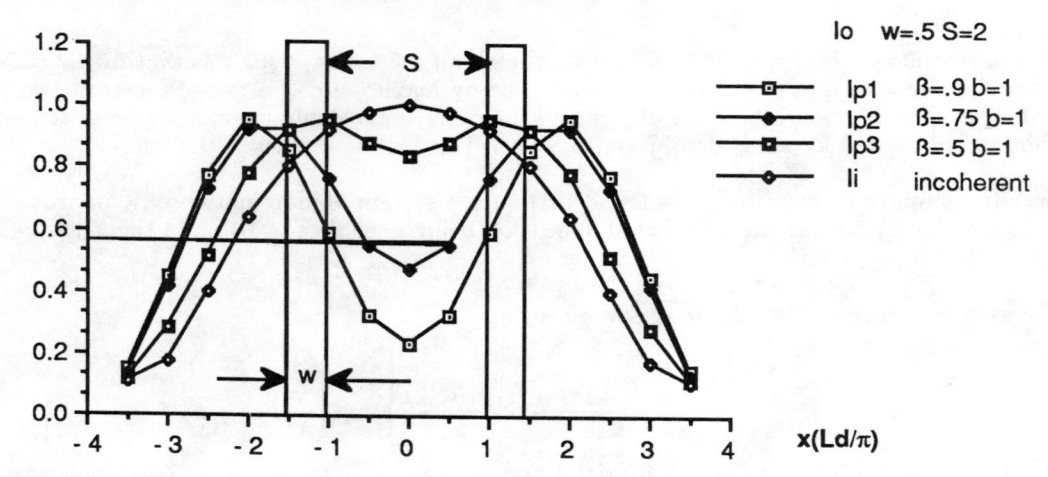

Fig.7 Image intensities for two-bar objects under different partially coherent illumination

Figure 7 shows the intensity distribution of a two-bar object in the image plane for the two-slit source with different β and b. For incoherent illumination the object is unresolvable. For partially coherent illumination, it is evident that the resolving power of the system is improved with the increase of β. If $\beta = 0.5$ the object is just resolved, and if $\beta = 0.9$ a significant improvement in resolution can be obtained.

In a optical microscope, a source of finite size is used so that absolutely incoherent illumination can not be achieved. For the case where the diameter of the image of a circular source is equal to that of the optical pupil (or quasi-incoherent illumination), the maximum resolution is the same as that for incoherent illumination, but the imaging of a two-point object with different separation differs from that for incoherent illumination. Fig.8 shows the effect of the illumination on the images of a two-bar object.

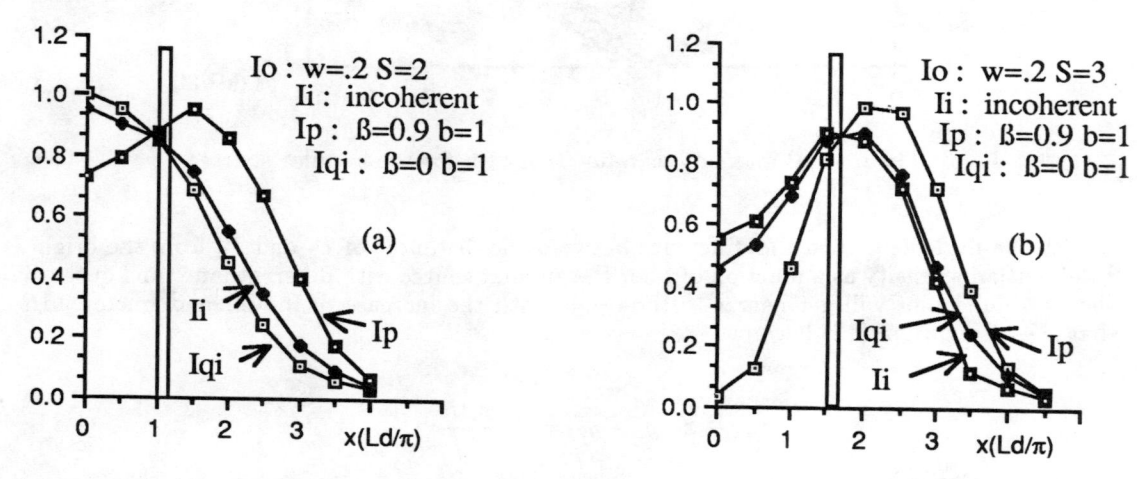

Fig.8 Image intensities for two-bar objects with gap (a) smaller than L_d (b) equal to L_d.

Figure 9 shows the intensity distribution of a two-bar object with different spaces between the bars. In Fig. 9(a), for incoherent illumination, the object cannot be resolved unless the space is bigger than the diffraction limit L_d. For

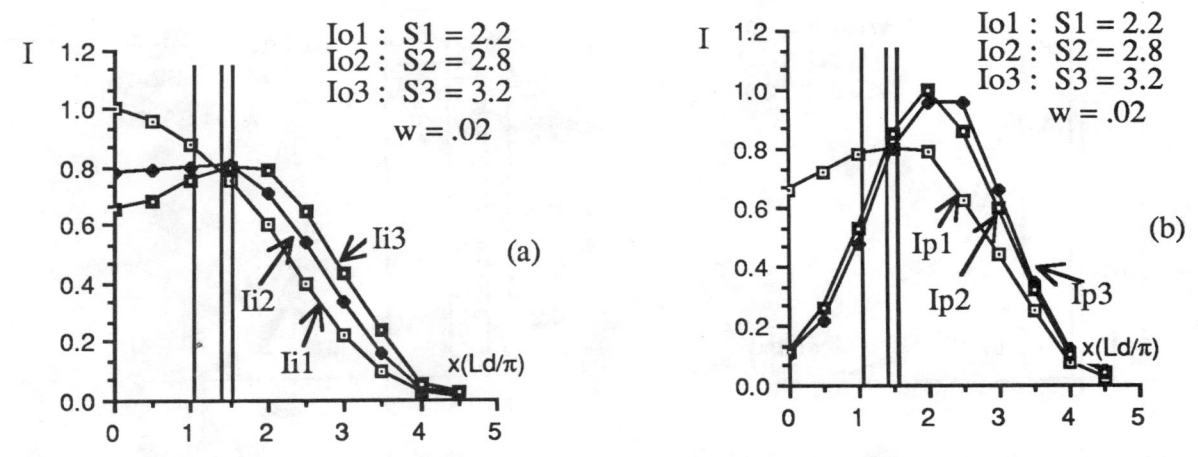

Fig.9 Image intensities for two-bar objects under (a) incoherent and (b) partially coherent illumination

partially coherent illumination formed by a two-slit source with $\beta = 0.9$, the maximum resolution is about 1.5 times as high as that in the case of incoherent illumination (Fig.9(b)).

It is found that there is a displacement of the Gaussian image point. It can be seen in Figs. (7)-(9) that, if the distance between a pair of points is smaller than L_d, although the two points can be distinguished from each other in the intensity distribution in the image plane, the separation of the intensity peaks is slightly greater than the separation of the object bars,. This probably means that these super-resolved images are not suitable for precision measurements. The same problem occurs in the optical scanning microscope[4]. That is why we emphasize the need for a priori information about the object in super-resolving imaging. For low spatial frequencies the proposed system gives the same performance as the traditional microscopes (Fig. 10).

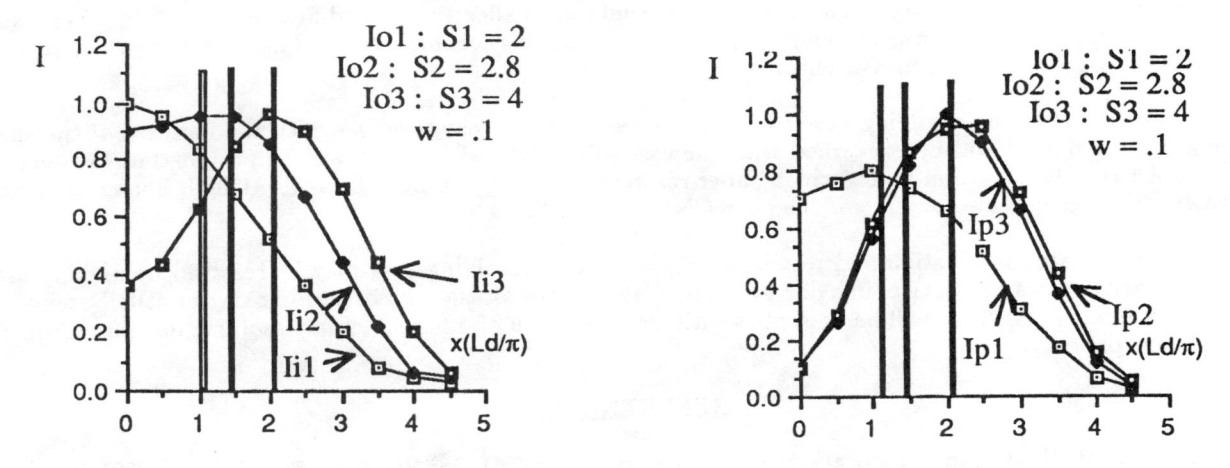

Fig.10 Image intensities for two-bar objects under (a) incoherent and (b) partially coherent illumination

Figs. 11(a) and 11(b) give further examples of partially coherent and incoherent imaging of two-bar objects with different spaces and widths, respectively. For partially coherent imaging, the smallest resolvable space is one fourth of that in the case of incoherent illumination. The side-lobes are still at a reasonable level. It can also be seen that with the increase in the distance between the pair of points these side-lobes are reduced as observed from the mutual intensity function in equation (9). It is by making use of this characterestic of partially coherent imaging that we can achieve a significant improvement in resolution without increasing the coherent noise level in the system.

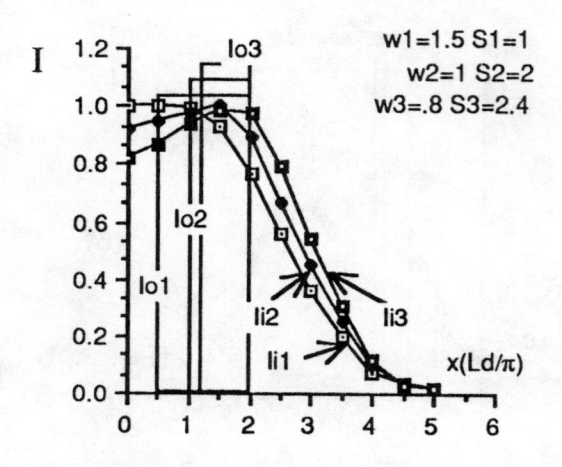

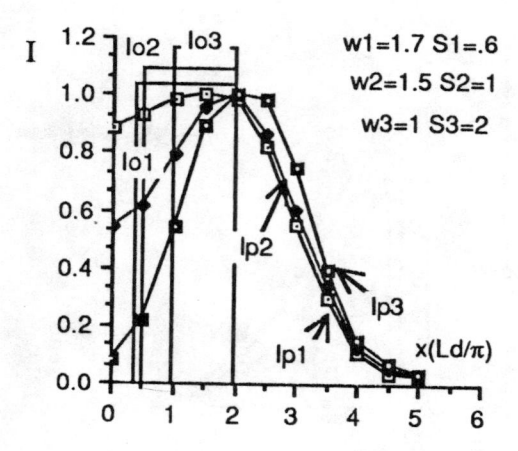

Fig.11 Image intensities for two-bar objects under (a) incoherent and (b) partially coherent illumination

5. CONCLUSIONS

In this paper, we have reported our research on super-resolution in optical microscopy. Threoretical analysis and numerical computing results have been given and a significant improvement in resolution has been achieved by using partially coherent illumination. Under partially coherent illumination, the image intensity distribution depends on the mutual intensity between all pairs of points in the object plane. It is possible to design an annular source with a small differece in outer and inner diameter so that we can get a mutual intensity with negative value at a pair of points separated by a distance which is smaller than the conventional resolution limit. Under such conditions we get a super-resolved image of the two-point object. There are some problems associated with this technique. One problem is that if the distance between a pair of points is smaller than the diffraction limit L_d, there would appear a displacement of the Gaussian image point. We feel that this technique can be used only for the purpose of observation but not for precision measurement.

Another problem is that the resolving power of the proposed system may decrease with the increase of the size of the object. Future papers will be concerned with the case of periodic objects. It will be proved that low Shannon number is not the only condition under which super-resolving imaging can be achieved. At high Shannon numbers super-resolving imaging is also possible if the object is non-periodic.

Problems of inadequate illumination and restricted effective super-resolving area may be resolved by making use of specially designed sources and optical imaging systems. Calculations for the case of periodic object distributions have also been carried out , and these will be presented with a comparision of different super-resolving imaging techniques in a subsequent paper.

REFERENCES

1. C.K. Rushford, R.W. Harris; "Restoration, Resolution and Noise", J.Opt.Soc.Am **58**, 539-545 (1968)

2. M. Bertero, E.R. Pike; "Resolution in Diffraction-limited Imaging, a singular value analysis", Optica Acta **29** , 727-746m (1982)

3. M. Born, E. Wolf, "Principles of Optics", 514-530, Fourth Edition, Pergamon (1968)

4. C.M. Ma; "Optical Imaging and Superresolving in the Scanning Microscope", To be presented at the annual meeting of the O.S.A , California (1988)

ECO1
SCANNING IMAGING

Volume 1028

SESSION 2

Scanning Microscopy I

Chair
Zoltan S. Hegedus
CSIRO (Australia)

A scanning differential optical system for simultaneous phase and amplitude measurement

Roland K. Appel, Chung Wah See and Michael G. Somekh

Department of Electrical and Electronic Engineering, University College London, Torrington Place, London, WC1E 7JE, United Kingdom

Abstract

A scanning optical system has been developed which can simultaneously and independently measure the differential amplitude and phase of light reflected off an object surface. The central element of such a system is an acousto-optic modulator which splits the incoming light into two beams of different optical frequencies, amplitude modulated in phase quadrature. These are focused with close proximity on the sample surface. The theoretical sensitivity of this type of differential optical system is 1×10^{-3} mrad in phase and 1 in 10^5 in reflectivity variation. In this paper we will look at the first implementation of the technique and practical problems encountered with this system. A second system is presented in which two first order beams from the Bragg cell are used to interrogate the sample. An analysis is made of the interpretation of the detected signals when different types of amplitude modulation is used. This latter system is shown to have an improved differential amplitude response. A number of samples have been looked at to assess the performance of the system. These demonstrate the high sensitivity of this system and its capability in separating amplitude and phase information. The advantages of extracting differential amplitude as well as phase information is discussed using a simple layered model. Potential applications include film thickness metrology, surface profilometry and measurement of refractive index variation. In addition it can be used to image low contrast objects such as biological specimens and defects on crystalline structures.

1. Introduction

Currently available optical techniques for looking at materials are able to measure either the amplitude or the phase variation of reflected light. Phase may be broadly interpreted in terms of topographic variation and the amplitude into compositional variation. Different materials alone may however cause phase contrast due to differences in their complex refractive indices. If an optical technique is to be used to accurately perform topography measurements, in general phase information alone is insufficient. Recently we described a new technique[1] which may be used to measure the phase and amplitude variation independently and simultaneously. This system is furthermore differential in that it responds to changes in the phase and amplitude responses[2-5]. In an absolute phase measurement system, microphonics limits depth resolution to typically[6] $\lambda/100$ or 50 Å. Differential systems on the other hand have been shown to have a phase sensitivity[7] of 3×10^{-3} mrad (equivalent to 1.5×10^{-3} Å in the context of topography) and a reflectivity sensitivity[8] of 3 parts in 10^5, both measured in a 1 kHz bandwidth.

This paper reports two implementations of the technique. In both systems, a Bragg cell is used to produce two beams which are focussed onto two close adjacent areas on the object. The differential phase is the difference in optical path length of the two reflected beams and the differential amplitude is the difference in reflected intensity. Thus two measurements are made from one small area on the sample. To obtain a line scan or image, the sample must be scanned relative to the light. The results presented in this paper comprise line scans where the object was mechanically scanned with stationary optics.

The basis of the technique[1] is that of a heterodyne interferometer. In a homodyne interferometer, both amplitude and phase information are mixed. By modulating the light with two separate frequencies, phase and amplitude information may be measured simultaneously and independently from two different frequency components of the detected signal. For a Bragg cell, these two frequencies can be the natural Bragg frequency shift between the zeroth and first orders and secondly, a low frequency modulation achieved by switching the Bragg cell drive on and off. This results in the intensity in each beam being modulated in anti-phase.

In the following section the initial implementation of the system will be introduced. Here the zeroth and first order beams from the Bragg cell are used. Results from this system are shown and it will be seen that draw backs of this lie in the differential amplitude part. This is caused by non-identical modulation of each beam and section 3 analyses this situation. Work on the first system leads directly on to the second where two first order beams from the Bragg cell are used to interrogate the sample. This is introduced in section 4 where conditions for suitable types of modulation are derived. Results from this system are presented and demonstrate the improved differential amplitude response of this second system. In section 5 we use a simple layered model to demonstrate how extraction of amplitude data can eliminate some of the ambiguities inherent in phase only profilometers. The possibility of using absolute amplitude information is also discussed.

2. A first implementation of the technique - the zeroth/first order system

Referring to this first implementation (figure 1) it operates as follows. Light from the laser enters the Bragg cell where it is split into a 1st and a zeroth order. These are frequency shifted by the Bragg frequency (ω_B) and sinusoidally amplitude modulated in phase quadrature at frequency ω_a, by periodic variation of the Bragg cell drive. The two beams are focussed normally onto the object where they experience different reflection coefficients and phase lags due to complex refractive index and object height variations. The two reflected beams are detected by detector D2 and, after passing through the Bragg cell for a second time, by detector D1.

The frequency components of interest from D1 take the form,

i) $\quad A\left(r_1^2 - r_2^2\right)\cos 2\omega_a t$ (1)

ii) $\quad B r_1 r_2 \cos\left[2\omega_B t + 2\delta\theta + 2\delta\phi + 2\delta\alpha\right]$ (2)

where $\delta\theta$ is the phase difference imposed on the two beams by the object topography; r_1, r_2 are the moduli of the complex reflection coefficients $\underline{r_1}, \underline{r_2}$ experienced by each beam; $\delta\alpha$ is the phase due to variation in the phase angle of the reflection coefficient and A,B and $\delta\phi$ are constants. $\underline{r_1}, \underline{r_2}, r_1, r_2$, and $\delta\alpha$ are related by the following expression,

$$\underline{r_1} = r_1 \exp(i\alpha_1) \quad (3)$$

$$\underline{r_2} = r_2 \exp(i\alpha_2) \quad (4)$$

with $\delta\alpha = \alpha_1 - \alpha_2$.

It can be seen that in general, when measurements are performed on a sample, there are three variables, namely phase (θ), and the complex reflection coefficient (r and α). Using this type of technique, we are now able to measure two of these simultaneously. The differential amplitude result which is the magnitude of the $2\omega_a$ frequency component, is proportional to the difference in the square of the amplitude reflection coefficients. The differential phase measurement is achieved by detecting the phase of the $2\omega_B$ component and yields information on the phase variation due to both object topography (θ) and the phase angle of the reflection coefficient (α).

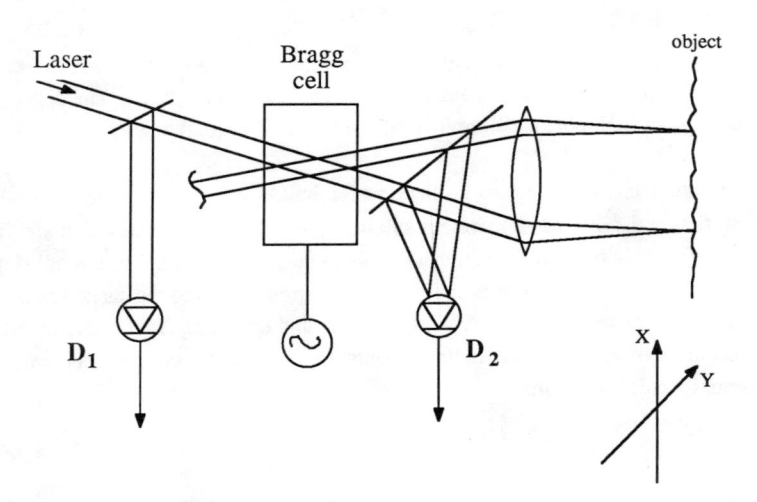

FIG. 1. Zeroth /first order system

2.1 Results from the zeroth /first order system

Two sets of results are presented which show the systems ability to measure differential phase and amplitude information simultaneously and independently. In this system, a HeNe laser of wavelength 632.8nm was used, the beam separation on the sample was 400µm and the focussed beam diameter was 40µm. Differential phase and differential amplitude measurements were made for reasons explained in section 3, using detector 1 and 2 respectively. The Bragg cell drive signal was triangularly modulated producing sinusoidal optical modulation.

Complementary samples were looked at, one with almost pure topography and the second with primarily reflectivity detail. Figure 2 shows line traces across a silicon wafer where the surface has been active plasma etched to produce a series of parallel tracks as shown in figure 2(a). The nominal step heights are 180, 300, 400, 550 and 700 Å which were confirmed to within ±10% by using a mechanical stylus probe.

The differential phase result is shown in figure 2(b). The slow underlying variation is due to warp of the wafer surface. As this is a differential result, the actual surface profile would be obtained by integration. The background phase variation has been subtracted to produce trace 2(c). The up and down peaks are the differential responses from 'up' and 'down' steps on the object. The size corresponds to the extra optical path length imposed by the steps. Since there is no compositional variation on this sample, the height of each peak can be converted into a step size. These compare well with measurements made by a mechanical stylus probe [see table 1].

TABLE 1. Comparison of step heights measured with stylus probe, zeroth /first order and two first order differential optical phase systems

Step	A	B	C	D	E	F	G	H
Stylus probe Å	180	310	315	430	440	530	540	700
zeroth /first order (Å)	163	326	302	424	490	579	609	713
two first order (Å)	187	283	265	452	465	573	563	

The differential amplitude result is shown in figure 2(d). Differential contrast on this sample is caused by scatterers, such as step discontinuities, scratches, surface roughness from the etching process and surface contamination (eg dust). The differential amplitude contrast from a point source is a dipole, the width of each pole is determined by the focussed beam diameter.

Figure 3 shows differential phase and amplitude line traces across a silicon wafer, half of which is implanted with As^+ at a doping level of 10^{13} ions/cm^2 [see figure 3(a)]. Looking first at the differential phase result in figure 3(b), it can be seen that any contrast caused by a change in the phase angle of the reflection coefficient is far smaller than that due to surface topography. The overall slope of the line trace is due to warp of the wafer surface. The differential amplitude results are shown in figure 3(c). Here the change in reflection coefficient is clearly visible. Also however, the dc levels on either side of the interface are different and non-zero. This is due to unequal modulation in each of the two beams and has proved to be a draw back of this system implementation. The next section is concerned with an analysis of how unequal modulation affects the differential amplitude results.

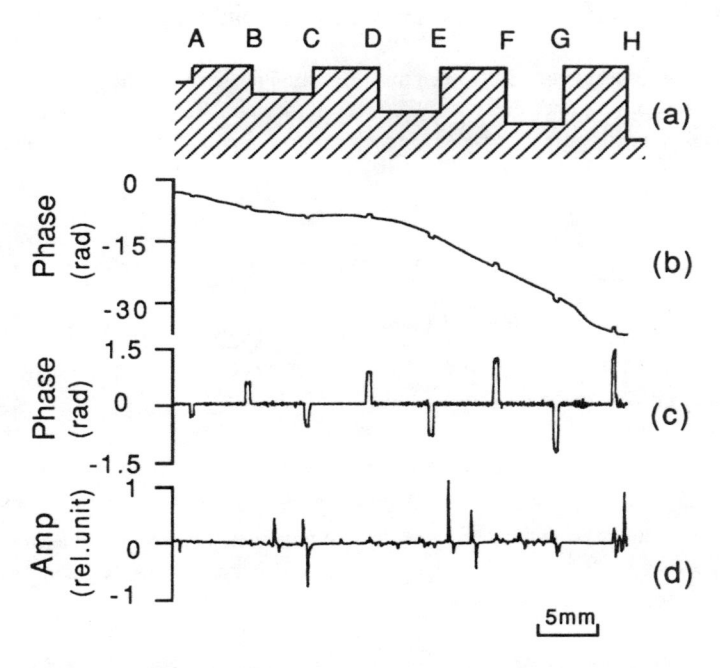

FIG. 2. Differential optical phase and amplitude line traces across an etched silicon wafer. (a) etched silicon surface, (b) differential phase, (c) differential phase (background variation subtracted), and (d) differential amplitude

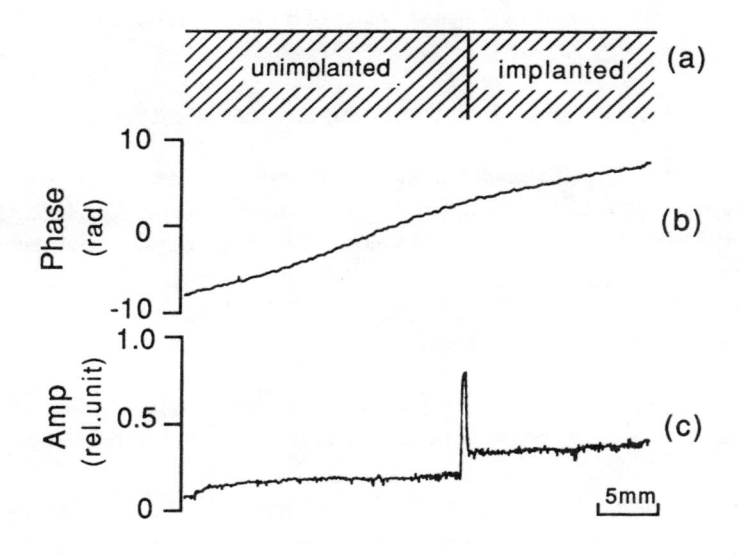

FIG. 3. Differential optical phase and amplitude line traces across a partly implanted silicon wafer. (a) silicon sample, (b) differential phase, and (c) differential amplitude

3. Analysis of a differential system with imperfect optical amplitude modulation

If the Bragg cell does not amplitude modulate each beam identically, then the differential phase result always remains valid, with a modification to the constant B in equation 2, whereas the differential amplitude result may not. To illustrate the dependence of the differential amplitude on the amplitude modulation, consider the case where the amplitude modulation per pass through the Bragg cell on the zeroth and first orders respectively is,

$$\left[1 - (1-\Delta)\cos^2\omega_a t\right]^{\frac{1}{2}} \tag{5}$$

$$\text{and} \qquad (1-\Delta-\beta)^{\frac{1}{2}}\cos\omega_a t \tag{6}$$

where Δ and β are constants. For identical modulation, Δ and β are zero in which case equations (5) and (6) reduce to $\sin\omega_a t$ and $\cos\omega_a t$ ie. identical modulation in phase quadrature and of equal amplitude. This was the condition assumed in deriving equation (1), which is the differential amplitude response, if all were perfect.

The form of the modulations in equations 5 and 6 have been chosen so as to model the physical situation. It was found that it was never possible to completely extinguish the zeroth order beam, hence the Δ term. Furthermore it was found that the Bragg cell produced other order beams besides the first, thus reducing the maximum intensity of the first order beam beam. The β term takes this into account. Using the amplitude modulations of equations 5 and 6, the differential amplitude signal detected at detector 2 and 1 respectively is,

$$\left(r_1^2 - r_2^2\right).(1-\Delta) - r_1^2\beta \tag{7}$$

$$\text{and} \qquad \tfrac{1}{2}\left(r_1^2 - r_2^2\right) - r_1^2(\Delta+\beta) + \tfrac{1}{2}\left\{\Delta^2 r_2^2 + (\beta+\Delta)^2 r_1^2\right\} \tag{8}$$

If we put in values of $\Delta=0.1$ and $\beta=0.05$, then the percentage error in the measurement of $(r_1^2 - r_2^2)$ from detector D2 is 5%, and from D1 is 30%. Hence with this example it can be seen that if there is any degree of imperfect modulation, then it is always better to user detector D2 for the differential amplitude measurement.

It can be seen in figure 1 that the two beams incident on detector 2 are not co-linear. Overlap of the two beams is not important because the differential amplitude signal does not rely on interference. The detector however must be large enough to detect all of both incoming beams. In contrast, the differential phase is an interference signal and a good overlap of the two beams is essential. Therefore the differential phase measurement is best made by detector 1.

4. A second implementation of the technique - the two first order system

Problems in achieving adequate amplitude modulation in the zeroth /first order system have led to a second implementation of the system (see figure 4). The Bragg cell is now driven by two r.f. signals. These are individually amplitude modulated with two identical but quadrature phase shifted signals. It is the two first order beams generated which are used to interrogate the sample. With this system it is in principle possible to compensate for any non-ideal characteristics of the Bragg cell.

4.1. Theory of two first order system

In this section, we will present a general analysis of the two first order systems. This analysis is valid for other implementations of the technique[1] such as the zeroth /first order system, but in this section it is discussed in the context of the two first order system.

Referring to figure 4, light from the laser (frequency ω_o) enters the Bragg cell, where it is divided into two first orders, upshifted in frequency by the two Bragg frequencies, ω_1, ω_2. The two first order beams can be written as,

$$E_1 = A_1 g_1(\omega_a t).exp\left(i\left[(\omega_o + \omega_1)t + \phi_1\right]\right) \tag{9}$$
$$E_2 = A_2 g_2(\omega_a t).exp\left(i\left[(\omega_o + \omega_2)t + \phi_2\right]\right) \tag{10}$$

where g_1, g_2 are the amplitude modulations of each first order per single pass of the Bragg cell and A_1, A_2, ϕ_1 and ϕ_2 are constants. g_1, g_2 could for example be equal to $\cos\omega_a t$ and $\sin\omega_a t$.

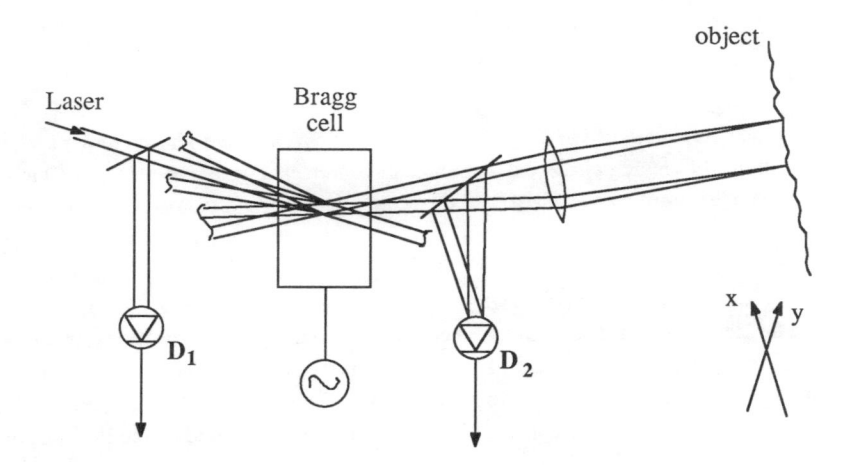

FIG. 4. Two first order system

The two beams are reflected off the sample where they experience two complex reflection coefficients $\underline{r_1}$, $\underline{r}$ and phase delays θ_1, θ_1 due to topography. After a second pass back through the Bragg cell the field at detector 1 is,

$$E = E_a + E_b \tag{11a}$$

where
$$E_a = \underline{r_1} A'_1 g_1^2(\omega_a t).exp\left(i\left[(\omega_b + 2\omega_1)t + 2\theta_1 + \phi'_1\right]\right) \tag{11b}$$

$$E_b = \underline{r_2} A'_2 g_2^2(\omega_a t).exp\left(i\left[(\omega_b + 2\omega_2)t + 2\theta_2 + \phi'_2\right]\right) \tag{11c}$$

and where $A'_1, A'_2, \phi'_1, \phi'_2$ are constants.

The light detected at detector 1 is,

$$I = (E_a + E_b).(E_a + E_b)^* \tag{12}$$

By decomposing this into its Fourier components, the differential phase signal is

$$B r_1 r_2 \cos\left[2(\omega_2 - \omega_1)t + 2\delta\theta + 2\delta\phi + 2\delta\alpha\right] \tag{13}$$

where r_1, r_2 and $\delta\alpha$ are related to the complex reflection coefficients according to equations 3 and 4 and B is a constant. If the following conditions are met:

(i) $g_2(\omega_a t) = \pm g_2(\omega_a t - \pi)$

(ii) $g_1(\omega_a t) = \pm g_2(\omega_a t - \pi/2)$

(iii) $A'_1 = \pm A'_2$

then the differential amplitude signal is,

$$C\left(r_1^2 - r_2^2\right)\cos 2\omega_a t \tag{14}$$

where C is a constant.

We will look at the above conditions. Conditions (ii) and (iii) are that the two modulations must be in phase quadrature and of identical form and that the depth of each modulation must be equal. Condition (i) tells us that we no longer need exact sinusoidal amplitude modulation. The amplitude modulation may be any function providing that it is even or odd. This result is important because the difficulties of exact sinusoidal modulation is greater in a two first order system than in the zeroth /first order system

If the modulation on each beam is not exactly the same, or the amplitude modulation is not in exact phase quadrature, then as with the zeroth / first order system, only the differential amplitude result is affected. In this case the analysis of section 3 still holds and it is better to use detector 2 for such measurement.

4.2 Results from the two first order system

Initial results from the two first order system are presented and compared with results from the zeroth /first order system. As in the previous system, a HeNe laser was used and the differential phase and amplitude measurements were made using detectors 1 and 2 respectively. The Bragg cell drive signals were triangularly modulated and the beam separation and focussed spot size on the sample was 260µm and 350µm.

We will look first at the results from the partly implanted silicon wafer as described in section 2.1 [Fig. 3 (a)]. The differential amplitude result shown in figure 5(b) can be seen to be a great improvement compared to that obtained with the zeroth /first order system [Fig. 3(c)]. The dc levels on both sides of the interface are now equal. This improvement can be attributed solely to there being identical modulation in each beam. The differential phase results from each system are shown in figures 3(b) and 5(a). These were taken with different lateral scales and over different areas and again show that any differential phase contrast caused by the ion implantation is smaller than that due to surface topography.

Figures 5(c,d,e) show results from the plasma etched silicon sample described in section 2.1. The differential phase result [figure 5(c)] again has an underlying variation due to warp of the wafer, and this has been subtracted away to reveal trace 5 (d). It can be seen that the sides of each peak from the two first order system are sloped compared to those from the zeroth /first order system. This is due to overlapping focussed beams being employed in the two first order system. The heights of these peaks have been converted into step heights and are shown in Table 1. They correspond very well with step height measurements from the zeroth /first order system and stylus probe. It can also be seen that the values all tend to be slightly less than those from the zeroth/first order system. This can be accounted for by the two beams being overlapped in the two first order system yet widely separated in the former system.

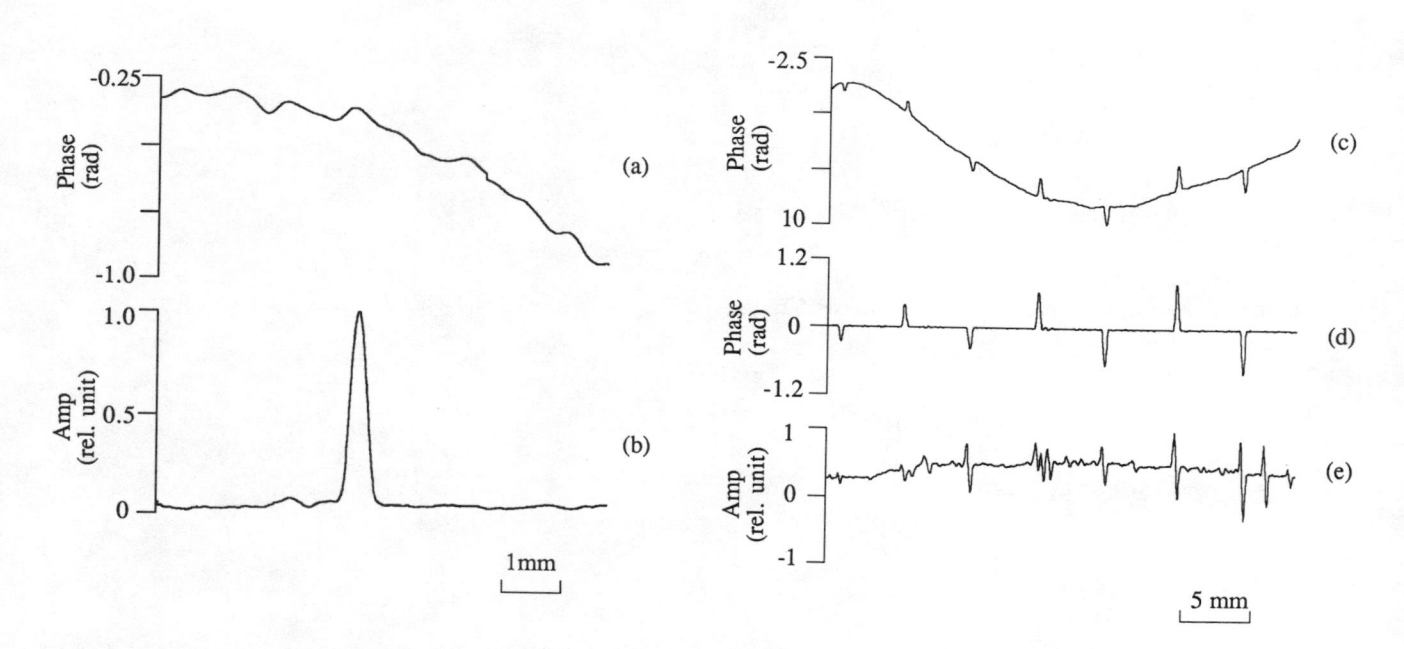

FIG. 5. Differential optical phase and amplitude line traces made by the two first order system across:
  (i) a partly implanted Si wafer. (a) differential phase, and (b) differential amplitude;
  (ii) an etched Si wafer. (c) differential phase, (d) differential phase (background variation subtracted away), and (e) differential amplitude

The differential amplitude result is shown in figure 5 (e). It can be seen however that the dc level is still non-zero and fluctuates about a mean level. There are two prime reasons for this. Firstly whilst the sample was mechanically scanned, due to wafer warp, the light moved across detector 2. This caused fluctuations in the dc level and may be overcome by using a larger area detector or measuring the differential amplitude from detector 1. Secondly, the mean dc level is due to the amplitude modulation on each first order not being in exact phase quadrature. This is a purely electronics problem and may be easily improved in future work.

5. Topographic interpretation of phase and amplitude information

This section will discuss some of the advantages of simultaneous amplitude and phase acquisition in optical profilometry and will give an idealised theoretical illustrative example.

One of the most pressing problems with optical metrology is that the phase information detected may arise either from topographical or structural variations of the sample. For instance a highly polished alloy with no topographical variation will give rise to a phase signal. Amplitude information can assist in determining whether the phase signal is topographic or due to compositional variation. A change in the amplitude of the reflected signal will generally indicate a compositional change. This change may however be causing some phase response. Monitoring of the amplitude can, at the very least, be used as a warning suggesting when phase results are likely to give an unreliable measure of surface profile.

In general, as shown in section 2, even if we have amplitude and phase information we cannot expect to perform a unique inversion to yield topography. Amplitude and phase information on its own cannot eliminate all the ambiguity present in the data but does reduce the number of independent additional assumptions that need to be made in order to characterise the sample surface.

In order to illustrate this point we give an idealised example, which nevertheless, does show the potential of this type of system. We consider a layered structure as shown in figure 6 below.

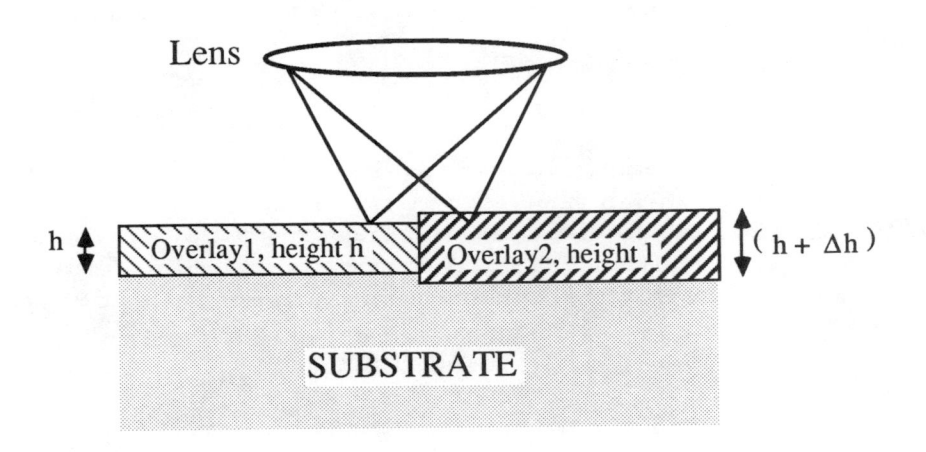

FIG. 6. Schematic diagram showing geometry of "sample" used in illustrative example

This structure may be taken to represent, for instance, two overlays of SiO_2 on silicon. A scan with a conventional phase profilometer would reveal a phase change across the interface which could arise from either topographical differences or differences in the thickness of the layer.

We make several simplifying assumptions for this particular example. The lens has a very low numerical aperture and the two beams are sufficiently well separated such that each beam interrogates one side of the interface only. We also look at the case where the layer thickness is small compared to the optical wavelength. Figure 7 shows the differential amplitude and phase terms for an overlay1 thickness of 50 Å and an

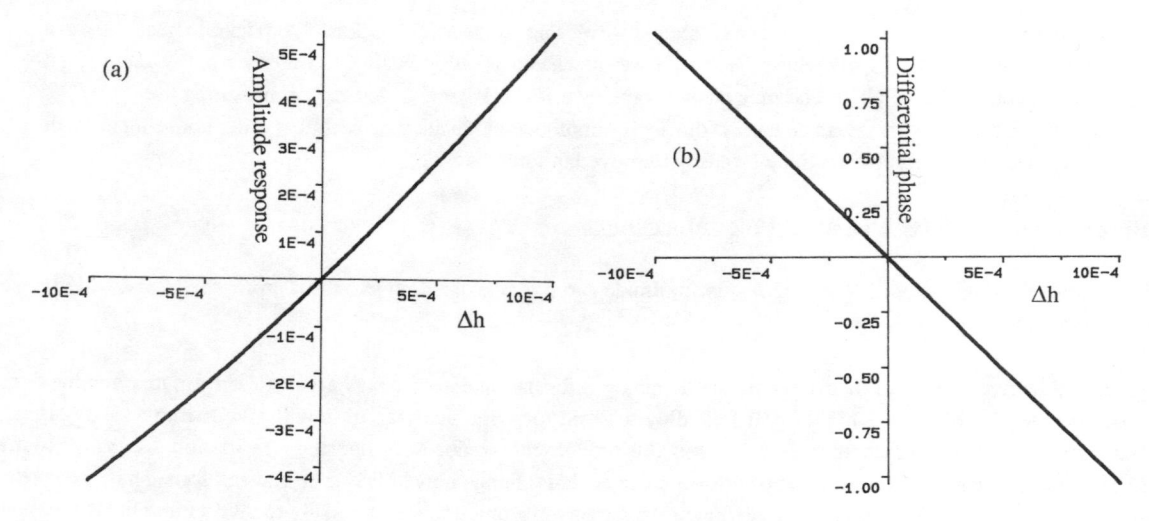

FIG. 7. Theoretical plot of (a) differential amplitude response and (b) differential phase response (degrees) versus difference in layer thickness, Δh (µm)

overlay2 of thickness, $(50+\Delta h)$ Å, where Δh is small compared to 50 Å. The abscissa represents the difference in response of the two overlays. The values of optical parameters used correspond to those of silicon dioxide and silicon. We can see that the differential amplitude signal allows us to determine the thickness difference, Δh, of the overlays. Hence the real height difference of the surface of the overlays is the deviation of the phase measurement from the value expected for the overlay thickness difference, Δh.

In this purely differential mode the combined system can be used to determine the topography and the variation of the layer thickness from a prescribed value. If in addition the absolute amplitude of the reflected signal is recorded, which may be readily measured in our system by taking the signal output at $4w_a$, the absolute layer thickness may be also be obtained. This value need not then be assumed when interpreting the differential results.

In order to further increase the information extracted from the system additional amplitude and phase information may be obtained using different laser wavelengths.

6. Conclusions

The second implementation using two first order beams has several advantages over the original system. Firstly the conditions for perfect differential amplitude modulation can be more closely met as the modulation of each beam can be controlled by external electronics rather than the physics of the Bragg cell. The theory presented in this paper shows that the modulation need not be sinusoidal which is particularly advantageous in the implementation of the new system where sinusoidal modulation is more difficult to attain. The results show that the differential amplitude signal is indeed superior for the new implementation when compared to our previous results.

7. Acknowledgments

The authors would like to thank the Science and Engineering Research Council and GEC for a CASE studentship for R.K.A., and SERC -Alvey for financial assistance for C.W.S..

8. References

1. C.W. See, R.K. Appel and M.G. Somekh, Appl. Phys. Lett. **53**,1 (1988)
2. C.W. See, M. Vaez Iravani and H.K. Wickramasinghe, Appl. Opt. **24**, 2373 (1985)
3. C.W. See and M. Vaez Iravani, Electron. Lett. **22**,1079 (1986).
4. C.C. Huang, Opt. Eng. **23**, 365 (1984).
5. G.E.Sommargren, Appl. Opt. **20**, 610 (1981).
6. G.E.Sommargren and B.J. Thompson, Appl. Opt. **12**, 2130 (1973).
7. M. Vaez Iravani and C.W. See, in Proceedings of SPIE conference on 'Scanning Microscopy Technologies and Applications', Los Angeles, CA, 1988 p.897.
8. C.W. See and M. Vaez Iravani, Electron. Lett. **22**,961 (1986).

Excentration errors combined with wavefront aberration
in a coherent scanning microscope

A.M. Hamed

Physics Department, Faculty of Science, Ain Shams University,
Cairo, Egypt

ABSTRACT

The Coherent Scanning Microscope (CSM) is mainly composed of two objectives arranged in tandem combined with the scanned object. The synchronized scanning of both of the object and the electron beam emitted from the cathod ray monitor will assure the construction of the temporal image. Since it is easy to aligne the first objective of the CSM, while it is not the case for the second objective, hence we assume that the later objective is subjected to tilting and to a lateral shift with respect to the first aligned objective. Another case assumes that the second objective is influenced by coma of third order in addition to the excentration errors. In both cases, the resulting impulse response $h_r = h_1 h_2$ is calculated. The above mentioned problems of excentration errors combined with wavefront aberration, which are both harmfull for the microscope resolution, are discussed.

1. INTRODUCTION

The basical concept of Coherent Scanning Microscope (CSM) has been given by Minsky[1] in his patent application, while Davidovits et al.[2] have built this microscope using a direct laser illumination, however, spatial coherence conditions have not been considered in this early work.

Nomarski[3] has given a right formula for the depth of field of CSM, following him, Sheppard and his collaborators have contributed by numerous papers[4-9] to the development of this new technique of microscopic imaging (named Confocal imaging). Hamed et al.[10-14] have proposed novel methods, using amplitude modulation technique, in order to improve the resulting impulse response of the CSM.

A pinhole detector is necessary in order to obtain a coherent image. Practically all imaging systems exhibit a certain degree of coherence so the illumination is always partial coherent. Hence, it is necessary to choose the pinhole diameter small enough in order to gain sufficient output energy i.e. compromising between the illuminating conditions and the pinhole diameter of the detector must be found in order to enhance the image contrast.

The confocal arrangement of the objectives L_1 and L_2 combined with the scanned object are the basic elements of the CSM as shown in fig. 1. In this schematic diagram, the scanning of both of the object and electron beam emitted from the cathod ray monitor simultaneously will assure the construction of the temporal image. Unfortunately, the CSM either working in transmission or at reflection will give weak output signal which needs amplification before its detection. For example the heterodyne technique of detection may be useful in this case in order to improve signal to noise ration S/N. We pay attention that the weak output signal simultaneously with the required high precision of object mechanical scanning will make this microscope much more difficult for its realization than the well known classical optical microscopes.

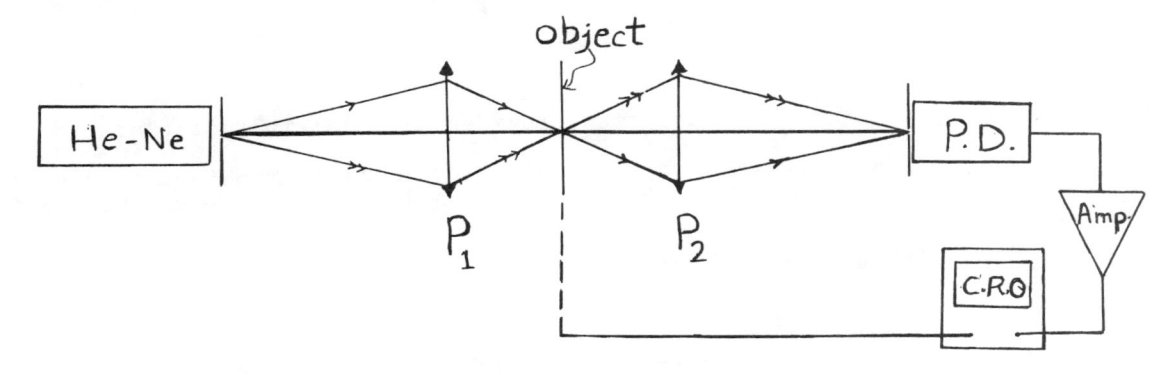

Figure 1. Figure 1. Schematic diagram of the Coherent Scanning Microscope.

The purpose of this study is to show the influence of excentration of the objectives on the resulting impulse response obtained by the CSM. A disaligned optical system of the CSM assuming that only the second objective exhibits a lateral shift is shown in fig. 2. Another disaligned optical system in which L_2 is tilted and shifted with respect to the optical axis of the aligned objective L_1 is shown in fig. 3. These serious problems of excentration will be analyzed in the following section taking in consideration the classical wavefront aberration.

2. THEORETICAL ANALYSIS

It is commode to calculate the intensity distribution of a Coherent Scanning Microscope (CSM) from the modulus square of the following convolution integral[4] :

$$I(w) = |g(w) * h_r(w)|^2 \qquad (1)$$

where $g(w)$ is the complex amplitude transmitted from the object and $h_r(w) = h_1 \times h_2$ is the resulting impulse response. h_1 is given for the first objective lens while h_2 is for the second objective.

2.1. Effect of excentration errors

Since it is easy to aligne the first objective of the CSM while it is not the case for the second objective, hence we assume that the later objective is subjected to tilting and to a lateral shift with respect to the first aligned objective. Under the above conditions the complex amplitude transmitted from the first objective for whatever oblique ray is :

$$P_1(\rho;\alpha) = P_{o1} \exp jk\rho \sin(\alpha) \quad , \quad |\rho/\rho_{o1}| \leq 1 \qquad j = -1 \qquad (2)$$

and that corresponding to the second shifted objective is :

$$P_2(\rho; \beta, \rho_d) = P_{o2}(\rho; \rho_d) \exp[jk\rho \sin(\beta)] \; ; \; |(\rho \pm \rho_d)/\rho_{o2}| \leq 1 \qquad (3)$$

where α and β are the tilting angles of the corresponding objectives L_1 and L_2 and ρ_d is the lateral shift introduced w.r.t. the optical axis. The sign plus or minus appearing in eq. (3) determine the direction of the displacement given to L_2 w.r.t. the optical axis as shown in fig. 2.

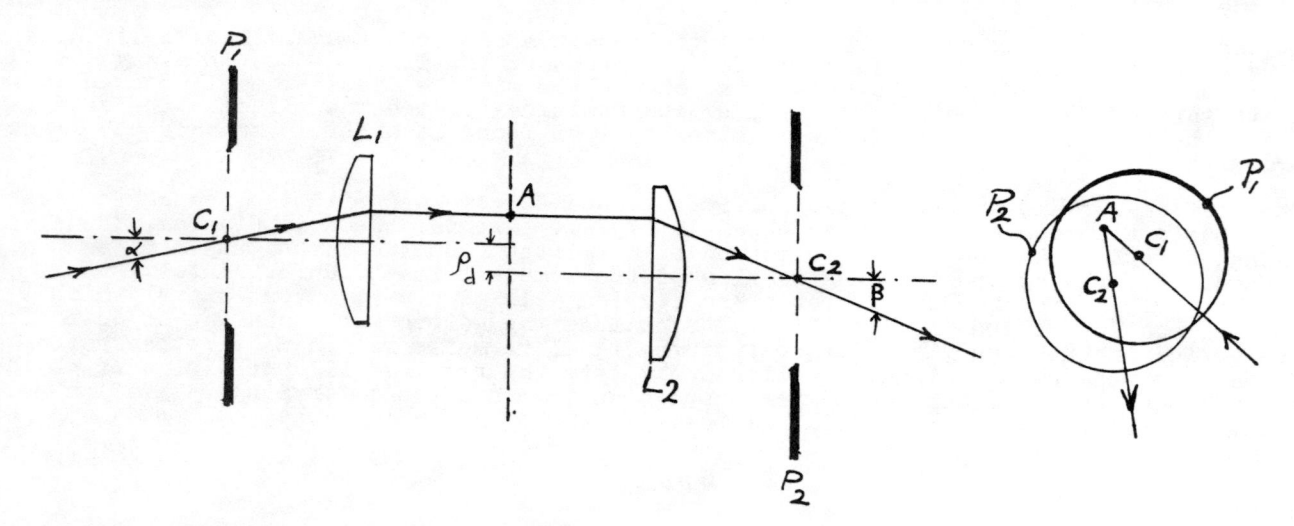

Figure 2. Set-up of disaligned optical system provided with lateraly displaced objective

The point spread function is obtained by operating the Fourier transformation upon eq. (2). After making use of convolution operations, we get this result for the PSF for the first objective as :

$$h_1(w; \alpha) = 2\pi\rho_{o1}^2 \Lambda_1(w) * \delta(r - f \sin\alpha) \qquad (4)$$

It is rewritten as :

$$h_1(w; \alpha) = 2\pi\rho_{o1}^2 \Lambda_1(w'); \quad w' = \frac{2\pi\rho_{o1}}{\lambda f}(r - f \sin\alpha)$$

where $\Lambda_n(w) = J_n(w) / w^n$ stands for the nth impulse response.

The new reduced coordinate w' is related to the ordinary reduced coordinate w through this equality:

$$w' = w \left(1 - \frac{f}{r} \sin\alpha\right) \quad ; \quad w = (2\pi/\lambda f)\rho_{o1} r$$

It is clear that the relative amount of shift $\Delta w/w = \frac{f}{r} \sin\alpha$.

The calculation of the PSF for the 2nd objective is not straightforward. Hence, before operating the transformation, it is helpful to change the radial limits of integration to become $\rho_{min} = \pm \rho_d$ and $\rho_{max} = \rho_{o2} \pm \rho_d$ instead of $(0, \rho_{o2})$. We have got this result:

$$h_2(w; \beta, \rho_d) = 2\pi (\rho_{o2} \pm \rho_d)^2 \left[\Lambda_1(w_u) - \tau^2 \Lambda_1(w_L) \right] \quad ; \quad \tau = \rho_d/(\rho_{o2} \pm \rho_d) \tag{5}$$

where $w_u = \frac{2\pi}{\lambda f}(\rho_{o2} \pm \rho_d)(r - f \sin\beta)$ and $w_L = \frac{2\pi}{\lambda f} \rho_d (r - f \sin\beta)$

It is clear that the upper limit of integration w_u is less than w by $(1 \pm \frac{\rho_d}{\rho_{o2}})(1 - \frac{f}{r}\sin\beta)$ and the lower limit w_L is less than w by $\frac{\rho_d}{\rho_{o2}}(1 - \frac{f}{r}\sin\beta)$.

The resulting impulse response is deduced from the product of eq. (4) and eq. (5) as:

$$h_r(w; \alpha, \beta, \rho_d) = 4\pi^2 \rho_{o1}^2 (\rho_{o2} \pm \rho_d)^2 \Lambda_1(w; \alpha) \left[\Lambda_1(w_u; \beta, \rho_d) - \tau^2 \Lambda_1(w_L; \beta, \rho_d) \right] \tag{6}$$

2.2. Combination of excentration and aberration

We calculate the intensity of the impulse response in the general case where the objectives of the CSM are influenced by coma of third order in addition to the excentration errors. These defects are both harmfull for the resolution of the microscope. We assume that the second objective is influenced by a tilting and lateral shift combined with a wavefront aberration. The transmitted amplitude is:

$$P_2(\rho; \rho_d, \beta, c_1) = \underbrace{P_{o2}(\rho; \rho_d)}_{\text{shift}} \cdot \underbrace{\exp(jk\rho \sin\beta)}_{\text{tilting}} \underbrace{\exp jk\Delta(\rho, \phi)}_{\text{aberration}} \tag{7}$$

In the case of small phase $K\Delta$ we can develop it in a series form[15] as:

$$\Delta(\rho, \phi) = d\rho^2 + \sum_{n=2}^{N} s_n \rho^{2n} + \sum_{n=1}^{N} c_n \rho^{2n+1} \cos\phi + a\rho^2 \cos 2\phi \tag{8}$$

where d: defect of focus coefficient, s: spherical aberration, c: coefficient of coma and a: coefficient of astigmatism. Retaining only the 3rd order come, eq.(8) becomes:

$$\Delta(\rho, \phi) = c_1 \rho^3 \cos\phi, \text{ where } \phi \text{ azimuthal coordinate in the objective plane.}$$

If the deviation of the wavefront does not exceed $\lambda/4$, then we retain from eq. (7) the terms up to the second order:

$$\exp jk\Delta(\rho, \phi) \simeq 1 + jk\Delta(\rho, \phi) - \frac{k^2}{2}\Delta^2(\rho, \phi)$$
$$= 1 + jkc_1\rho^3 \cos\phi - \frac{k^2}{2} c_1^2 \rho^6 \cos^2\phi \tag{9}$$

From eq. (7) and eq. (9), we write the aberrated pupil function as:

$$P_2(\rho; \rho_d, \beta, c_1) = P_{o2}(\rho; \rho_d) \exp(jk\rho \sin\beta) \times$$
$$\left[1 + jkc_1 \rho^3 \cos\phi - \frac{k^2}{2} c_1^2 \rho^6 \cos^2\phi \right] \tag{10}$$

The Fourier transformation is operated upon eq. (10) to obtain the PSF as:

$$h_2(w; \rho_d, \beta, c_1) = \left\{ F.T.[P_{o2}(\rho; \rho_d)] \right\} * \delta(r - f \sin\beta)$$
$$+ jkc_1 \left\{ F.T.[P_{o2}(\rho; \rho_d) \rho^3 \cos\phi] \right\} * \delta(r - f \sin\beta)$$
$$- \frac{k^2 c_1^2}{4} \left\{ F.T.[P_{o2}(\rho; \rho_d) \rho^6 (1 + \cos 2\phi)] \right\} * \delta(r - f \sin\beta)$$

Making use of the properties of Bessel function

$$J_n(x) = \frac{j^{-n}}{2\pi} \int_0^{2\pi} \exp(jx\cos\phi)\exp(jn\phi)\,d\phi$$

then we find that $F.T.[P_{o2}(\rho;\rho_d)\rho^3\cos\phi] = 0$

and $F.T.[P_{o2}(\rho;\rho_d)\rho^6\cos 2\phi] = 0$

We get finally this result for the PSF as :

$$h_2(w;\rho_d,\beta,c_1) = 2\pi(\rho_{o2}\pm\rho_d)^2 \cdot A(w;\beta,\rho_d) - 2\pi(\rho_{o2}\pm\rho_d)^8 [B(w;\beta,\rho_d,c_1)$$
$$- 6C(w;\beta,\rho_d,c_1) + 24D(w;\rho_d,\beta,c_1) - 48E(w;\rho_d,\beta,c_1)] \quad (11)$$

where

$$A = \Lambda_1(w_u) - \tau^2\Lambda_1(w_L) \quad , \quad B = (kc_1/2)^2[\Lambda_1(w_u) - \tau^8\Lambda_1(w_L)]$$

$$C = (kc_1/2)^2[\Lambda_2(w_u) - \tau^8\Lambda_2(w_L)], \quad D = (kc_1/2)^2[\Lambda_3(w_u) - \tau^8\Lambda_3(w)]$$

$$E = (kc_1/2)^2[\Lambda_4(w_u) - \tau^8\Lambda_4(w)]$$

The resulting impulse response and its intensity are calculated from eq. (4) and eq. (11) as : $h_r = h_1 \cdot h_2$ and $I = h_r(w;\beta,\rho_d,c_1)^2$ (12)

In the case of perfect alignment of the optical system ($\rho_d = 0, \beta = 0$) we obtain a result valid for an objective suffering only from simple coma :

$$h_2(w;c_1) = 2\pi(1 - k^2c_1^2/4)\Lambda_1(w) + 3\pi k^2c_1^2[\Lambda_2(w) - 4\Lambda_3(w) + 8\Lambda_4(w)] \quad (13)$$

(the pupil radius is taken to be a unity $\rho_{o2} = 1$).

In this case, the intensity response is deduced from eq. (4) and eq. (13). For free aberration and perfectly aligned confocal system, one get the result given by Sheppard[4] :

$$I(w) = const.\Lambda_1^4(w) \; ; \; c_1 = 0 \quad (14)$$

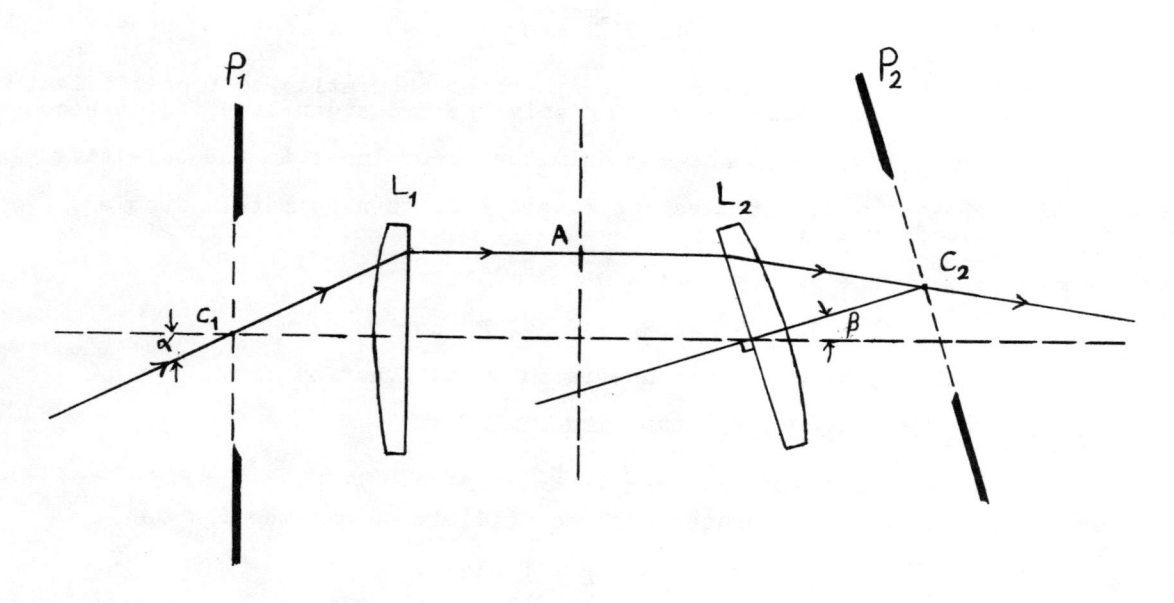

Figure 3. Set-up of disaligned optical system provided with objectives possessing both of an orthogonal tilting combined with lateral shift.

3. RESULTS AND DISCUSSION

The intensity distribution of a point object (resulting impulse response) is calculated in the case where the second objective of the CSM in subjected to a lateral shift ρ_d due to failure of alignment. The computational results of eq. (6) are graphically represented as shown in fig. (4 a,b). A comparative curve representing the case of misalignment in a classical optical microscope is drawn. The numerical apertures of both of the objectives of the CSM are chosen to be 0.5. It is shown, from the obtained results, that :

- the central intensity peak of the resulting impulse response is nearly unaffected for lateral shift not exceeding 500 μm.
- the cut-off spatial frequency of the classical optical microscope r_c = 0.75 μm is greater than that obtained for the CSM r_c = 0.6 μm where both of the microscopes are subjected to equal lateral shift.
- as the lateral shift greatly increases, the numerical aperture of the misaligned objective will decrease from NA = ρ_o/f to NA' = NA - (ρ_d/f). Consequently, according to Rayleigh criterion of resolution the cut-off spatial frequency may be displaced toward the greater values.

Another set of curves for the intensity response is calculated from eq. (13) and is drawn as shown in fig. 5 and fig. 6 where the second objective is affected by a coma of third order. The comatic coefficient is chosen to be 0.05, 0.1 and 0.15. It is shown that the cut-off spatial frequency r_c is shifted towards the smaller values at the expense of the enhancement of the secondary peaks when the coefficient of coma exceeds 0.1. Refering to the results, presented graphically in fig. 5, 6; it is shown that the maximum value of the intensity distribution is much degraded with increasing coma, e.g. a reduction of the maximum intensity estimated by about 43% is obtained for c_1 = 0.2 while only 12% is obtained for c_1 = 0.1.

We conclude from the obtained result that the excentration errors combined with wavefront aberration will modify the resulting response and will degrade the image contrast with respect to perfectly aligned optical system.

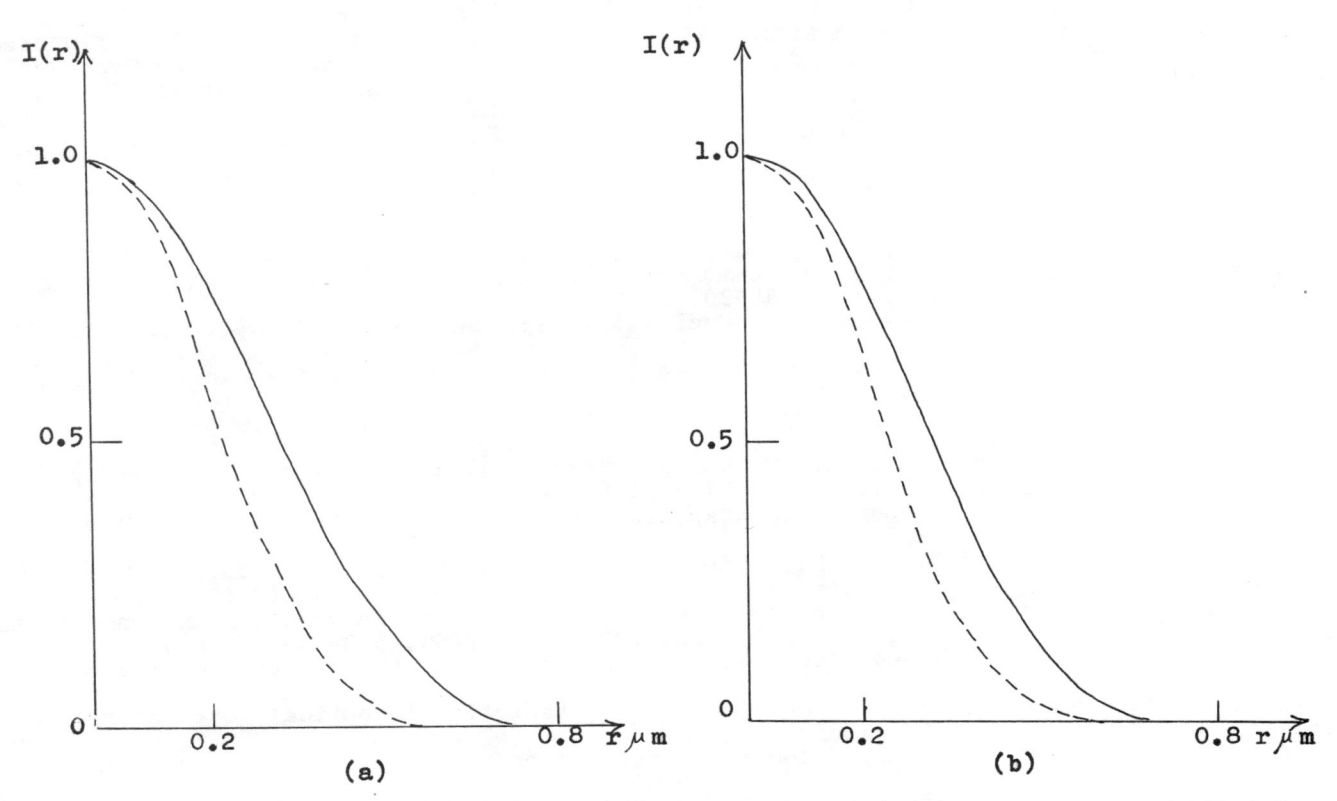

Figure 4. Intensity response of a CSM where the second objective is laterally shifted by an amount ρ_d, where ρ_d = 100 μm fig. 4a and ρ_d = 500 μm fig. 4b.

——— : for classical optical microscope
- - - - : for coherent scanning microscope.

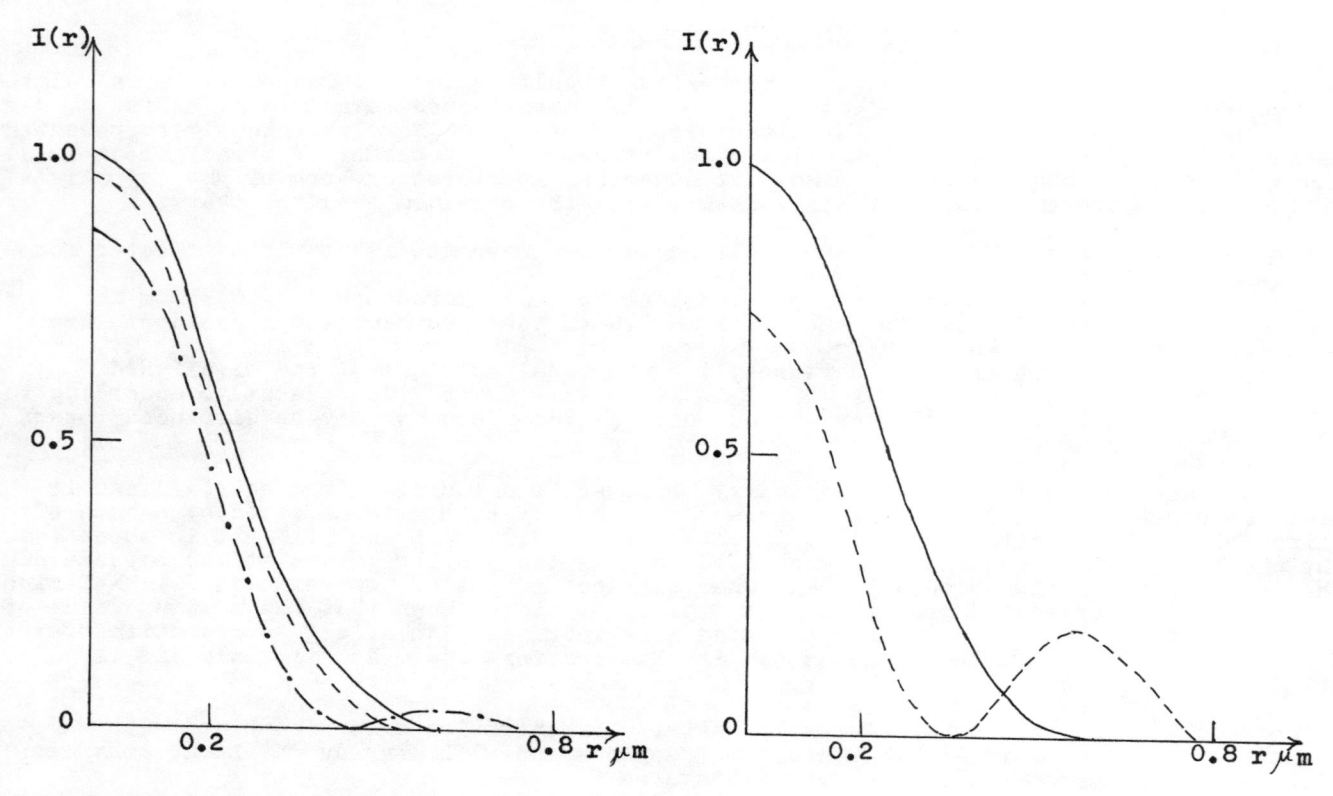

Figure 5. Intensity response of a CSM where the second objective is influenced by a coma of third order.
— : coefficient of coma $c_1 = 0$
--- : $c_1 = 0.05$
-.- : $c_1 = 0.1$

Figure 6. Intensity response of a CSM where the second objective is influenced by coma of $c_1 = 0.15$ represented by the discontinuous line while the solid line is given for $c_1 = 0$ for comparison.

4. REFERENCES

1. M. Minsky, U.S. Patent 3013467, Microscopic Apparatus, Dec. 19, 1961.
2. P. Davidovits and M.D. Egger, "Scanning laser microscope for biological investigation", Appl. Opt., 10(7), 1615-1619 (1971).
3. G. Nomarski, "Simple method for reducing the depth of focus", J. Opt. Soc. Am. 21, 1166 (1975).
4. C.J.R. Sheppard and A. Coudhury, "Image formation in the scanning microscope", Opt. Acta 24(10), 1051-1073(1977).
5. T. Wilson and J.N. Gannaway, "The imaging of periodic structures in partially coherent and confocal scanning microscopes", Optik 54(3), 201-210 (1979).
6. I.J. Cox, C.J.R. Sheppard, and T. Wilson, "Improvement in resolution by nearly confocal microscopy", Appl. Opt. 21(5), 778-781 (1982).
7. T. Wilson, "Imaging properties and application of scanning optical microscopes", Appl. Phys. 22, 119-128 (1980).
8. C.J.R. Sheppard and T. Wilson, "Effect of spherical aberration on the imaging properties of scanning optical microscopes", Appl. Opt. 18(7), 1058-1063 (1979).
9. C.J.R. Sheppard and T. Wilson, "Fourier imaging of phase information in scanning and conventional optical microscopes", Phil. Trans. Ray Soc. A295, 513-536 (1980).
10. J.J. Clair and A.M. Hamed, "Theoretical studies on optical coherent microscopes", Optik 64(2), 133-141 (1983).
11. A.M. Hamed and J.J. Clair, "Image and super resolution in optical coherent microscopes", Optik 64(4), 277-284 (1983).
12. A.M. Hamed and J.J. Clair, "Studies on optical properties of confocal scanning optical microscope" Optik 65(3), 209-218 (1983).
13. A.M. Hamed, "Aberration studies utilising an optoelectronic coherent microscope", Optik 67(3), 279-290 (1984).
14. A.M. Hamed, "Resolution and contrast in confocal optical scanning microscopes", Opt. & Laser Tech. 16(2), 93-96 (1984).
15. A. Boivin, "Theorie et calcul des figures de diffraction de revolution, pp. 92-97, Laval U.P., Quebec (1964).

OPTIMIZATION OF RECORDING CONDITIONS IN LASER SCANNING MICROSCOPY

Tobias Damm, Michael Kaschke, and Uwe Stamm

Friedrich-Schiller University of Jena
Max-Wien-Platz 1, Jena, 6900, GDR

ABSTRACT

The measuring process of laser scanning microscopy is analyzed. The distortion of the recorded image due to finite focus diameter and the temporal response of the electronic recording device is estimated. The influence of noise on the accuracy of the recorded image is considered. Optimum recording conditions such as choice of scan velocity and time constant of the registration device are discussed for minimal total error.

1. INTRODUCTION

In laser scanning microscopy (LSM) the image is achieved sequentially by scanning a diffraction-limited laser spot relative to the surface of an object in a raster type scan. Thus the image of the object is built up pointwise by recording a signal arising from the illuminated spot and storing it into a digital memory the content of which may be displayed on a screen /1/.

In general the recorded image differs from the original due to the finite spot size, the finite raster step width, the limited response time of the electronic registration system, and due to the influence of noise and laser fluctuations.

In this paper we discuss optimum recording conditions for LSM with respect to the condition that the total difference between the recorded image and the original should be a minimum.

2. CHARACTERIZATION OF THE RECORDING PROCESS

Let us first investigate of the build-up of the recorded signal. The laser spot is scanned line by line along the object. Due to this operation mode we want to restrict our considerations to a one dimensional scan along one line. Since the illuminating spot is of finite diameter the detector receives a signal which corresponds to the convolution of the object given by the object function $S(x')$ with the focal function $F(x')$, as sketched in Fig. 1. The detector (e.g. a Photomultiplier, an amplifier for OBIC-signals) has a conversion sensitivity E and generates a noise $r(t)$. The output of this detector is digitized and stored into the frame memory as a function of the spatial position of the laser spot. In order to compare the signal $D(x)$ with the object $S(x')$ we introduce a scaling factor $1/E$ (describing overall conversion efficiency) into our signal chain. Because the laser spot is scanned with the velocity $v = dx/dt$ we must take into account the finite response time of the detection and digitization device represented by $R(t)$. Thus we have an output signal

$$D(x) = S(x) + (D_d(x) - S(x)) + D_n(x) \tag{1}$$

which differs due to the noise (D_n) and the distortions (D_d-S) from the original property or structure of the object. At a given spatial position x the corresponding memory contains the value:

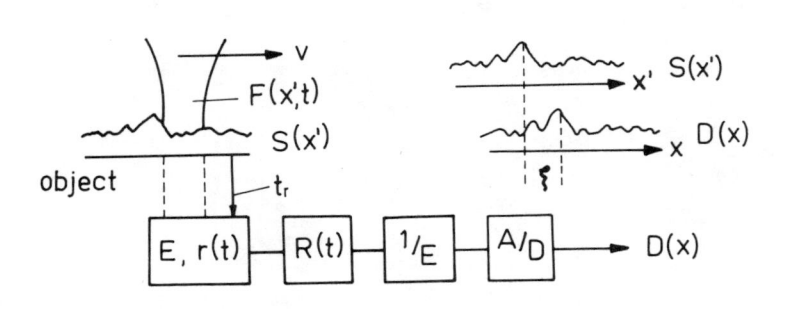

Figure 1. Signal chain in a laser scanning microscope.

$$D(x) = \frac{1}{E}\int_0^\infty dt' \int_{-\infty}^{+\infty} dx' E\, R(t')\, S(x-x'-vt'+\delta)\, F(x') + \frac{1}{E}\int_0^\infty dt'\, R(t')r(x/v-t') \qquad (2)$$

where δ is an additional parameter which is introduced to compensate for a possible constant shift ξ between any spatial position in the object and its counterpart in the image. This shift can result from the running time t_r of the signal from the object to the detector (see below).

The formula for $D(x)$ is similar to that of the output signal in cw spectroscopy /2/ or to the recorded signal in continuously working picosecond pump and probe pulse spectrometers /3/. Therefore we can apply a procedure similar to that given in /2/ and /3/. In order to have a measure for the distortions and the influence of noise on the recording process we will define a distortion error F_d and a noise error F_n.

2.1. Error due to distortions

As a measure for the distortions due to the final focal width and the finite response time of the electronic device we will use the normalized mean square deviation F_d between the original S and the distorted signal D_d.

$$F_d = \overline{(D_d(x) - S(x))^2}^x \Big/ \overline{S^2(x)}^x \qquad (3)$$

The spatial averaging is performed over the length of at least one total line scan. Substituting of (2) into (3) yields:

$$F_d = \frac{1}{\overline{S^2(x)}^x}\int_0^\infty dt' \int_0^\infty dt'' \int_{-\infty}^\infty dx' \int_{-\infty}^\infty dx''\, R(t')\, R(t'')\, F(x')\, F(x'') \cdot$$

$$\overline{S(x+x'-vt'+\delta)S(x+x''-vt''+\delta)}^x$$

$$- \frac{2}{\overline{S^2(x)}^x}\int_0^\infty dt' \int_{-\infty}^\infty dx'\, R(t')\, F(x')\, \overline{S(x+x'-vt'+\delta)S(x)}^x + 1 \qquad (4)$$

The products $\overline{S(x_i)S(x_i-\varepsilon)}^x$ can be expressed by the autocorrelation function $\Phi(\varepsilon)$ of the object function S

$$\Phi(\varepsilon') = \int_{-\infty}^\infty S(x_1)S(x_1-\varepsilon')\, dx_1 \Big/ \int_{-\infty}^\infty S^2(x')\, dx' \qquad (5)$$

Hence, for the distortion error holds

$$F_d = \int_0^\infty dt' \int_0^\infty dt'' \int_{-\infty}^\infty dx' \int_{-\infty}^\infty dx''\, R(t') \cdot R(t'') \cdot F(x') \cdot F(x'') \cdot \Phi(x''-x'-v(t''-t'))$$

$$- 2 \cdot \int_0^\infty dt' \int_{-\infty}^\infty dx'\, R(t') \cdot F(x') \cdot \Phi(x'-vt'+\delta) + 1 \quad . \qquad (6)$$

Let us now discuss some properties of the elements which influence the distortions. For the object function it should only be assumed that it has an upper limit of the cut off frequency f_o of the corresponding power spectrum of spatial frequencies, which physically means that the object structure has only variations with lengths larger than $1/2f_o$. (In agreement with the sampling theorem the cut off frequency can be expressed by $f_o = 1/2l_o$ where l_o is the sampling point distance.)

In order to simplify the calculations we will further assume that this power spectrum L(f) is constant below f_o. This is not a strong restriction because there are no demands concerning the phase distributions over spatial frequencies. Therefore an infinite number of functions S(x) can be ascribed to such a spectrum. For the autocorrelation function $\phi(x)$ then holds

$$\Phi(x) = \frac{1}{2f_o} \int_{-f_o}^{f_o} df\, L(f)\, \exp(2\pi i f x) \quad , \quad L(f) = \begin{cases} 1 & \text{if } |f| \le f_o \\ 0 & \text{if } |f| > f_o \end{cases} \quad (7) \tag{7}$$

The electronic registration system is described by the response function R(t) of a low pass filter having a characteristic time constant. This behaviour results e. g. from the time constant of the detector, from the finite sample time of the analog-digital converter, and from the digitization time. One digitized value D(x) is representative for the interval Δx to the previous digitized value $D(x-\Delta x)$. Consequently, higher spatial frequencies than $1/\Delta x$ can not be recorded and so a low pass behaviour occurs. Although the low pass behaviour leads to an increase of the distortions due to smearing out the signal it is, however, of advantage concerning the suppression of noise.

We want to treat two cases of low pass filters which should be representative for the complete registration system.

I   ideal integrator:

$$R^I(t) = 1/\tau \quad \text{if } t_r \le t < \tau + t_r \qquad \text{otherwise } R(t) = 0 \tag{8a}$$

II RC-filter:

$$R^{RC}(t) = (1/z)\exp(t_r - t)/z \quad \text{for } t > t_r. \tag{8b}$$

The value t_r represents the running time of the signal from the laser spot to the detector and the digitizer. Distortions of the signal along this path are neglected. It should be noted that this running time t_r can take countable values if one thinks of OBIC and photothermal laser scanning microscopy /1/.

In order to simplify the calculations we approximate the focal function by $F(x) = (1/r_o\sqrt{\pi})\exp(-x^2/r_o^2)$. The value r_o is connected with the FWHM-diameter d_o of the focus by $d_o = 2r_o\sqrt{\ln 2}$.

With (7) and (8) into eq. (6) we obtain the distortion

$$F_d^I = 1 + (1/2f_o)\cdot \int_{-f_o}^{f_o} df\cdot \exp(-2\pi^2 r_o^2 f^2)\cdot [\sin^2(\pi f v \tau)/(\pi f v \tau)^2]$$

$$- (1/f_o)\cdot \int_{-f_o}^{f_o} df\cdot \exp(-\pi^2 r_o^2 f^2)\cdot [\sin(\pi f v \tau)/(\pi f v \tau)]\cdot \exp\{-2\pi i f[(vt_r - \delta) - \tfrac{1}{2}v\tau]\} \tag{9a}$$

and

$$F_d^{RC} = 1 + (1/2f_o)\cdot \int_{-f_o}^{f_o} df\cdot \exp(-2\pi^2 r_o^2 f^2)/[1+(2\pi f v z)^2]$$

$$- (1/f_o)\cdot \int_{-f_o}^{f_o} df\cdot \exp[-\pi^2 r_o^2 f^2 - 2\pi i f(vt_r - \delta)]/[1+2\pi i f v z]\cdot \tag{9b}$$

for the integrator and the RC-filter, respectively.

Now we want to discuss the shift between a position at the object and the corresponding position in the image which is a result of the time constant of the low pass filter and of the running time t_r, as mentioned above. Due to this shift the correlation of the image with the object will be rather small. This correlation can be improved (i. e. the distortion error can be minimized) by variation of the parameter δ. The optimal value of δ derived from eq. (9) by differentiation with respect to δ is

$$\delta^I = vt_r + v\tau/2 \quad \text{and} \quad \delta^{RC} = vt_r + vz \tag{10}$$

for the integrator and the RC-filter, respectively. Since the error due to noise is independent of the shift between image and object we will also minimize the total error. For sake of simplicity we expand the distortion error into a power series of the space frequency f. Using the optimum values of δ we obtain in the lowest order

$$F_d^I{}_{opt} = (\pi f_o)^4 [v^4 \tau^4/180 + r_o^2 v^2 \tau^2/15 + r_o^4/5] \tag{11a}$$

for the integrator and

$$F_d^{RC}{}_{opt} = (\pi f_o)^4 (16 v^4 z^4/15 + r_o^4/5 + 4 r_o^2 v^2 z^2) \tag{11b}$$

for the RC-filter, respectively.

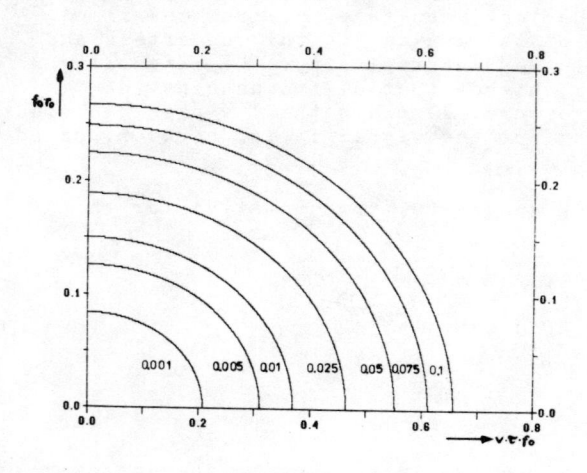

Figure 2.
Minimum distortion error as a function of scan parameters v and r_o (normalized with respect to the upper limit of the object spatial frequencies f_o).

2.2. Error due to noise

In order to have a similar expression for the deviations between the original and the image due to influence of noise we define a noise error F_n^x by

$$F_n = \overline{D_n^2}^x / \overline{S^2}^x \tag{12}$$

D_n is the output of the recording system arising from the fluctuations of the illuminating laser and from the noise of the detector device. From the signal chain we derive the expression

$$D_n(x) = \frac{1}{E} \cdot \int_0^\infty dt' \cdot R(t') \cdot r(x/v - t') \tag{13}$$

According to eq. (1) we will assume a noise amplitude r(t) which is additive and independent of the signal. It should be noted that this assumption is in general not valid for the fluctuations of the illumination laser since they are monitored via the structure of the object (see discussion below).

For the noise error the correlation $\Theta(t') = \overline{r(t) r(t-t')}$ of the noise amplitude is of importance only. Under the condition that this correlation has a constant power spectrum within the bandwidth of the low pass filter the noise can be approximated to be δ-correlated $\Theta(t') = \overline{r^2} \cdot \delta(t')$. Thus, we can derive for the noise error F_n in general

$$F_n = \frac{\overline{r^2}}{E^2 \cdot S^2} \cdot \int_0^\infty dt'' \int_0^\infty dt' \cdot R(t'') \cdot R(t') \cdot \delta(t''-t') \tag{14}$$

and in particular for the integrator and the RC-filter

$$F_n^I = N/[f_o v \tau] \quad \text{and} \quad F_n^{RC} = N/[2 f_o v z]. \tag{15}$$

respectively. N represents the noise to signal ratio given by constant values

$$N = (f_o v^2/E^2) \cdot [\overline{r^2}/S^2(x)]. \tag{16}$$

3. OPTIMUM SCAN PARAMETERS

The total error $F = F_d + F_n$ results by adding eq. (9) and eq. (15). It depends on the scan parameters, the sensitivity and the noise of the detector, and on the object. Applying the approximated distortion error of eq. (11) an optimum time constant τ of the electronic recording device as a function of scan parameters can be derived by minimizing the total error. This procedure results in algebraic equation of fifth order in :

$$(v\tau)^5 + a \cdot (v\tau)^3 - b = 0 \tag{17}$$

with $a^I = 6 \cdot r_o^2 v^3$, $b^I = 90 \cdot N/f_o^5 \pi^4$ \hfill (17a)

and $a^{RC} = 15 \cdot r_o^2 v^3 / 8$, $b^{RC} = 15 \cdot N/64 \cdot f_o^5 \pi^4$ \hfill (17b)

for the integrator and the RC-filter, respectively.

Here we do not want to discuss the exact solution of eq. (17) but will give instructive results of numerical calculations.

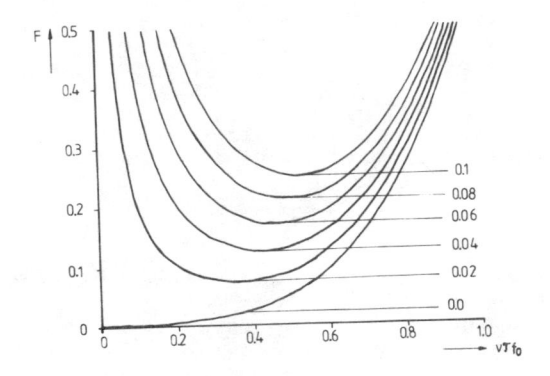

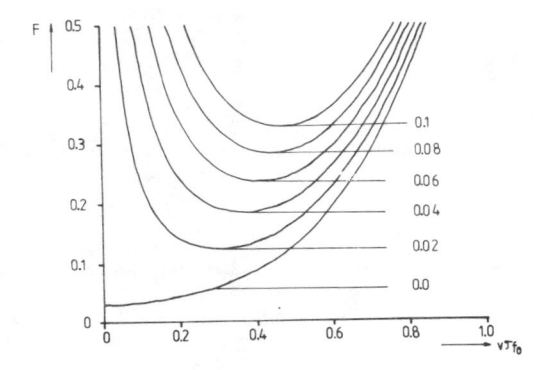

Figure 3. Total error versus normalized integration time constant $\tau v f_o$ and as function of parameter $N/\tau v f_o$.
 a) $r_o f_o = 0.1$ \hspace{2cm} b) $r_o f_o = 0.2$

Fig. 3 shows the total error versus the normalized time constant $\tau v f_o$ and as function of the reduced noise to signal ratio N (eq.(16)). It is evident that for increasing N (given in units of $\tau v f_o$) the total error increases considerably. The minimum of F is due to the balance between distortion error (large values of $\tau v f_o$, smearing out of the signal) and influence of noise (small values of $\tau v f_o$).

4. GENERALIZATION OF THE APPROXIMATIONS

In order to derive the optimal scan parameters we approximated the object function to have a rectangular power spectrum. This approximation should be dropped now and we want to demand the existence of an upper cut off frequency f_o of the spectrum of spatial frequencies only. L(f) of eq. (7) is than to be replaced by an arbitrary normalized function L'(f) according to

$$L(f) = 2f_o L'(f) \quad \text{with} \quad \int df\, L'(f) = 1 \quad \text{and} \quad L'(f > f_o) = 0. \tag{18}$$

The distortion error of eq. (11) is then

$$F'_{d\,opt} = \mathcal{L} \cdot F_{d\,opt} \quad \text{where} \quad \mathcal{L} = \int df\, L'(f)(\alpha f)^4 \tag{19}$$

with $\alpha^I = \pi v \tau$ and $\alpha^{RC} = 2\pi v z$, respectively. The total error can than be written as $F = \mathcal{L} * (F_d + F_n/\mathcal{L})$. The total error and the noise to signal ratio are only modified by a factor depending on $\mathcal{L}$. From this relation the same optimal scan parameters as given in eq. (17) can be derived.

Next, we would like to discuss the case when the noise depend on the structure function, i.e. we have a signal dependent noise. This is e.g. the situation when the fluctuations of the laser intensity are much larger than the noise of the detector. For the output due to these fluctuations we have the value

$$D_n(x) = \frac{1}{E} \int_0^\infty dt' \int_{-\infty}^\infty dx' \cdot R(t') \cdot r'(x/v - t') \cdot S(x-x'-vt'+\delta) \cdot F(x') \qquad (20)$$

The integration along x' leads to a structure function $S'(x-vt')$ which is averaged over the focus width. $r'(t')$ represents now <u>the fluctuations of the laser intensity</u>. The noise error F_n again depend on the correlation $r'(x-vt') \cdot S'(x-vt') \cdot r'(x-vt'') \cdot S'(x-vt'')$ only, where r' is the noise transformed from the time coordinate to the space coordinate by $t = x/v$. The power spectrum of this correlation extends to infinity but it is not necessary constant at low spatial frequencies because it is modified by the spectrum of S'. Usually microscopic objects have very small structures which should be investigated, but due to the averaging over the focus area the width of the spectrum of S' is approximately $1/r_o$. However, at very low frequencies this spectrum can also be assumed to be constant. As an approximation the same noise error as given in eq. (15) can be derived if the mean squared noise value of eq. (16) is replaced by $r^2 \cdot S'^2$.

Since the image is stored at discrete points only it is of interest to find an optimum step width x_{opt}. We want to restrict the discussion to the case of the integrator which should be realized by the sample and hold unit of the A/D-converter. The integration of the signal in the sample unit (integration time constant τ) is performed over the time t_c between two digitizations only. Consequently, the normalized electronic response function $R(t)$ is

$$R(t) = 1/t_c \quad \text{for} \quad t_r < t < t_r + t_c \qquad (21)$$

and we can derive the same optimum time constant from eq. (17) if τ is replaced by t_c. Since t_c is the temporal distance between two recorded points we obtain then for the optimum step width from eq. (17) and Fig. 3, respectively:

$$x_{opt} = (vt_c)_{opt} \qquad (22)$$

In Figures 4 and 5 there are shown images of a step-like object recorded under optimized and non-optimized scan conditions with a laser scanning microscope in comparison with the corresponding computer calculated images (eq. (2)). Clearly, an increase of the integration time constant t_c leads to a reduction of noise, but at the expense of spatial resolution if it exceeds the optimum value for a given noise to signal ratio.

5. ACKNOWLEDGEMENT

We are greatly indebted to Prof. B. Wilhelmi for valuable help and stimulating discussions.

6. REFERENCES

/1/ T. Wilson and C.J.R. Sheppard, <u>Theory and Practice of Scanning Optical Microscopy</u> (Academic, London, 1984)

/2/ M. Schubert, "Über die unvermeidlichen Fehler bei der Registrierung von Spektren", Experim. Techn. Physik 6, 203 (1958)

/3/ T. Damm, W. Triebel, and B. Wilhelmi, "Measurement Conditions for Continously Recording Picosecond Pump-and-Probe Spectrometers", IEEE Journ. Quant. Electr. <u>QE-19</u>, 635 (1983)

/4/ B. Wilhelmi, "Untersuchung optimaler Messbedingungen beim spektroskopischen Abtastverfahren", Experim. Techn. Physik (25), 141 (1967)

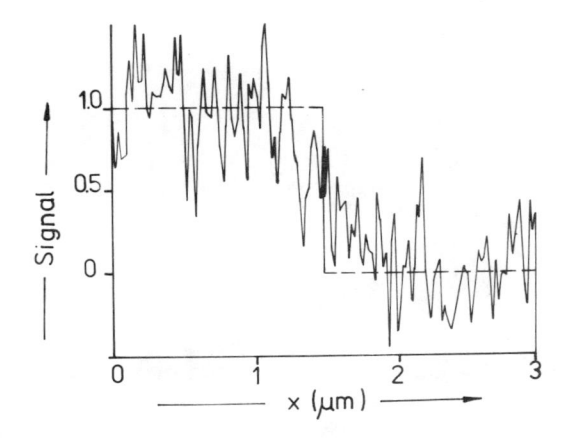

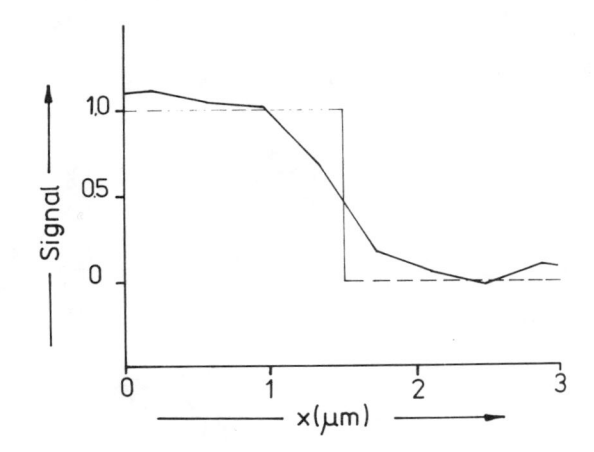

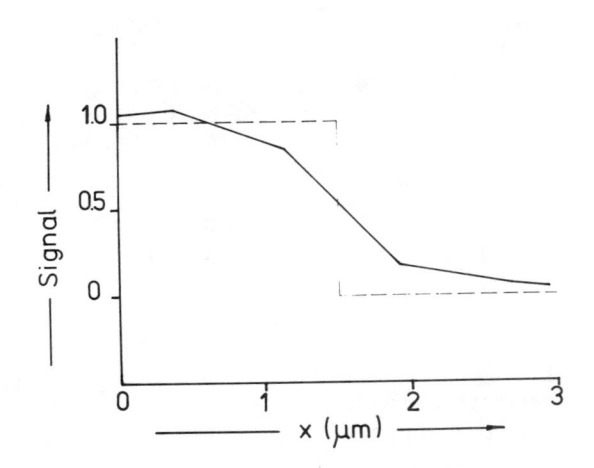

Figure 4. Calculated image of a step-like object for different recording conditions.
a) $v\tau/r_o = 0.06$ b) $v\tau/r_o = 1.0$
c) $v\tau/r_o = 3.0$

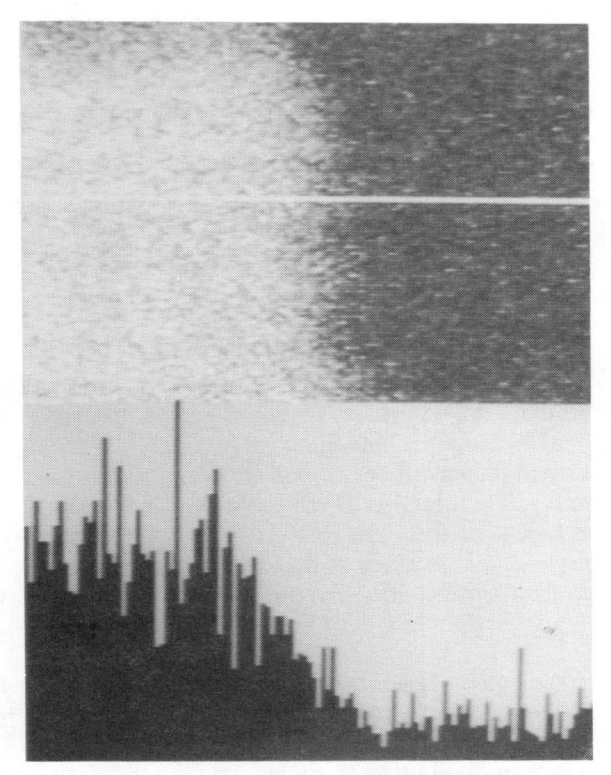

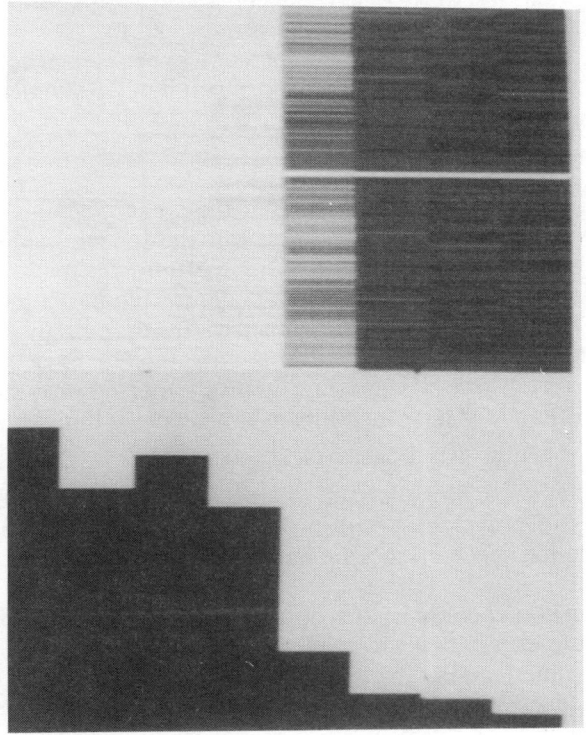

Figure 5. Recorded LSM-images of a step-like object for different recording conditions.
a) $v\tau/r_o = 0.06$
   $v = 0.0806$ µm/µs, $\tau = 40$ µs, $r_o = 0.4$ µm
b) $v\tau/r_o = 1.0$
   $v = 0.0824$ µm/µs, $\tau = 160$ µs, $r_o = 0.4$ µm
Note the distortion due to noise in case a) of too small step width recording.

Phase-shifting and Fourier transforming for sub-micron linewidth measurement

Yiping Xu, Evelyn Hu, Glen Wade

Center for Robotic Systems in Microelectronics, University of California
Santa Barbara, CA 93106, USA

ABSTRACT

By using a translating phase mask, a Fourier-transforming lens and a spatial filter, we can process laser light reflected from a surface in such a way as to avoid the diffraction effects common in conventional imaging. This optical technique is thus inherently capable of measuring minute surface features such as the width of deposited or inscribed lines with better than Rayleigh resolution and should therefore be applicable for metrology of sub-micron features. The novelty of the technique is to transfer linewidth information into the zero-order spatial frequency component of the light reflected from the surface. We present analyses and computer simulations to detail the effects of such features of the system as the sharpness of a step edge in a phase-shifting mask, the magnitude of the phase shift introduced by the mask, the variation in reflectivity and height of various regions of the surface structure, and the effect of instrumental noise on the determination of linewidth.

Experimental measurements were performed on specimens with large feature dimensions to verify the inherent capability of the technique. The results agree well with theoretical predictions.

It is hard to validate this technique at smaller dimensions because of the necessity for precise lateral translation of the mask with respect to the surface and the sensitivity of the system to the mask-to-surface distance. We discuss modifications in the next-generation experimental set-up that will address both these issues. Current results indicate that this technique will be viable well into the submicron range.

INTRODUCTION

Present advances in state-of-the-art semiconductor integrated circuits include the continued increase in device density and chip sophistication with the concomitant shrinkage of minimum device dimensions to below one micron in size. An important task in microelectronics manufacturing is to automatically inspect small surface structures on fabricated semiconductor components such as photoresist-patterned lines. Since the dimensions of line geometries determine the electrical characteristics of a circuit, we have been particularly interested in how to measure such critical dimensions as width and depth. The most-widely used instruments for determining linewidth have been based on optical imaging methods that employ a microscope in conjunction with some type of measuring attachment. These instruments are severely limited by diffraction and are adequate only for circuit geometries of one micron and above.[1] Alternative means of metrology are being developed, such as in-line inspection by low-voltage scanning electron microscopy. This technique provides the requisite resolution, but its limitations include possible damage resulting from the interaction between electron beam and substrate material, as well as inconvenience in carrying out the inspection in a vacuum.

The accuracy of the optical imaging methods depend upon the characteristics of the optical image of the microscope and is affected by diffraction and aberrations in the optical system as well as the degree of coherence of the illumination. The images produced by such systems are at best diffraction limited. A major problem in obtaining reliable linewidth measurements using these systems is therefore how to determine where the edge of the line actually is. An optical threshold for locating line edges is commonly used in these systems.[2-6] The proper value for the threshold, however, strongly depends on reflectance variations across the surface and the thickness of the line being measured. Accurate measurements are theoretically possible only when one or both of these parameters are known, which is not always the case in practical situations.

We have proposed a novel phase-shifting technique for linewidth measurement that does not involve an optical microscope and does not require any knowledge of relative reflectances and phase differences.[7-8] In this paper we present further analyses and computer simulations to detail the effects of such features of the system as the sharpness of a step edge in a phase-shifting mask, the magnitude of the phase shift introduced by the mask, the variation in reflectivity of various regions of the surface structure, and the effect of instrumental noise on the determination of linewidth.

Experimental measurements were performed on specimens with large feature dimensions to verify the inherent capability of this technique. The results agree well with theoretical prediction.

It is hard to validate this technique at smaller dimensions because of the necessity for precise lateral translation of the mask with respect to the surface and the sensitivity of the system to the mask-to-surface distance. We discuss modifications in the next-generation experimental set-up that will address both these issues. Current results indicate that this technique will be viable well into the submicron range.

PHASE-SHIFTING AND FOURIER-TRANSFORMING TECHNIQUE

To accurately measure submicron linewidths on photomasks and wafers, we proposed a new technique using a phase-shifting mask, a Fourier-transforming lens and a spatial filter (pinhole) as shown in Figure 1. The specimen is located at the front focal plane of the lens. The mask, placed close to the line specimen, consists of two different sections, one with 90° of phase shift more than that of the other. At the back focal plane of the lens, a photodetector with a pinhole is aligned and fixed at the focal center. Since this plane is the lens's Fourier-transform plane, only the zero-order spatial frequency component of the light reflected from the specimen passes through the pinhole and is detected. As the mask is translated laterally with respect to the line being measured, the intensity is detected as a function of the mask position.

The novelty of the technique is the transfer of linewidth information into the zero-order spatial frequency component of the reflected light by accurate translation of the phase mask. Unlike conventional optical imaging systems, only the zero-order frequency intensity is detected as the output signal. This technique is therefore not limited by the numerical aperture of the system, invariably a critical factor in all diffraction-limited optical imaging systems. The linewidth measured by this technique is relatively independent of variations in the optical reflectance and in the height between the patterned feature and the substrate. Calibration for the system is not necessary. Current analysis indicates that this technique should be viable well into the submicron range using an optical laser.

THEORETICAL ANALYSIS AND SIMULATIONS

Mathematic representation

For simplicity, we consider a single, rectangular-shaped line on a wafer characterized by the complex reflectance function as follows

$$r(x) = \begin{cases} r_1 e^{j\Delta\phi} & \text{for } 0 \leq |x| \leq \frac{W}{2} \\ r_2 & \text{for } \frac{W}{2} < |x| \leq \frac{P}{2} \end{cases} \quad (1)$$

where r_1 and r_2 are the reflectivities of the line and substrate respectively, W is the width of the line, and P is the width of the incident light spot. A relative phase difference $\Delta\phi$ for light reflected from the wafer surface is introduced by the thickness h of the line and is equal to $4\pi h/\lambda$.

A phase-shifting mask consisting of two precisely fabricated sections separated by a steep edge running parallel to the line is placed close to the line and translated laterally with respect to it. Let x=0 designate the center of the line being measured. When the edge between the two sections is translated from x=0 by a distance d, the complex transmittance of the phase-shifting mask can be expressed as

$$t(x) = \begin{cases} 1 & \text{for } x < d \\ e^{j\frac{\pi}{2}} & \text{for } x \geq d \end{cases} \quad (2)$$

Since the incident light passes through the phase mask, reflects from specimen, and passes through the mask again, the difference in the phase shift between sections due to the round-trip propagation is 180°. We can take into account the phase shift by multiplying the complex reflectivity function of Eq. (1) by t(x) from Eq. (2). By taking the Fourier transform of the product, we obtain the spatial spectrum of the reflected light as follows

$$R(f_x) = \begin{cases} \frac{j}{\pi f_x}\left\{r_1 e^{j\Delta\phi}\left[e^{j2\pi f_x d} - \cos(\pi f_x W)\right] + r_2\left[\cos(\pi f_x W) - \cos(\pi f_x P)\right]\right\} & \text{for } 0 \le |d| \le \frac{W}{2} \\ \frac{1}{\pi f_x}\left\{r_1 \sin(\pi f_x W) e^{j\Delta\phi} - r_2\left[\sin(\pi f_x W) - j(e^{-j2\pi f_x d} - \cos(\pi f_x P))\right]\right\} & \text{for } \frac{W}{2} < |d| \le \frac{P}{2} \end{cases} \quad (3)$$

Eq. (3) shows that the spatial spectrum strongly depends on the edge position of the phase mask. For the case when d=0 (that is, when the phase mask is co-centered with the line) the spectrum is as shown in Figure 2. It is very different in shape from the shape of the function sinx/x which is the shape of the spectrum of light that would be reflected from the wafer without the phase mask. One marked difference is that the zero-order spatial frequency component, at a maximum value without the mask, becomes zero with the mask because of complete phase cancellation. We now consider only the change in the zero-frequency component as the phase mask is moved. The zero-frequency intensity can be written as

$$I(0) = \begin{cases} 4r_1^2 d^2 & \text{for } 0 \le |d| \le \frac{W}{2} \\ r_1^2 W^2 + r_2^2(2d-W)^2 + 2r_1 r_2 W(2d-W)\cos(\Delta\phi) & \text{for } \frac{W}{2} < |d| \le \frac{P}{2} \end{cases} \quad (4)$$

Inspection of Eq. (4) shows that the output signal is a quadratic function of the displacement of the center-edge boundary of the mask from the center of the line feature when the displacement is within the line width. As the boundary of the phase mask is translated out of the line region so that the entire line is covered by only one section of the mask, the output signal depends on the relative reflectance r_2/r_1 and the relative phase difference $\Delta\phi$. As a special case when r_2 is zero, the output signal will be constant when the displacement is greater than a half linewidth.

Effects of variations in reflectivity and physical structure

We use computer simulations to examine the effects of variations in reflectivity and physical structure of various portions of the surface. For convenience we assumed the linewidth to be one micron wide so that the edges of the line were located at |x|=0.5 micron. Figure 3 shows the theoretical zero-order intensity as a function of phase mask position for several different values of the relative reflectances involved. Such differences could result from different material combinations of feature and substrate. The height of line feature was fixed the same. Figure 4 shows the zero-order intensity as a function of phase mask position for several different values of relative phase difference, $\Delta\phi$ representing variations of line height. For these calculations, the relative reflectance r_2/r_1 was fixed at 0.5.

From Figures 3 and 4 we see some changes in the curves caused by variations in relative reflectivity and relative phase difference. The changes appear only when the central boundary of the mask is translated beyond the region of the line. The curves are all identical within the line region and correspond to a well-shaped quadratic function because of the phase cancellation of the light reflected from the substrate. All curves are zero valued when the mask is co-centered with the line which causes complete phase cancellation and they exhibit break-points (for $\Delta\phi \le 90°$) or peaks (for $90° < \Delta\phi \le 180°$) at the line edges because of the discontinuities of reflectivity and geometrical structure. The linewidth is determined by reading the distance between the break-points. It is obvious that the linewidth determination is independent of both relative reflectance and relative phase difference.

Effects of phase-mask quality

As we have previously stated, this technique is not diffraction limited. Its accuracy, however, depends upon the precision of phase-mask fabrication, the accuracy of the translator and the minimum detectable signal of the photodetector. Since the above results are derived from an ideal case where the phase mask can provide an exact

180° phase shift and has a perfect step edge, it is necessary to consider the situation where these conditions are not satisfied. The actual phase shift introduced by the mask can be written as $180° - \Delta(x)$ or $180° + \Delta(x)$ where $\Delta(x)$ is a phase error caused by the mask not being optically flat. Several effects will take place. First, the zero-order intensity will no longer go to zero even when the phase mask is co-centered with the line being measured because of incomplete phase cancellation. Second, the curve is not a pure quadratic function of the displacement of the mask and depends on the phase error $\Delta(x)$ when the mask boundary is within the line region. However, this phase error will not cause any change in the position of the break points appearing in the curve which correspond to the actual edges of the line since at these points the entire line is covered by one section of the mask. Theoretical analysis indicates that all the above-mentioned effects are negligible when the maximum phase error $\Delta(x)$ is less than 30°. Curves (a) and (b) in Figure 5 show a comparison when the actual phase shift is 150° and 180°.

Unlike the effects caused by phase error, the sharpness of the step edge of the mask will not affect the center point of the curve. However, a sloping edge will make the determination of the break-points more difficult. This reduces the accuracy of linewidth determination. This indicates that the sharpness of the step edge of the mask is the most important factor for the fabrication of the phase-shifting mask.

Noise considerations

In practice, one can use curve-fitting methods to process the acquired data and determine the break points from the interceptions of the fitted curves. But for low signal-to-noise ratios we would expect errors to reduce the accuracy. Systematic noise can be expected from the detection system, i.e. photomultiplier, in the form of dark noise. Noise will also be introduced if there is any variation in the laser light intensity used to illuminate the sample.

We have employed the Monte-Carlo method to study the effect of instrumental noise on the determination of linewidth. The simulated signal data are computer generated. All data are then corrupted by random noise. Two kinds of noise were taken into consideration. One is the additive noise such as dark noise which is independent of incident light and the other is associative noise such as quantum noise which depends on the magnitude of the illumination. These kinds of noise are alternatively added to the signal data in order to examine which one of them is the main factor that reduces measurement accuracy. We used the least-square-error curve-fitting method and calculated the typical quadratic central curve section with its two outer horizontal lines from the noisy data. The linewidth is determined by finding the distance between the two point of intersection between the quadratic section and the two horizontal lines. Our studies show that additive noise is the major source which affects the accuracy of the linewidth determination. For example, assume that the additive noise intensity is less than or equal to 10% of the maximum signal intensity (that is, the signal-to-noise ratio = 10). Nine points were used for fitting the quadratic section and ten points for fitting two horizontal lines. The statistical relative-error distribution based on 2000 trials is shown in Figure 6. This calculation indicates that the relative error is less than 4% for a signal-to-noise ratio of 10. Further analysis shows that the relative error will be halved if the signal-to-noise ratio is doubled.

As we expect, the relative error can be reduced by using a small step size of the translator, that is, measuring more data points to fit the curve. But this will also slow down measurement speed and increase computation. There is a trade-off between them. Computation simulations have been done to examine how the relative error decreases as the number of measurements increases. Assume that N is the number of points used to fit the curve. In other words, it is the number of measurements within the line. For the given N, 2000 sets of data are computed generated and corrupted by random noise. We use each set of data to calculate the linewidth and its relative error. An average relative error and its standard deviation are calculated based on 2000 relative errors. This calculation procedure is repeated by increasing N. Results are shown in Figure 7 where the mean (a) and the standard deviation (b) of relative errors decrease as N increases. We can see that the improvement of the relative error increases a little after N is greater than 20.

Since the mask position data are recorded for fitting the curves, the position error will also affect the accuracy of the linewidth determination. We again used the Monte-Carlo method for studying this kind of error. In addition we took the effect of error in the phase mask position into account by adding the appropriate noise to the signal data. The results from this analysis lead to the same conclusion stated above.

Considerable effect must be expended to increase the signal and reduce the noise sufficiently to permit high measurement accuracy. In a practical situation it may be difficult. By adding one more photodetector to provide a reference signal, the noise caused by variations in laser light intensity can be removed by normalization. Since most of the energy in the reflected light is concentrated in the zero and low orders, the proposed system should be capable of achieving good signal-to-noise ratio.

PRELIMINARY EXPERIMENTAL VERIFICATION

Initial experiments were performed to verify the theory described above.[8] By using a phase mask that provides a single-pass phase shift of 180° between its two sections, the phase-masking technique may be tested in the transmission mode. This is diagrammed in Figure 8. The object was a slit in an opaque material. Such an object in the transmission mode is obviously analogous to a bright line on a substrate with zero reflectance in the reflection mode. The phase mask was fabricated by etching a sharp step into a piece of SiO_2 with hydrofluouric acid to provide a single-pass phase shift of 180° difference in phase between etched and unetched sections. A micrometer-driven translation stage with one thousandth of an inch resolution was used to translate the phase mask. An achromatic lens of approximately 8 inches focal length performed the spatial Fourier transformation. A photomultiplier was placed behind a spatial filter which was a pinhole 50 microns in diameter fixed at the center of the Fourier-transform plane using a two-axis positioner. A He-Ne laser was used to illuminate the specimen. The output signal, proportional to the intensity of the zero-order, was read using a digital voltmeter.

As one example, Figure 9 shows a plot of experimental data and the corresponding theoretical curve. The data show quadratic behavior within the line region and relatively constant behavior beyond that region. This agrees with theoretical prediction. The uneven value of two horizontal lines was caused by impefect optical flatness differences between the etched and unetched sections of the phase mask. We used the least-square error method to process the experimental line segments. This gives the solid-line curve of Figure 9. Although there were rounding effects associated with the experimental data near the line edge positions, which was due to the sloped edge of the phase mask, the linewidth could be accurately determined by measuring the distance between the intersection points of the quadratic curve and the horizontal lines. In this example, the linewidth was calculated as being 631.2 microns by this method whereas it was measured as being approximately 630 microns by using an optical microscope.

TECHNICAL CHALLENGES FOR FUTURE EXPERIMENTATION

Although the preliminary experimentation has demonstrated the validity of the concept, it will be hard to validate this technique at smaller dimensions because of the necessity for precise fabrication of the phase mask with a very steep, straight edge, the necessity for precise lateral translation of the mask with respect to the substrate, and the high sensitivity of the system to the mask-to-substrate distance. Methods exist for attacking all these problems. We are currently investigating the construction of the next-generation experimental set-up that will consider these issues.

Phase-mask fabrication. As stated above, the steepness and straightness of the step edge of the mask are very important. High resolution lithography and dry etch pattern transfer should be able to provide a mask with sufficiently steep and straight edges.

Micro-translation. For submicron linewidth measurement, one factor which limits the resolution of the system is the size of the steps taken while scanning. In particular, the step size should be much smaller than the linewidth. Hence, a system designed for 0.5 micron resolution must use a translator with step size < 500 Å in order to achieve a good accuracy. A piezoelectric drive can be built to achieve such high resolution. For example, Binnig and Rohrer[9] have used piezoelectric stacks to achieve a practical positioning accuracy of about 0.2 Å in scanning tunneling microscopes. We intend to employ such apparatus in future experiments to accomplish the phase-mask translation for submicron linewidth measurements.

Micro-alignment. In order to maintain the diffraction of the light emerging from the mask, the phase mask should be placed as close as possible to the specimen. This might cause difficulties for practical alignment. For example, the translating mask may scratch the specimen if the mask plate physically contacts it. This is essentially the same problem as is encountered in near-field optical scanning microscopy where the gap between the probe and sample at its closest point is a few angstroms.[10] By monitoring tunneling current, direct mechanical contact between probe and sample can be avoided while maintaining an extremely small gap during scanning. Our theoretical study has shown that the distance between the phase mask and the surface structure should be less than 500 Å for measuring linewidths below one micron.

SUMMARY

A new optical technique has been described for measuring submicron linewidths on integrated-circuit masks and wafers. Theoretical analysis and computer simulations have been made to determine expected limitations of the method. Preliminary experimental results show good agreement with theoretical predication. Future experimental modifications have been discussed. This new technique has the advantages of system simplicity and high lateral resolution. Its potential applications include lithography, submicron alignment and mask-defect inspection.

ACKNOWLEDGMENT

The authors would like to acknowledge their appreciation of the Center for Robotic System in Microelectronic for its support and to thank Alan Mar for his assistantce in doing experiment. This work was supported by the National Science foundation under contract number 08421415.

REFERENCES

1. P.H. Singer, "Linewidth Measurement: Approaching the Submicron Dimension," Semiconductor International, pp. 48-54, March 1983.
2. D. Nyyssonen, "Linewidth measurement with an optical microscope: the effect of operating conditions on the image profile," Applied Optics, vol.16, No.8, pp. 2223-2230, Aug. 1977.
3. D. Nyyssonen, "Theory of optical edge detection and imaging of thick layers," J. Opt. Soc. Am., vol.72, No.10, pp. 1425-1436, Oct. 1982.
4. D. Nyyssonen, "Calibration of optical systems for linewidth measurements on wafers," Optical Engineering, vol.21, No.5, Sept./Oct. 1982.
5. D. Nyyssonen, "Practical method for edge detection and focusing for linewidth measurements on wafers," Optical Engineering, vol.26, No.1, pp.081-085, Jan. 1987.
6. W.M. Bullis and D. Nyyssonen, "Optical linewidth measurements on photomasks and wafers," in VLSI Electronics: Microstructure science, N.G.Einspruch, ed., vol.3, pp. 301-346, Academic Press, New York,1982.
7. Y. Xu, A. Mar, G. Wade, E.L. Hu, "New technique for submicron linewidth measurement," Proceedings of the SPIE, Vol. 849, pp. 87-91, 1987.
8. A. Mar, Y. Xu, G. Wade, E.L. Hu, "Linewidth measurement using a translating phase mask," Proceedings of the IEEE Third International Electronic Manufacturing Symposium, pp. 70-75, 1987.
9. G. Binnig and H. Rohrer, "The scanning tunneling microscope," Sci. Am. vol. 253, pp. 50-56, August, 1985.
10. E. Betizig, A. Lewis, A. Harootunian, M Isaacson, and E. Kratschmer, "Near-field scanning optical microscopy (NSOM)," Biophys. J., vol.49, pp. 265-279, 1986.
11. J.W. Goodman, Introduction to Fourier Optics, Chapter 6, McGraw-Hill, New York, 1968.

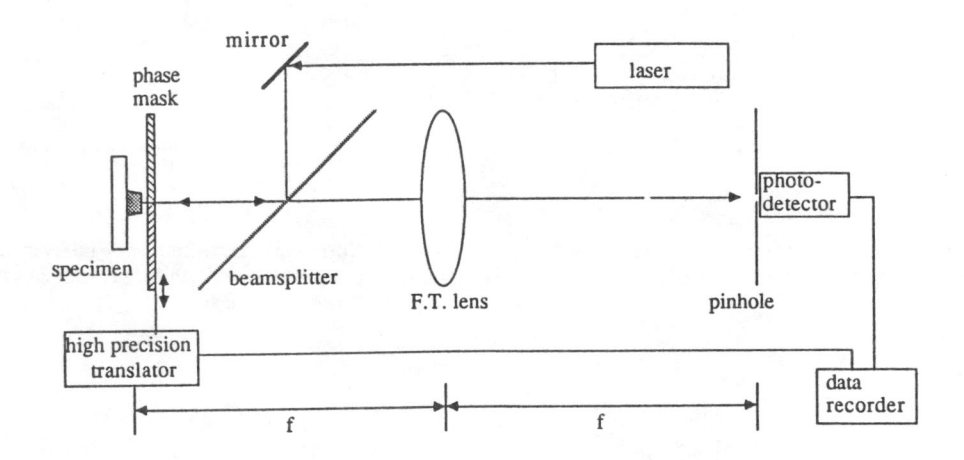

Fig. 1 Basic system set up for measuring linewidth with high lateral resolution

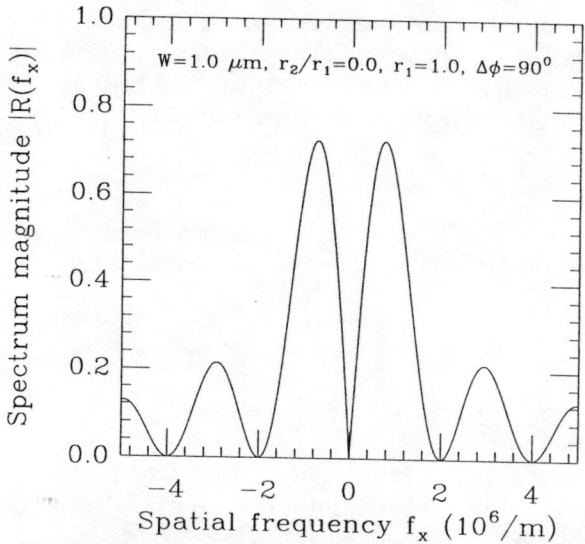

Fig. 2 The magnitude of the spectrum of light reflecting from the wafer and passing through a phase mask centered on the line.

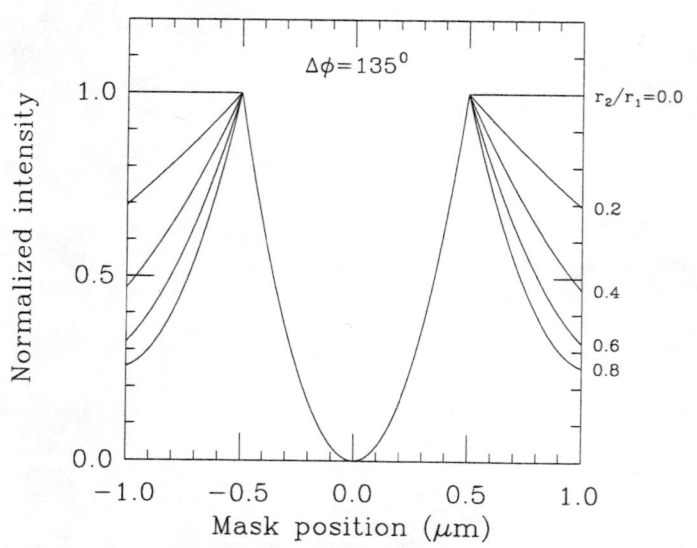

Fig. 3 Normalized zero-frequency intensity as a function of phase-mask position for different line-substrate reflectivity ratios.

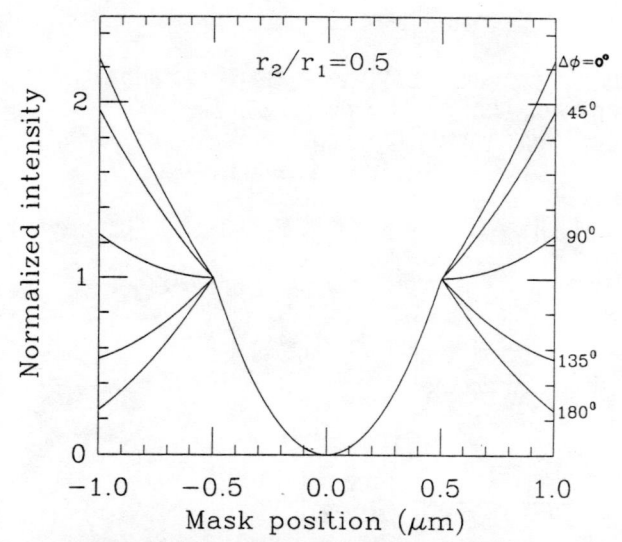

Fig. 4 Normalized zero-frequency intensity as a function of phase-mask position for different value of $\Delta\phi$.

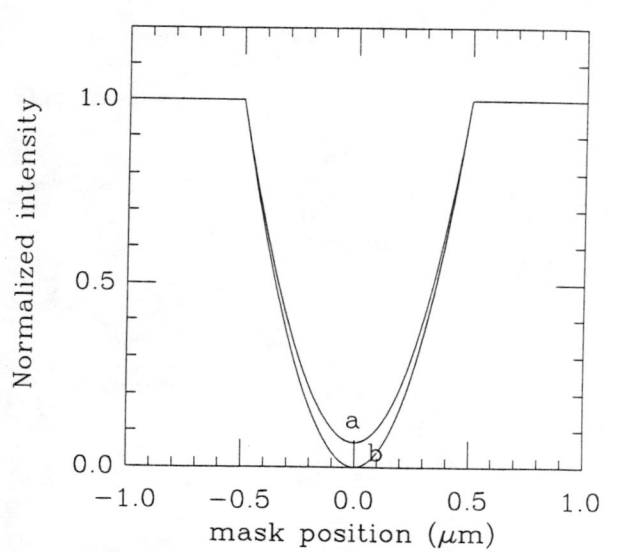

Fig. 5 Normalized zero-frequency intensity as a function of phase-mask position for (a) 150^0 phase shift (b) 180^0 phase shift.

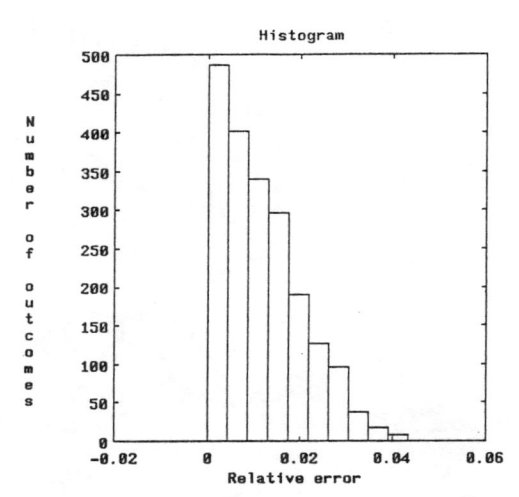

Fig. 6   Statistical relative error distribution for $\frac{S}{N}=10$.

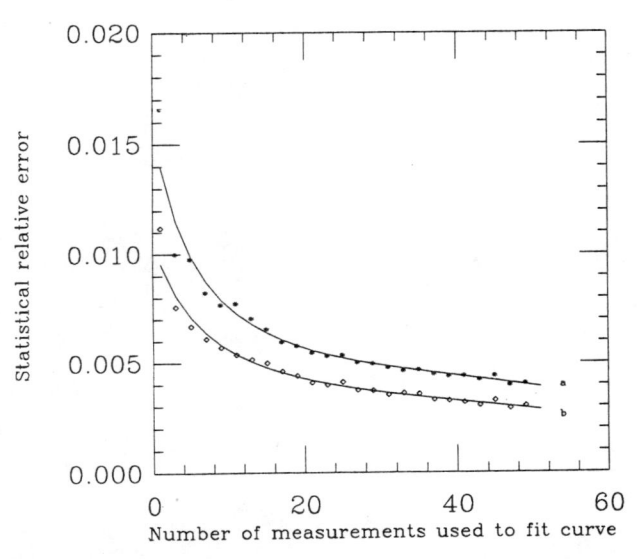

Fig. 7   The mean (a) and standard deviation (b) of relative error as a function of measurements based on 2000 simulations and $\frac{S}{N}=10$.

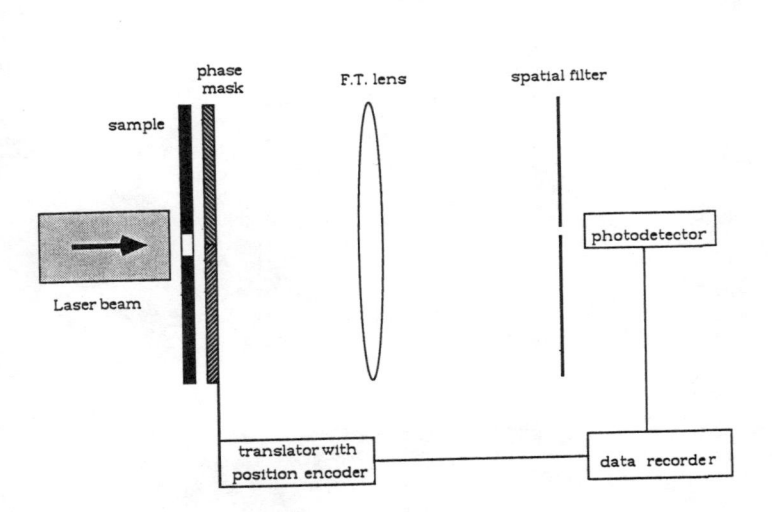

Fig. 8   Experimental set-up for transmission mode.

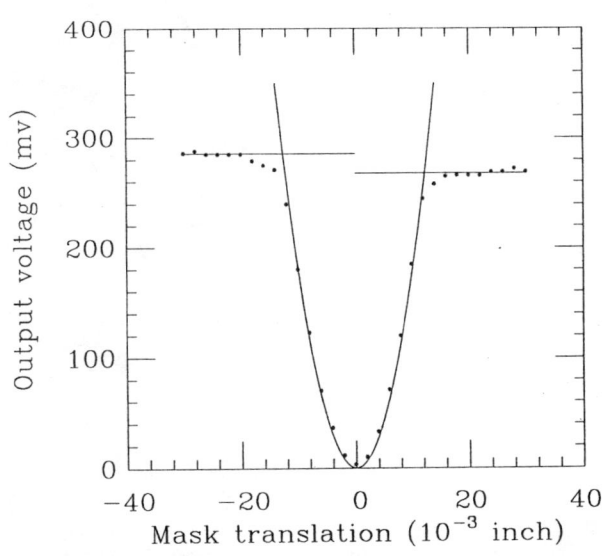

Fig. 9   Preliminary experimental data (dotted line) and corresponding theoretical curve (solid line).

Measurement of the degree of coherence in conventional microscopes

Andreas Glindemann

Optisches Institut der TU Berlin,
Str.d.17.Juni 135, 1000 Berlin 12, FRG

ABSTRACT

The spatial degree of coherence affects the quality of imaging systems. Especially in microlithography and microscopy the image quality is optimized by choosing the "right" degree of coherence. Hence it is necessary to know how to adjust the desired degree of coherence. The illumination system of a microscope allows one to vary several parameters : field stop, aperture stop and Köhler/critical illumination. The purpose of this paper is to determine the influence of these parameters on the degree of coherence in the object plane of a usual illumination system. The degree of coherence is measured as variation of modulation of interference fringes, produced by a shear interferometer, with the shear distance. For any kind of illuminating aperture (circular, i.e. bright-field, annular, i.e. dark-field, rectangular and double slit) and for both Köhler and critical illumination we found a very good agreement with the concept of Hopkins' effective source, predicting a Fourier relationship between the intensity distribution in the pupil of the condenser and the degree of coherence in the object plane.

1. INTRODUCTION

Since the fundamental experiment by Thompson and Wolf /1/ in 1957 to verify the coherence theory and namely the van Cittert-Zernike theorem several arrangements for the determination of the degree of coherence have been presented (e.g. /2,3/). Limited interest has been shown however for the application of these schemes of measurement to optical instruments, although the quality of imaging systems is affected by the degree of coherence and is often optimized by choosing the appropriate degree of coherence. Hence it is necessary to know how to choose and to adjust an illumination system to get the desired degree of coherence.

The purpose of this paper is to investigate which parameters of a usual illumination system (field stop, aperture stop, Köhler/critical illumination) govern the degree of coherence in the object plane of a microscope.

A theoretical prediction was given by Hopkins /4/ when he introduced the concept of the effective source. According to this concept the degree of coherence in the object plane depends on the intensity distribution in the pupil of the condenser system. The relation between the intensity distribution in the pupil and the degree of coherence in the object plane of the microscope is given by the van Cittert-Zernike theorem, relating the two quantities by a Fourier transform.

Using a microscope setup with a usual condenser and a usual light source (a halogen lamp) we were performing measurements of the degree of coherence in the image plane of the microscope and the pupil of the condenser by means of a shear interferometer that is inserted between the objective and the image plane. The degree of coherence is determined by measuring the variation of modulation of the interference fringes with the shear distance. The connection between the degree of coherence in the object plane and that in the image plane was given by Hopkins /4/ and Carpenter and Pask /5/ assuming an incoherent light distribution in the pupil. We shall give a slightly different approach dealing with the wave front aberration function of the microscope objective.

2. THEORY

In this section we shall give a brief summary of the Hopkins' effective source model as far as it will be needed for the interpretation of our experiments. Then the imaging of the coherence properties will be discussed and the measured quantity, i.e. the modulation of the interference fringes will be related to the degree of coherence.

All derivations and formulae sketched in this section can be read in more detail in books on coherence theory, e.g. Marathay /6/.

Figure 1 shows the layout of a microscope with the minimal number of elements. The investigation is performed for a uniform refractive index n=1 and for quasimonochromatic illumination. When regarding the propagation of partial coherent light through a system the mutual coherence function (MCF) $\Gamma(r_1,r_2)$ is used, from which the degree of coherence $\gamma(r_1,r_2)$ derives by normalization. It is

$$\gamma(r_1,r_2) = \frac{\Gamma(r_1,r_2)}{\sqrt{\Gamma(r_1,r_1)\,\Gamma(r_2,r_2)}} \quad . \tag{1}$$

r_1 and r_2 are two points in any plane. For simplicity all formulae in this section are written one-dimensional. There is no restriction for the extension to two dimensions.

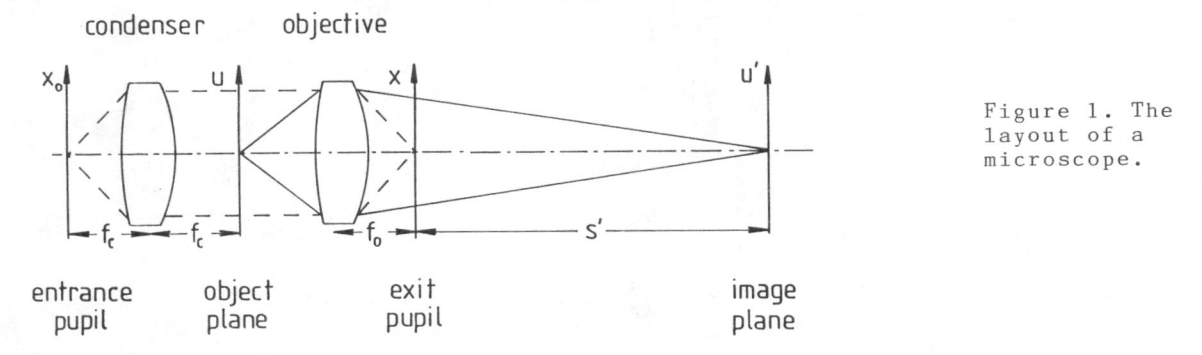

Figure 1. The layout of a microscope.

2.1. Hopkins' effective source

Starting in the entrance pupil of the condenser the MCF $\Gamma_{ent}(x_{o1},x_{o2})$ contains all information about the elements of the illumination system lying in front of it. With the entrance pupil in the front focal plane of the condenser the exit pupil lies at infinity and angle coordinates $\alpha_i = x_{oi}/f_c$ (i=1, 2) are required. The MCF in the exit pupil reads as follows

$$\Gamma_{ex}(\alpha_1,\alpha_2) = \Gamma_{ent}(\alpha_1 f_c, \alpha_2 f_c)\, P_c(\alpha_1)\, P_c^*(\alpha_2)\;, \qquad (2)$$

with $P_c(\alpha_i)$ being the complex pupil function of the condenser which includes the physical size of the exit pupil, and its phase part incorporates the wave front aberration terms.

The connection with the MCF in the object plane is given by the generalized van Cittert-Zernike theorem, yielding

$$\Gamma_{ob}(u_1,u_2) = \iint \Gamma_{ex}(\alpha_1,\alpha_2)\, \exp(-ik(u_1\alpha_1 - u_2\alpha_2))\, d\alpha_1 d\alpha_2\;, \qquad (3)$$

neglecting a constant and with $k=2\pi/\lambda$. Integrations are extended from $-\infty$ to $+\infty$ unless otherwise noted.

As the integral cannot be solved normally, Hopkins approximated the MCF in the entrance pupil by an incoherent, constant light distribution. Then the MCF in the entrance pupil becomes

$$\Gamma_{ent}(x_{o1},x_{o2}) = I_o \delta(x_{o1} - x_{o2})\;. \qquad (4)$$

I_o is the constant intensity.

Now the MCF in the object plane can be written as

$$\Gamma_{ob}(u_1,u_2) = \iint I_o P_c(\alpha_1) P_c^*(\alpha_2)\, \delta(\alpha_1-\alpha_2)\, \exp(-ik(u_1\alpha_1-u_2\alpha_2))\, d\alpha_1 d\alpha_2\;,$$
$$\Gamma_{ob}(u_1-u_2) = \int I_o |P_c(\alpha)|^2\, \exp(-ik\alpha(u_1-u_2))\, d\alpha\;, \qquad (5)$$

where $I_o |P_c(\alpha)|^2$ is called Hopkins' effective source. The MCF in the object plane, which is a function of the coordinate difference only, and the intensity distribution in the exit pupil form a Fourier pair now. $|P_c(\alpha)|^2$ is a real function. The wave front aberrations of the condenser do not affect the MCF in the object plane. Only the size of the pupil, i.e. the numerical aperture in a real system, controls the state of coherence in the object plane. With $I_o' = I_o/\Gamma(0)$ the degree of coherence becomes

$$\gamma_{ob}(u_1-u_2) = \int I_o' |P_c(\alpha)|^2\, \exp(-ik\alpha(u_1-u_2))\, d\alpha\;. \qquad (6)$$

This formula will be taken to relate the measured degree of coherence to the illuminating aperture.

2.2. Imaging of the MCF

With the general form of the MCF in the object plane $\Gamma_{ob}(u_1,u_2)$ (Eq.(3)) the image is obtained by a double Fourier transform, yielding

$$\Gamma_{im}(u_1',u_2') = \iint \Gamma_{ob}(u_1,u_2) \, F(u_1'-\beta u_1) \, F^*(u_2'-\beta u_2) \, du_1 du_2 \, , \qquad (7)$$

where $F(u_i'-\beta u_i)$ is the amplitude impulse response of the microscope objective and β is the Gaussian magnification.

To solve this double convolution the equation is transferred into the frequency space with $R_i = x_i/(s'\lambda)$ and becomes

$$\tilde{\Gamma}_{im}(R_1,R_2) = \tilde{\Gamma}_{ob}(\beta R_1, \beta R_2) \, P(\lambda s' R_1) \, P^*(\lambda s' R_2) \, . \qquad (8)$$

$\tilde{\Gamma}_{im}$ and $\tilde{\Gamma}_{ob}$ are the two-dimensional, spatial Fourier-transforms of Γ_{im} and Γ_{ob} respectively and P is the pupil function of the microscope objective.

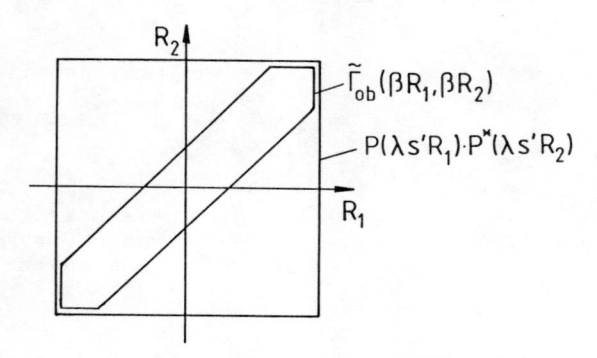

Figure 2. Top view on the plane R_1, R_2 in frequency space (Eq.(8)). See the text for further explanation.

The presentation of $\tilde{\Gamma}_{ob}$ in Fig.2 displays the first zero of the function obtained by the Fourier transform, if the illuminating aperture is smaller than the imaging aperture and if the size of the image is of the same order of magnitude as the size of the pupil. The extension of $\tilde{\Gamma}_{ob}$ in the direction of the difference of the coordinates R_1-R_2 is rather narrow compared with the extension of the product of the pupil functions. If the phase part of $P(\lambda s' R_i)$, the wave front aberration function, varies much slower than $\tilde{\Gamma}_{ob}$ goes to zero with respect to the difference R_1-R_2, than the following approximation may be performed:

$$\tilde{\Gamma}_{im}(R_1,R_2) \simeq \tilde{\Gamma}_{ob}(\beta R_1, \beta R_2) \, |P(\lambda s' \frac{R_1+R_2}{2})|^2 \, . \qquad (9)$$

This means that as long as $\tilde{\Gamma}_{ob}$ fits into the square PP* (see Fig.2), an objective with a rather smooth wave front aberration function does not affect the imaging of the MCF. This result is very similar to that of Hopkins /4/ and Carpenter and Pask /5/ who require incoherence for the light distribution in the frequency space however.

The factor β indicates that the imaged MCF is enlarged like a normal object.

Under the given condition the measurement of the degree of coherence in the image plane yields the scaled version of the degree of coherence in the object plane.

2.3. Measurement of the degree of coherence

The quantity to be measured in our experiment is the modulation of the interference fringes in a shear interferometer. The interference pattern of two plane, quasimonochromatic, sheared and tilted wavefronts in the image plane is

$$\begin{aligned} I(u') = &\langle |V(u'-\frac{\Delta}{2}) \exp(-ik\frac{\delta}{2}u') + V(u'+\frac{\Delta}{2}) \exp(ik\frac{\delta}{2}u')|^2 \rangle \, , \\ &\langle |V(u'-\frac{\Delta}{2})|^2 + |V(u'+\frac{\Delta}{2})|^2 + V(u'-\frac{\Delta}{2})V^*(u'+\frac{\Delta}{2})\exp(-ik\delta u') + \\ &+ V^*(u'-\frac{\Delta}{2})V(u'+\frac{\Delta}{2})\exp(ik\delta u') \rangle \, , \end{aligned} \qquad (10)$$

where V is the analytic signal, Δ is the amount of shear and $+\delta/2$ and $-\delta/2$ are the tilt angles against the image coordinate axis u'. The statistical average has to be taken, indicated by $\langle\ \rangle$. Using the definition of the MCF

$$\Gamma_{im}(u'-\tfrac{\Delta}{2},u'+\tfrac{\Delta}{2}) = \langle V(u'-\tfrac{\Delta}{2})V^*(u'+\tfrac{\Delta}{2})\rangle \qquad (11a)$$
$$= \langle V^*(u'-\tfrac{\Delta}{2})V(u'+\tfrac{\Delta}{2})\rangle ,$$

and the relationships

$$I_o(u'\pm\tfrac{\Delta}{2}) = \langle |V(u'\pm\tfrac{\Delta}{2})|^2\rangle \quad \text{and} \quad I_o(u') = I_o(u'\pm\tfrac{\Delta}{2}) , \qquad (11b)$$

the interference pattern takes the form

$$I(u') = 2I_o(u')\left(1+\gamma_{im}(u'-\tfrac{\Delta}{2},u'+\tfrac{\Delta}{2})\cos(k\delta u')\right) . \qquad (12)$$

The modulation of $I(u')$ equals the amount of the degree of coherence. If the wavefront is spherical of otherwise aberrated the form of the fringes is affected but not the modulation /2/.
The conclusion of this section is that measuring the modulation of the interference fringes in the image plane yields the amount of the degree of coherence in the object plane if a good objective is used.

3. EXPERIMENT

The experimental setup is shown in Fig.3. The light source is a 12 V 100 W halogen lamp imaged by the lens L1 to infinity and by the lens L2 in the aperture stop of the condenser for realizing Köhler illumination. For critical illumination L2 is removed and the position of L1 is corrected in order to image the filament in the object plane. The field stop is imaged by the condenser, a Zeiss condenser with NA = 0.9, in the object plane.

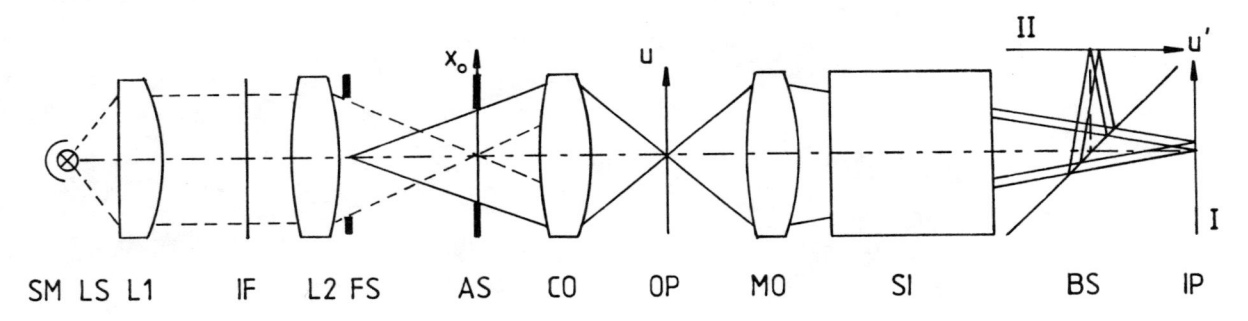

Figure 3. Scheme of the experimental setup utilized for the investigation. It is: SM - spherical mirror, LS - light source, L1 and L2 - lenses, IF - interference filter, FS - field stop, AS - aperture stop, CO - condenser, OP - object plane, MO - micro objective, SI - shear interferometer, BS - beam splitter, IP - image plane I and II.

The object plane is imaged by a microscope objective in the image plane passing through a shear interferometer. The beam splitter divides the image into two planes. That allows one to control all modifications with a CCD camera in plane I and to measure the interference pattern with a photo element behind a slit (4mm x 10μm) in plane II. Applying the principle of differential scanning /7/ to this setup gives a spatial resolution of 2μm with the 10μm slit. By an interference filter quasimonochromatic light is produced. Micro objectives with apertures and magnifications ranging from 10/0.25 up to 50/0.85 were used. A spherical mirror behind the lamp doubles the image of the filament to improve the uniformity of the illumination.

The shear interferometer is a modified Michelson type presented by Kelsall /8/. It is shown in Fig.4. A mechanical displacement of amount d causes a shear in the image of amount 2d. The displacement is realized with a micrometer screw and controlled with a linear transducer of Heidenhain. The minimal adjustable displacement is 1μm. The phase is adjusted in the other arm of the interferometer with the help of a Piezo element. In that arm the two mirrors are single adjustable to correct a possible roof error of the two fixed mirrors in the other arm. This scheme allows an independent control of phase shift and shear of the wavefront.

The interferometer used for this investigation was built by Fa.Studio S, Berlin and is displayed schematically in Fig.5 . With this arrangement the optical paths through the beam splitter is only half of that in Fig.4 . The optical path length is approximately 150mm, the maximal beam diameter is 12mm, resulting in a numerical aperture of 0.1 .

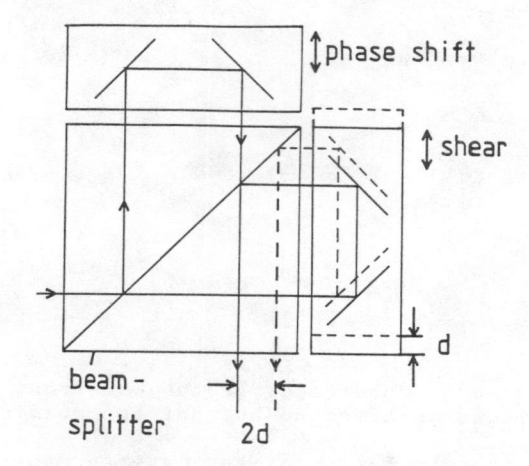

Figure 4. A Michelson type shear interferometer.

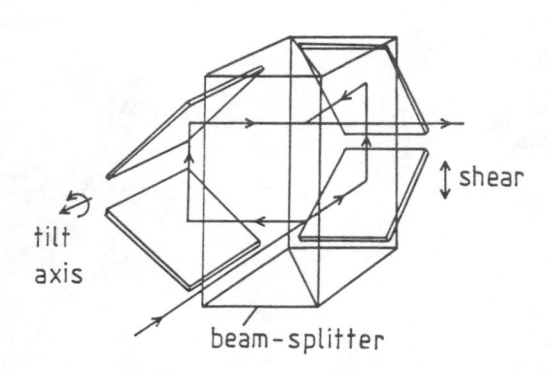

Figure 5. A variation of Fig.4 with a smaller beam splitter.

With these technical data the instrument can be inserted between the micro objective and the image plane. The final tube length is 250mm because the additional beam splitter and mechanical elements do not allow a closer approach.

A tilt between the two arms causes a tilt of the sheared wavefronts, i.e. interference fringes. The distance of two interference fringes is adjusted to 0.3mm .

The maximal modulation in the image plane was measured to be C=0.89 . For a very similar Michelson type interferometer, where the plane mirrors are replaced by triple mirrors, a theoretical investigation was presented by Leonhardt /9/. For the same angle of incidence at the beam splitter he obtains a maximal modulation of C=0.89 , too. For our interferometer the maximal modulation should be between the case discussed by Leonhardt and the Michelson with plane mirrors, where C=1 .

4. RESULTS

All results, except the last (Fig.9), to be presented are given in form of the degree of coherence in the object plane as a function of the difference coordinate u_1-u_2 , which is obtained by multiplying the shear distance in the image by the Gaussian magnification, according to Sec.2 . For the presented results two different objectives were used: a 10/ 0.25 Spindler&Hoyer and a 40/0.85 Leitz objective. For illuminating apertures 0.1 and 0.2 the Spindler&Hoyer objective was used, for all other the Leitz objective. For comparison with Hopkins' effective source model the theoretical curves (Eq.(6)) are displayed in all figures.

The first series of measurement (Figs.6 and 7) was performed with the bright field Zeiss condenser with a maximal numerical aperture of 0.9 . The aperture stop is adjustable from 0.1 to 0.9 in steps of 0.1 after calibration with an apertometer.

The results in Fig.6 were obtained with Köhler illumination, $\lambda = 545\pm21$ nm and an illumination aperture of 0.1 . The object size determined by the field stop is 0.4mm for Fig. 6a and 0.8mm for Fig.6b .

The difference between the two curves is the higher maximum for the smaller object size (Fig.6a). The reason for this is the reduction of straylight in the interferometer with reduced beam diameter.

Applying Eq.(6) for the two-dimensional case to the bright-field condenser with NA=0.1 yields as degree of coherence the Bessel function of first order divided by its argument, which we call Besinc function. The best curve fit is obtained if the Besinc function is normalized to the measured maximum and for an illuminating aperture of NA=0.08 for both Figs.6a and 6b. If the light distribution in the pupil of the condenser was not perfectly incoherent the result would be the same (see Marathay /6/ pp.95-99, where he calculated the propagation of the MCF starting with a partial coherent source with Gaussian intensity distribution and Gaussian degree of coherence.). However, the degree of coherence in the pupil of the condenser is affected by the size of the field stop what will be discussed in more detail at the end of this section. If the difference between the measured illuminating aperture of 0.08 and the adjusted one of 0.1 was caused by the partial coherence in the pupil of the condenser then a difference should appear between the results for object size 0.4mm (Fig.6a) and 0.8mm (Fig.6b). As there is no difference between the two cases there is no influence of the size of the field stop on the degree of coherence. The reason for the smaller illuminating aperture probably is an error in calibrating the aperture stop.

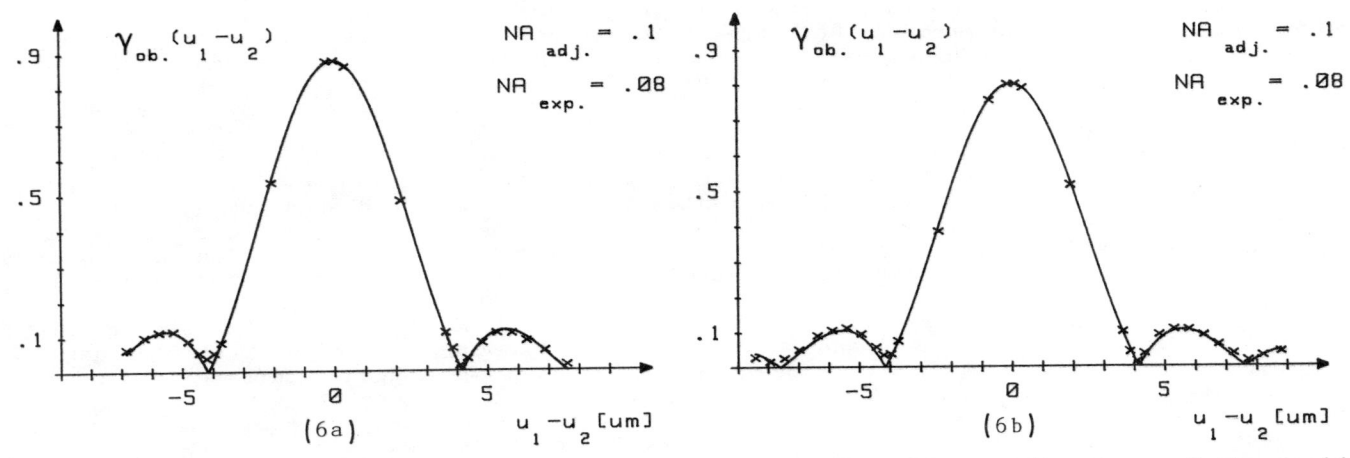

Figure 6. The degree of coherence in the object plane for different diameters of the field stop. The adjusted numerical aperture (NA_{adj}) and the numerical aperture resulting from the best fit of the Besinc function on the experimental data (NA_{exp}) are displayed. There is no influence of the size of the field stop on the degree of coherence measurable.

Other parameters being varied during this series are the focal length of L2, resulting in a different magnification of the filament in the aperture stop, the region of measurement in the object plane, i.e. on axis and in the field, different interference filters and the change between Köhler and critical illumination as described in Sec.3 .

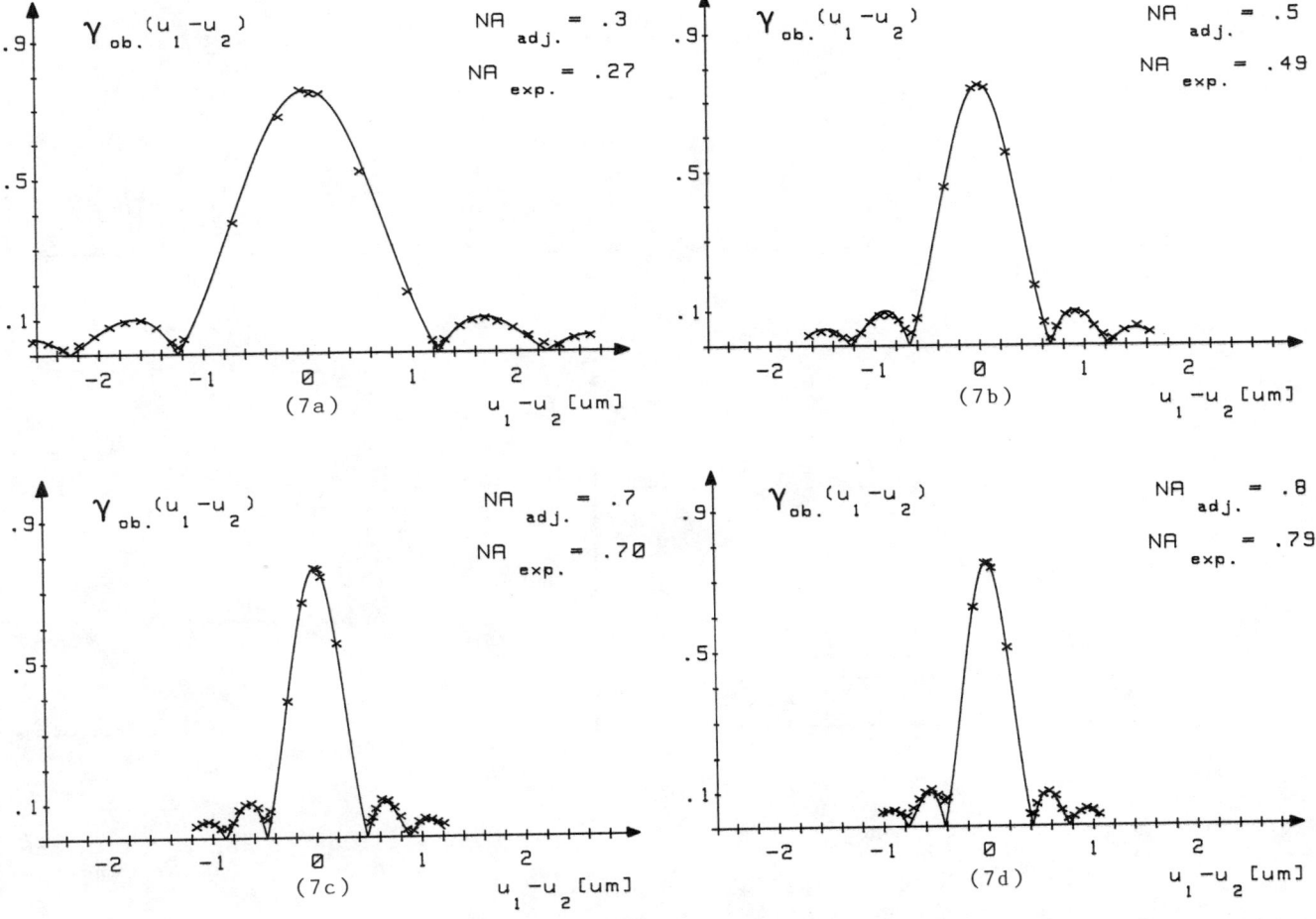

Figure 7. The degree of coherence in the object plane for different illuminating apertures The adjusted numerical aperture (NA_{adj}) and the measured aperture (NA_{exp}) are displayed.

However, none of these parameters affects the degree of coherence measurably. Only the illuminating aperture determines the degree of coherence in the object plane of an illuminating system. Hopkins' effective source model can be applied to usual condenser systems without restrictions, if the diameter of the image is not smaller than 4mm. Below this limit a slightly increasing influence of the field stop on the degree of coherence is possible.

The results in Fig.7 were obtained with Köhler illumination, the 40/0.85 Leitz objective and an object size of 0.45mm corresponding to an image size of 18mm for the standard tube length of 160mm. We consider this arrangement to be a standard arrangement in brightfield microscopy

It can be seen that the difference between the measured illuminating aperture, resulting from the best curve fit, and the adjusted one becomes smaller for greater values. The Besinc functions fit very well on the experimental data of each experiment.

The curves displayed in Fig.8 represent the degree of coherence for dark-field illumination (Fig.8a), double slit (b) and rectangular (c). (It should be clear that the annular aperture of the dark-field condenser must be smaller than the imaging aperture to perform the experiment.) As the degree of coherence and the illuminating aperture form a Fourier pair exactly like the amplitude impulse response in the image and the imaging aperture, the annular illuminating aperture causes a degree of coherence whose form is well known from the amplitude impulse response of an annular imaging aperture. The formula is given e.g. in the book of Wilson and Sheppard /10/ and its graph is displayed in Fig.8a with appropriate parameters.

The Fourier transform of a double slit is a cosine function multiplied by a broad sinc function, displayed in Fig.8b . In Fig.8c the degree of coherence has the form of a sinc function for the rectangular illuminating aperture.

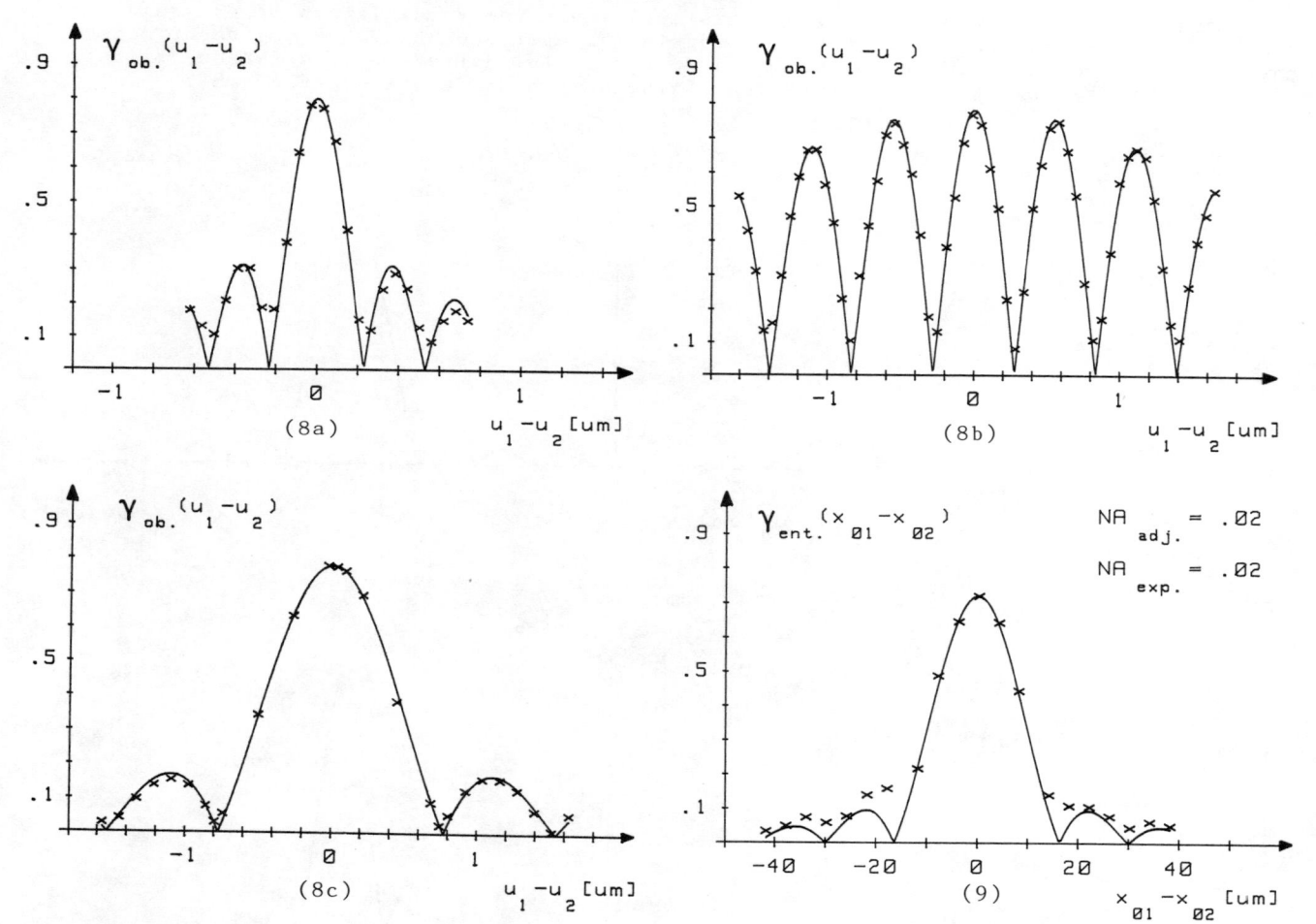

Figure 8. The degree of coherence in the object plane for different forms of the illuminating aperture.

Figure 9. The degree of coherence in the entrance pupil of the condenser

The result shown in Fig.9 was obtained when the shear interferometer was inserted between the field stop and the aperture stop plane of the condenser, measuring the degree of coherence in the entrance pupil of the condenser. According to the van Cittert-Zernike theorem the experimental data fit on a Besinc function with an illuminating aperture of 0.02 , which corresponds exactly to the adjusted radius of the field stop of 5mm and the distance between the plane of the field stop and the plane of measurement of 250mm. The deviations between the experimental data and the theoretical curve are very large here. Different from the preceding results the difference coordinate $x_{o1}-x_{o2}$ equals the shear of the interferometer. For the preceding results the shear was multiplied by the Gaussian magnification to obtain the difference coordinate in the object plane. As the curve is very steep in the vicinity of the zeros the shear must be extremly uniform for the whole area of measurement. Any deviation from this condition causes values of modulation that belong to the neighborhood of the adjusted shear. Another reason for the deviations is the lack of uniformity of illumination in the pupil of the condenser, which is the image of the filament (see Fig.3) .

The adjusted radius of the field stop of 5mm corresponds with an object size of 0.4mm . For the experiment leading to the results in Fig.6a this object size was adjusted, together with an illuminating aperture of NA=0.1 corresponding to the radius of the aperture stop of 0.5mm. If the coordinate of the first zero of the degree of coherence is considered to be a typical value of the function (sometimes it is called coherence width), we have an amount of 16μm for this typical value in the aperture stop plane (see Fig.9). As 16μm is very small compared to the radius of the aperture stop of 0.5mm, Hopkins' assumption is that the illumination of the entrance pupil is approximately incoherent. The conclusion from this assumption is confirmed by all our experiments.

5.CONCLUSIONS

The concept of the effective source is applicable to usual illumination systems. For different arrangements (Köhler and critical illumination) and variation of the parameters field stop, aperture stop and wavelength we found a very good agreement of the measured degree of coherence with Hopkins' concept predicting a Fourier relationship between the intensity distribution in the illuminating aperture and the degree of coherence in the object plane.

6.ACKNOWLEDGMENTS

I would like to thank Prof.J.Kross for giving me the possibility to perform this investigation and for many helpful discussions and H.Gerloff, G.Harth and G.Ehlke for expert technical help.

7.REFERENCES

1. B.J.Thompson and E.Wolf, "Two-Beam Interference with Partially Coherent Light", J.Opt. Soc.Am.47(10),895-902(1957).
2. D.N.Grimes,"Measurement of the Second-Order Degree of Coherence...", Appl.Opt.10(7), 1567-1570(1971).
3. W.H.Carter,"Measurement of second-order coherence in a light beam...", Appl.Opt.16(3), 558-563(1977).
4. H.H.Hopkins,"Applications of Coherence Theory in Microscopy and Interferometry", j.Opt. Soc.Am. 47(6), 508-526(1957).
5. D.J.Carpenter and C.Pask,"Coherence properties in the image of a partially coherent object", J.Opt.Soc.Am. 67(1), 115-117(1977).
6. A.S.Marathay, Elements of Optical Coherence Theory, J.Wiley&Sons, New York (1982).
7. A.Glindemann, "Superresolution by Differential-Scanning",Proc.SPIE 813, 513-514(1987).
8. D.Kelsall, "Optical Frequency Response Characteristcs in the presence of Spherical Aberration..." Proc.Phys.Soc.73, 465-479(1959).
9. K.Leonhardt, "Kontrast, Helligkeit und Phasenverschiebung der Interferenzstreifen...", Optik 35(5), 509-523(1972).
10. T.Wilson and C.Sheppard, Scanning Optical Microscopy, p.16, Academic Press, London(1984).

Confocal Interference Microscopy

Colin J.R. Sheppard, Douglas K. Hamilton, Hubert J. Matthews

Oxford University, Department of Engineering Science
Parks Road, Oxford. OX1 3PJ England

ABSTRACT

Confocal imaging is a coherent technique so that the amplitude and phase information can be extracted using interference techniques. A confocal interference microscope has been constructed based on a Michaelson interferometer. Phase changes of 25 mrad., corresponding to height changes of 1 nm. can be detected. The interferometer can be used to measure surface profiles, and also to measure aberrations in the optical system.

1. INTRODUCTION

In a confocal microscope[1] the object is illuminated with a focused light spot and the reflected or transmitted radiation collected and focused on to a small pinhole placed in front of a photodetector. An image is generated by scanning the object relative to the focused spot. Confocal microscopy has an optical sectioning property[2] which allows the investigation of thick objects including the measurement of surface topography. However the sensitivity is limited to perhaps 10 nm. Confocal imaging is a coherent process. Thus by constructing a confocal interference microscope it is possible to extract the amplitude and phase of the image signal.

2. CONFOCAL INTERFERENCE MICROSCOPY

A confocal interference microscope has been constructed[3] based on a Michaelson interferometer. In this novel arrangement a light beam is split into two with a beam splitter: the object beam is focused on to the specimen with a microscope objective and the reference beam reflected from a plane mirror. The object and reference beams are combined by the beam splitter and focused on to the point detector. It thus differs from the Linnik interferometer which has objectives in both arms. This system has the great advantages over conventional interference microscopes that there is no need to match optics, and alignment is also simplified. This is because the complex amplitude of the reference beam only at the point detector is important. A similar transmission instrument based on the Mach-Zehnder interferometer[4] also has lenses in one arm only.

The intensity signal from the detector may be written

$$I = |S|^2 + |R|^2 + 2\mathrm{Re}\{SR^*\} \qquad (1)$$

where S, R are the complex amplitudes of the signal and reference beams. The interference term can be isolated by obtaining signals with two different reference beam phases. This can be achieved by using two detectors[3], one in each output arm of the interferometer, or by directly altering the reference beam phase[5]. In order to extract both the amplitude and phase of the signal beam three measurements are necessary.

Fig.1 shows four images of a TEM grid. In Fig.1(a) a confocal image is shown, whilst Fig.1(b) shows a confocal interference image from a single detector. In Fig.1(d) the interference term has been extracted using two detectors. Fig.1(c) shows that the original confocal image can also be extracted from the two detector signals.

3. SURFACE PROFILING

The interference term can also be used to lock on to a dark fringe using a feedback system which controls either the axial position of the object[6] or the reference beam phase[5]. The former method has greater range but the latter is much faster. An example of a profile from the surface of an aluminium film produced using the second method is shown in Fig.2. The phase of the reference beam is altered by an electro-optic modulator. Resolution of the method is seen to be of the order of 1 nm., or $\lambda/600$. A complete frame is generated in 2s.

In order to ensure that the fringes are as strong as possible it is necessary to adjust the relative intensities of the signal and reference beams. This can be done using neutral density filters. Another method is to rotate the plane of polarisation of the laser, thereby altering the transmission properties of the beamsplitter[3]. However because of polarisation effects in the electro-optic modulator, object or high-aperture optics it is better to align the laser polarisation either in or perpendicular to the plane containing the normal to the beamsplitter[5]. In this case it is necessary to subtract a constant electronically in order to extract the interference term[5].

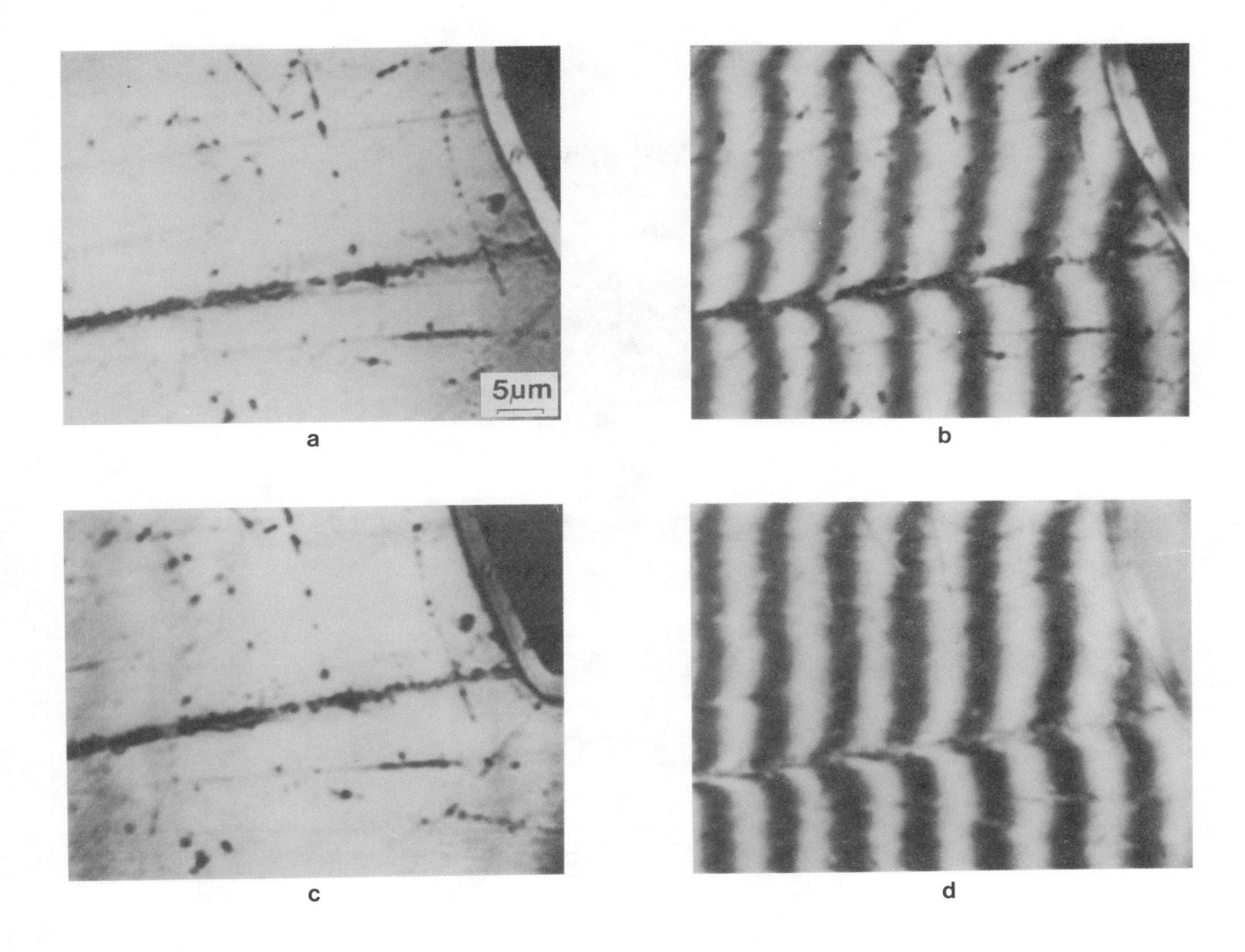

Figure 1. (a) confocal image of a TEM grid, (b) confocal interference image from a single detector, (c) the confocal image can be generated by adding two detector signals, (d) the interference term has been extracted by subtracting two detector signals.

4. ABERRATION MEASUREMENT

Interference methods can be used to extract the amplitude and phase of the signal beam. One application of such an approach is the measurement of the confocal defocus signal. If the object is a perfect reflector then the signal beam for a circularly symmetric system obeying the sine condition varies with defocus as[7]

$$S(z) = \int_0^\alpha P^2(\theta) \exp(2jkz\cos\theta) \sin\theta\cos\theta \, d\theta, \qquad (2)$$

where α is the angular aperture of the objective. Introducing the optical coordinate[8]

$$u = 4kz\sin^2(\alpha/2), \qquad (3)$$

we have for an aberration-free lens

$$S(u) = \exp\left\{\frac{ju}{2}\cot^2\left(\frac{\alpha}{2}\right)\right\} \left[\frac{\sin(u/2)}{u/2} + \frac{j\tan^2(\alpha/2)}{u/2}\left(\frac{\sin(u/2)}{u/2} - \cos(u/2)\right)\right]. \qquad (4)$$

The complex exponential term represents the fringe information, whereas the term in square brackets is an envelope. By observation in a confocal interference system the modulus and phase of $S(u)$ can be extracted. This process is simplified by two procedures:

(i) Scanning the objective rather than the reflector in z, a phase factor exp(2jkz) is repressed[9] and the phase of the resulting signal is constrained to lie in a range ±π.

(ii) Because P is band-limited S is analytic so that if we measure its real part its imaginary part can be determined directly from a Hilbert transform relationship. Thus in this case only two reference beam phases are necessary rather than three[10].

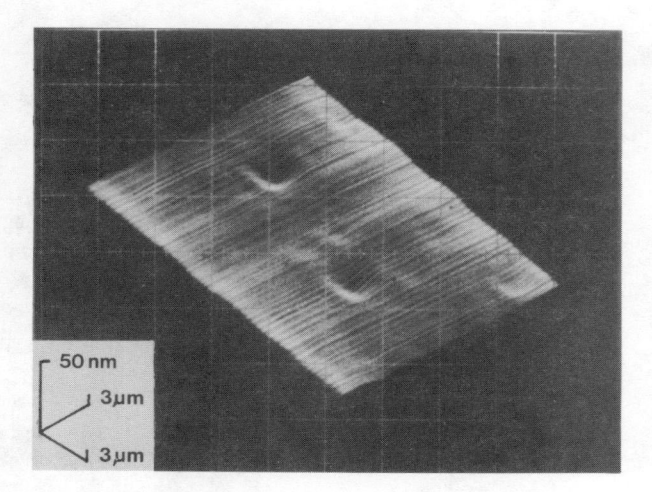

Figure 2. Surface profile of an aluminium film

Once we have obtained the complex defocus signal, from the Fourier transform relation (2) we can extract the amplitude and phase of the pupil function, corresponding to the apodization and wavefront aberration of the objective. These are conveniently plotted against the variable s given by

$$s = \sin^2(\theta/2)/\sin^2(\alpha/2). \tag{5}$$

Fig.3 shows some results for an objective of numerical aperture 0.85. The plane of best focus has been

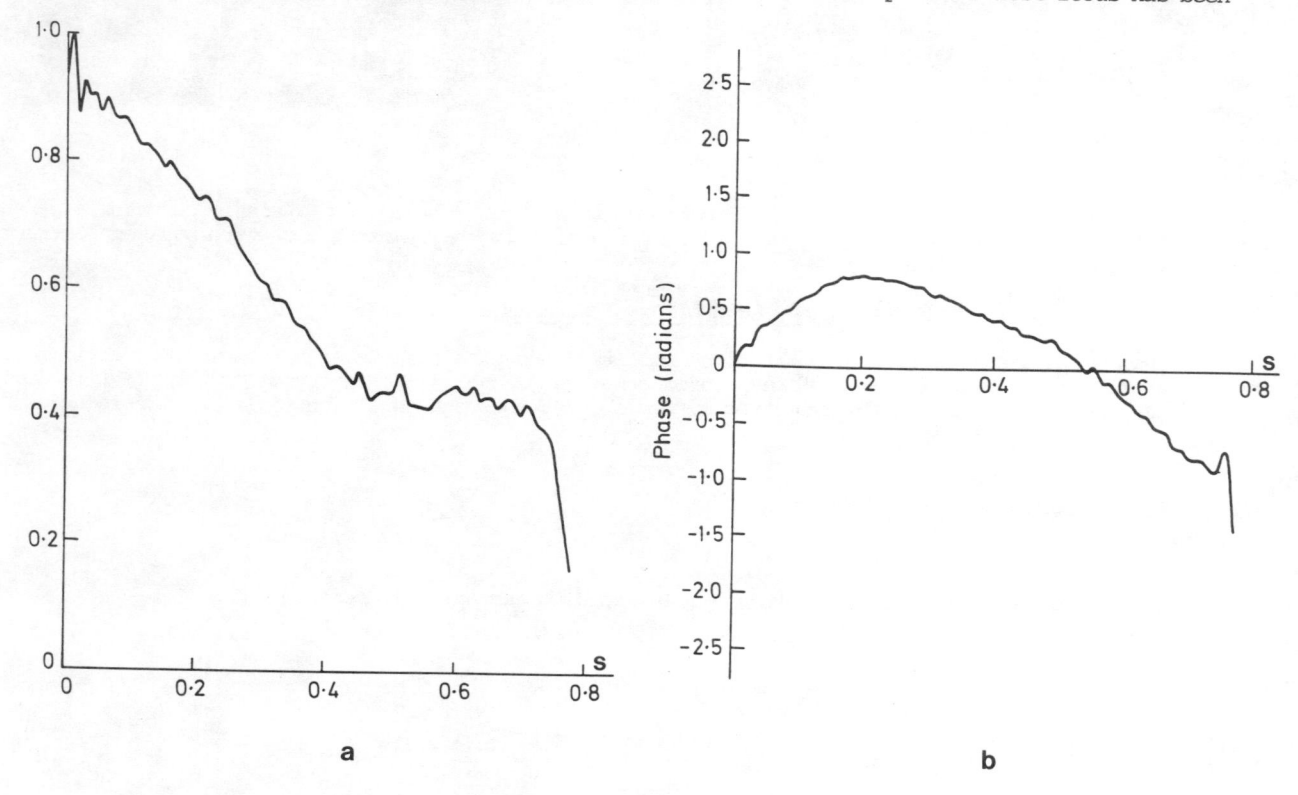

a

b

Figure 3. (a) Apodization and (b) wavefront aberration for an objective of numerical aperture 0.85

selected by optimization of the Strehl intensity. The spikes at the beginning and end of the curves are computational artifacts. The modulus falls sharply towards zero when is just less than 0.8 corresponding to the edge of the pupil, implying that the true aperture of the lens is 0.78. The falloff in transmissivity is caused by Fresnel losses at the surfaces of the lens elements. Calculated mean wavefront deformation for the lens was $\lambda/18$. This method allows measurement of objective lens aberrations <u>in situ</u> to a sensitivity of about $\lambda/100$.

5. CONCLUSIONS

A confocal interference microscope allows the phase of the confocal image to be retrieved. A phase-locked system can be used for surface profiling with a sensitivity of 1 nm. The aberrations of the optical system can be measured to about $\lambda/100$ by analysis of the defocus signal.

6. REFERENCES

1. C.J.R. Sheppard and A. Choudhury, "Imaging in the scanning microscope", Opt. Acta, 1051-1073 (1977).
2. C.J.R. Sheppard and T. Wilson, "Depth of Field in the scanning microscope", Opt. Lett. 3, 115-117 (1978).
3. D.K. Hamilton and C.J.R. Sheppard, "A confocal interference microscope", Opt. Acta, 29, 1573-1577 (1982).
4. C.J.R. Sheppard and T. Wilson, "Fourier imaging of phase information in conventional and scanning microscopes", Phil. Trans. Roy. Soc. Lond. A529, 513-536 (1980).
5. H.J. Matthews, D.K. Hamilton and C.J.R. Sheppard, "Surface profiling by phase-locked interferometry", Appl. Opt. 25, 2372-2374 (1986).
6. D.K. Hamilton and H.J. Matthews, "The confocal interference microscope as a surface profilometer", Optik 71, 31-34 (1985).
7. C.J.R. Sheppard and T. Wilson, "Effects of high angles of convergence on V(z) in the scanning acoustic microscope", Appl. Phys. Lett. 38, 858-859 (1981).
8. C.J.R. Sheppard and H.J. Matthews, "Imaging in high-aperture optical systems", J. Opt. Soc. Am. A, 4, 1354-1360 (1987).
9. D.K. Hamilton and C.J.R. Sheppard, "Interferometric measurements of the complex amplitude of the defocus signal V(z) in the confocal scanning optical microscope", J. Appl. Phys. 60, 2708-2712 (1986).
10. H.J. Matthews, D.K. Hamilton and C.J.R. Sheppard, "Aberration measurement by confocal interferometry", to be published.

ECO1
SCANNING IMAGING

Volume 1028

SESSION 3

Confocal Microscopy II

Chair
G. J. Brakenhoff
University of Amsterdam (Netherlands)

Confocal and conventional modes in tandem scanning reflected light microscopy

Alan Boyde

University College London, Department of Anatomy & Developmental Biology
Gower St., London WC1E 6BT, England

ABSTRACT

The addition of alternative, conventional non-confocal transmitted illumination to the tandem scanning reflected light microscope enables it to be used with all familiar and conventional modes of the microscope to discover a location of interest in a prepared sample, using the confocal modes when desired, or vice versa. Depth of field for the non-confocal transmitted light image can be controlled by the aperture of illumination. Conventional and confocal (coloured) images can be mixed in any proportion in the same frame.

INTRODUCTION

The first traceable invention of a confocal scanning light microscope (CSLM) is that of Minsky:[1] he wished to look at Golgi preparations of brain, in which silver is deposited as a reactive product in or on just a few nerve cells and the sections may be fourty times thicker than we would normally use in conventional light microscopy. The reason for using thick slices is that it is difficult to reconstruct a tissue ("the wiring diagram of the brain") from many thin slices. Minsky's idea was to slice the brain optically, seeing only a limited amount of 3-D complexity at once, yet to be able to put it back together by being able to focus through all the optical slices. The Tandem Scanning Microscope (TSM) is a direct view, real time confocal microscope: Petran invented it to look at neurons in live brain tissue slices.[2-6]

In a confocal microscope, only one very small volume in the defined focal plane within the specimen is brightly illuminated at one time. Light from that point is collected, strongly discriminating against scattered or fluorescent light from all points in front of and behind that point within the sample by means of an aperture placed in the intermediate image plane of the objective lens. In both the Minsky[1] and Petran[2-6] microscopes the definition of the illuminated spot is achieved by imaging an aperture in the intermediate image plane of the condensor-objective. Minsky scanned the sample mechanically. Petran scanned multiple conjugate apertures in both illumination and detection channels using a highly derived Nipkow disc.

The advantage of extracting information from only one plane in the object is offset by practical disadvantages: it is the purpose of this communication to describe new means of reversing this situation and to be able to enjoy the advantages of both conventional and confocal modes of operation in one CSLM - in our case, a TSM.

The problems of conventional light microscopy are contrast and resolution, particularly in the reflection mode which must be used with a thick object and in fluorescence microscopy. Contrast is low because of the high DC signal level due to backscattering of light (or fluorescence) from other layers in the sample and from optical surfaces in the microscope. Resolution is low because of confusion of detail from features lying in the out-of-focus halo zone. These are overcome by confocal operation.

The problems of confocal observation are that nothing is seen (there is no signal) if the specimen is out of focus. It is therefore difficult to find an area of interest by rapid searching of poorly focussed images as we do in routine biological microscopy. Further, the depth of field is so small that features of interest may be missed unless all the possible optical (Z) sections are scanned, as well as all the X,Y fields.

The time needed to scan large numbers of fields in Z, as well as X,Y, is greatly reduced in the TSM: because large numbers of apertures (light beams) are scanned simultaneously, the frame rate is much higher than standard TV and it is not noticed that it is a scanning device. To this is added the comfort of viewing a real optical image rather than a scanning display monitor.[2-6] The TSM, therefore, really requires no second or back up method of finding the field of view; it is a stand alone device. At the other extreme, the ideal, on-axis, minimum aberration, scanned specimen type of laser CSLM[7-11] presents unacceptably low frame rates, even if the specimen may be oscillated. By scanning the laser beam with mirrors, the frame rates may be accelerated to one per second at full display resolution, but with the disadvantage of working off axis shared with the TSM. However, this rate is not sufficient for locating fields in XYZ, and because unacceptable radiation damage may be done in finding a field before an image can be captured, a second method of searching is

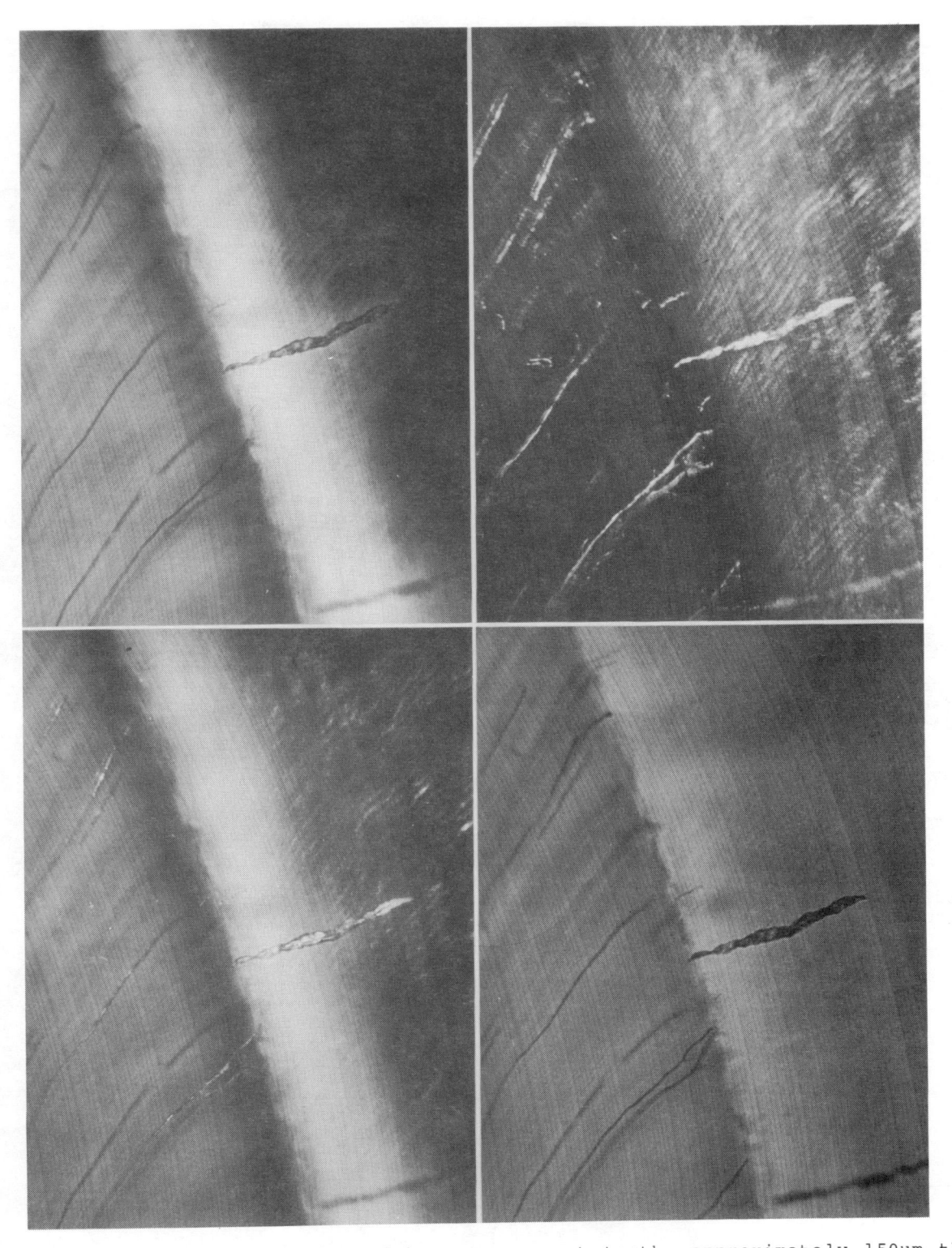

Figure 1. Longitudinal sawn section of human permanent tooth, approximately 150um thick, imaged in the Petran and Hadravsky TSM.[4,5] The base of a Zeiss polarising microscope was used to provide transmitted non-confocal illumination. In each field, enamel is right and dentine left. The prominent feature at centre is an enamel "spindle". Objective used was 40/1.3 oil, width of each field is 240um. (a) transmitted polarised light; extinction in enamel further from junction with dentine. (b) reflected TSM confocal image showing reflections from spindle, enamel prisms boundaries and dentine tubules where these lie in plane of focus. (c) combined transmitted polarised light and reflected TSM images taken with both illumination sources running. (d) ordinary transmitted light only.

virtually imperative. There is no doubt then that a great improvement for the single beam of laser origin CSLMs came with the "black box" configuration in which an existing light microscope is used.[12-15] The beam is scanned in the eye point of the photo-tube of a conventional upright microscope and hence in the focussed-on plane in the specimen. The reflected or fluorescent light is taken back through the photo tube and descanned with the same device. This approach has the advantage that one can use an ordinary specimen in the conventional microscope to locate a discrete region in the specimen.

In presenting the case for the black box CSLM,[14,15] it has been implied that such simple means of correlation are not available in other designs, but they are in the TSM. We can cross-correlate using an ordinary microscope, using no electronic devices and this approach may be much more attractive to the user than one dependent on video images.

2. EXPERIMENTAL ARRANGEMENTS

2.1. Conventional inverted light microscope under TSM

The light provided by the TSM is used to illuminate an inverted optical microscope. The same field of view is then seen in transmitted light, but via a different objective lens. This was very easy to set up on the Petran-Hadravsky Plzen TSM. A low magnification (long working distance) objective may be fully illuminated by defocussing the TSM.

2.2. Conventional light microscope base (illuminator and stage) under TSM

The bottom half of a Zeiss polarising microscope placed under the Plzen TSM was used to illuminate a conventional translucent specimen backwards. This arrangement allows the rotation of the specimen using the rotating stage of the polarising microscope to avoid the correspondence of linear structures in the specimen with the scanning lines. More importantly, the second source of illumination only passes through the disc on the observation side, so that the disc merely acts as a neutral density filter but has no other function.

2.3. Substage illuminator and condensor attached to a conventional LM stage on the TSM

A conventional LM specimen stage with substage condensor focussing and centration and a separate illumination source was built on to a TSM constructed by Tracor Northern (Middleton WI, USA). This arrangement was optically identical to the foregoing, but both more versatile and mechanically stable. This addition to the TSM makes it possible to use all the usual modes of LM directly in or with the TSM. Many of these possibilities have already been proven in our laboratory.

With a microscope in this configuration, it is now possible to observe the confocal reflected light fluorescence image, simultaneously with any transmitted light image (including all methods of bright-field and dark-field illumination). In addition, because it is easy to remove the disc from the Tracor TSM, it is also possible to use the TSM to provide epi-illumination for conventional reflected light microscopy or epifluorescence microscopy with no confocal function. The possible modes include:- confocal reflection (including interference reflection contrast IRC), confocal fluorescence, conventional reflection, conventional fluorescence, dark-field, bright-field, polarisation, epipolarisation, phase contrast, other interference methods, amplitude contrast from conventional stains, and the use of colour filters to code contrasting (non-coloured) conventional and confocal images.[16]

A field stop or aperture in either illuminating channel can be imaged in the "TSM" to make it possible to display only portions of the field of view thus illuminated, so that selected areas in the field of view can be imaged in one or the other or both of the trans-illumination, and reflected confocal modes.

3. APPLICATIONS

3.1. Colour

Many of the advantages presented by the possibility to perform simultaneous or alternate confocal and conventional imaging in the TSM can only be demonstrated in colour. The conventional stained or fluorescence image is coloured: the TSM works in full colour.

3.2. Autoradiograms

Autoradiograms consist of a developed silver grain containing emulsion applied to the surface of a conventional stained histological section. The two layers are separated by a finite distance and cannot be brought into sharp focus simultaneously. With the method described here, however, sharp resolution of the silver grains in the reflected confocal

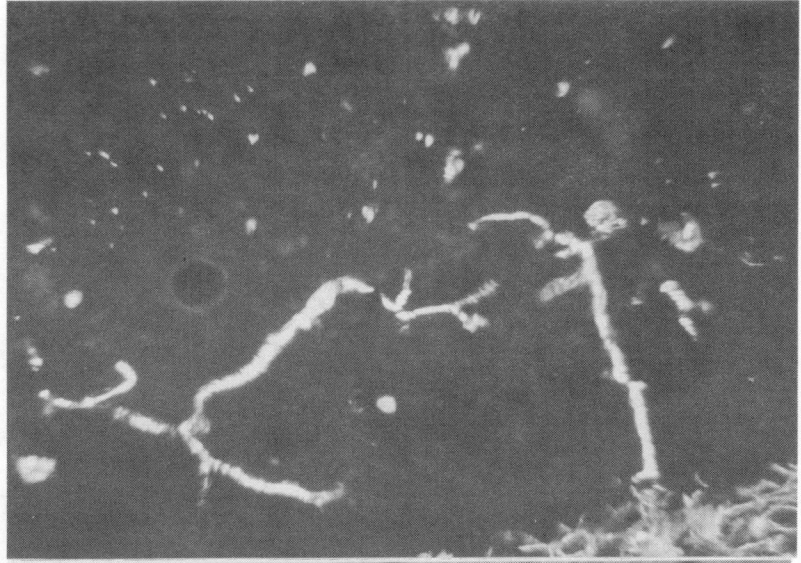

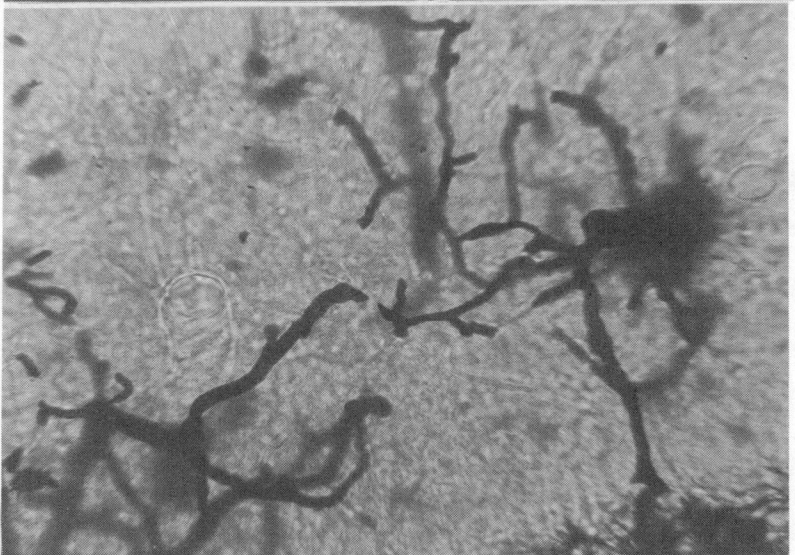

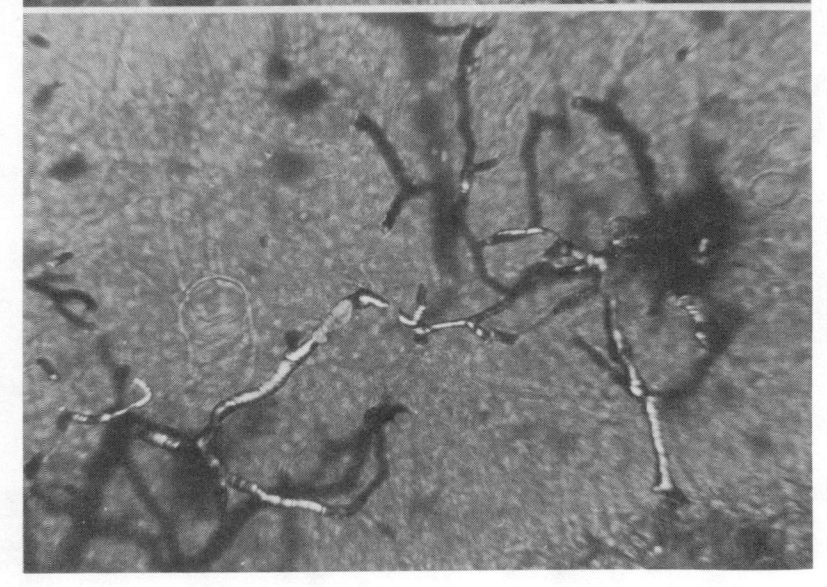

Figure 2. Thick section of hamster cortex stained by Golgi impregnation, here showing distribution of blood vessels, imaged in Petran-Hadravsky TSM[4,5] using 25/0.65 oil immersion objective. Width of each field is 400um. Use of automatic metering 35mm camera back (Olympus OM2) to record these images has led to considerable relative over-exposure of the TSM reflected confocal image shown in (a) at top.
(b) shows ordinary transmitted light image of same field with very large depth of field: less well focussed vessels are 110um distant from those shown in focus in (a).
(c) combined TSM confocal reflected image (bright features) and transmitted, non-confocal, large depth of field image. Note how greatly reduced exposure time leads to reduction in dimensions of features in the TSM component of the image. This effect is equivalent to reduction in gain in an electronic contrast controlled system like a laser CSLM.

TSM images can be combined with an enhanced depth of field image of the histological detail within the section, simply obtained by reducing the effective aperture of the objective for the transmitted light image by lowering the condenser or using a lower NA condenser. By adjusting lamp intensities and appropriate use of neutral density filters, we can arrange to mix the images in any balance whilst maintaining the colour balance.

3.3. Phase contrast

Many cells are studied in tissue or organ culture as sparse distributions on plastic dishes. The finding of cells may be slowed up in the confocal reflected mode if the dishes do not have flat bottoms. The identification of cell types is nearly always based upon well known phase contrast microscopical appearances. Thus it may be necessary to use "phase" to find them. If this is so, then why use TSM? The TSM provides a simple and effective way of forming high contrast interference reflection (IRC) images which map out the distribution of the very thin peripheral portions of cells which cannot be properly resolved by phase imaging. These images give information about the thickness of the thin parts of the cell as well as about the distribution of focal contacts (close encounters) where cells are closely apposed to the substrate to which they adhere. With a flat substrate, the experienced TSM user can find cells just as rapidly in the TSM-IRC image.

3.4. Polarised light (PLM)

Polarised light microscopy is widely used in the characterisation of both geological and biological mineralised tissue samples (Fig. 1a,b,c,d). It is therefore advantageous to be able to compare and match appearances in PLM and TSM. Confocal reflected polarised light microscopy can be achieved in the TSM, and it is anticipated that this may be particularly useful in mineralogical and geological applications.

3.5. Large and small depth of field

There are many reasons to want to increase the depth of field in biological microscopy, particularly to be able to have some 3-D impression of the distribution of complex features. Such increase in depth of field must be traded for resolution. In the modified TSM considered here, however, both large and small depths of field can be combined in the same image frame.

The main objects in Fig. 2 are blood vessels in a 250um thick Golgi stained section of brain tissue. Portions of these vessels are imaged at 110um out of focus features in the transmitted light (TL) image (Fig. 2b). The confocal TSM image of the identical field of view (Fig. 2a) shows only those features which were really in focus. Fig. 2c shows the combined TL and TSM.

4. CONCLUSIONS

The combination of confocal and non-confocal methods are very easily achieved in a standard TSM. The arrangement suggested is practical and convenient for the direct view comparison of all but low intensity fluorescence images in the TSM. To view the latter, it

is necessary to use a sensitive image intensification device (we have had the best results with a Peltier cooled CCD camera). Although it has not been our experience that it is necessary to use other than the usual TSM image to locate objects (for example, specific cell types) for fluorescence imaging, it is probable that the combination of imaging modalities described here will be useful in this regard, particularly for casual users and beginners.

In making a comparison of the enhanced TSM method described here with the existing mirror scanning off-axis laser CSLMs,[12-15] it should be noted that - because of their slow scanning speed - it is not just an advantage to have other modes of image formation available: it is a pure necessity. Finding an area of interest would otherwise be too difficult. Further, when operating in the fluorescence mode, the high powered laser beam may induce photo-bleaching so rapidly that an area has to be located by conventional means first.

The methods for cross-correlation considered here will be equally applicable in a one-sided TSM[16] and might make that configuration of instrument potentially more useful in biology than it could be at the present. For example, one could imagine a one-sided TSM working principally in fluorescence in which the field of view was selected in a normal light microscopic image.

5. ACKNOWLEDGEMENTS

This work has been supported by the Science and Engineering Research Council (U.K.). I would like to thank Tim Watson and Roy Radcliffe for various forms of assistance. Sheila Jones and Louise Taylor have focussed our attention on biological problem areas. Miguel Freire provided excellent Golgi preparations.

6. REFERENCES

1. M. Minsky, "Microscopy apparatus," United States Patent Office. Filed Nov. 7, 1957, Granted Dec. 19, 1961. Patent No. 3,013,467 (1961).
2. M. Petran, M. Hadravsky, M.D. Egger and R. Galambos, "Tandem-scanning reflected light microscope," J. opt. Soc. .Am. 58, 66-664 (1968).
3. M.D. Egger and M. Petran, "New reflected-light microscope for viewing unstained brain and ganglion cells," Science 157, 305-307 (1967).
4. M. Petran, M. Hadravsky, J. Benes, R. Kucera and A. Boyde, "The tandem scanning reflected light microscope, Part I - the principle, and its design," Proc. roy. microsc. Soc. 20, 125-129 (1985).
5. A. Boyde, "The tandem scanning reflected light microscope, Part II - Pre-Micro '84 applications at UCL," Proc. roy. microsc. Soc. 20, 131-139 (1985).
6. M. Petran, M. Hadravsky and A. Boyde, "The tandem scanning reflected light microscope," Scanning 7, 97-108 (1985).
7. G.J. Brakenhoff, P. Blom and P. Barends, "Confocal scanning light microscopy with high aperture immersion lenses," J. Microsc. 117, 219-232 (1979).
8. H.T.M. Van der Voort, G.J. Brakenhoff, J.A.C. Valkenburg and N. Nanninga, "Design and Use of a computer controlled confocal microscope for biological applications," Scanning 7, 66-78 (1985).
9. T. Wilson, "Imaging properties and applications of scanning optical microscopes," Appl. Phys. 22, 119-128 (1980).
10. T. Wilson and C. Sheppard, Theory and Practice of Scanning Optical Microscopy. Academic Press, London (1984).
11. R.W. Wijnaendts van Resandt, H.J.B. Marsman, R. Kaplan, J. Davoust, E.H.K. Stelzer and R. Stricker, "Optical fluorescence microscopy in three dimensions: microtomoscopy," J. Microsc. 138, 29-34 (1984).
12. K. Carlsson, P.E. Danielson, R. Lenz, A. Liljeborg, L. Majlof and N. Aslund, "Three-dimensional microscopy using a confocal laser scanning microscope," Optics Letters 10, 53-55 (1985).
13. K. Carlsson and N. Aslund, "Confocal imaging for 3-D digital microscopy", Applied Optics 26, 3232-3238 (1987).
14. J.G. White, W.B. Amos and M. Fordham, "An evaluation of confocal versus conventional imaging of biological structures by fluorescence light microscopy," J. cell Biol. 105, 41-48 (1987).
15. W.B. Amos, J.G. White and M. Fordham, "Use of confocal imaging in the study of biological structures," Applied Optics 26, 3239-3243 (1987).
16. A. Boyde, "Colour-coded stereo images from the tandem scanning reflected light microscope (TSRLM)," J. Microsc. 146, 137-142 (1987).
17. G.Q. Xiao, G.S. Kino, "A real-time confocal scanning optical microscope," Proc. 4th Int. Symp. on Optical and Optoelectronic App. Sci. and Eng. The Hague (1987).

Imaging theory for the scanning optical microscope

G. S. Kino, C-H. Chou, and G. Q. Xiao

Edward L. Ginzton Laboratory, Stanford University
Stanford, California 94305

I. INTRODUCTION

The confocal scanning optical microscope (CSOM) has the major advantages over the standard microscope that it has extremely good range resolution and cross-sectioning capability, somewhat better transverse resolution, and is well adapted to quantitative measurements.[1,2] The basic reason for the cross-sectioning capability is that the objective lens is illuminated from a collimated beam, as shown in Fig. 1a, which focuses the beam to a small spot on the object. The light reflected from the object then returns back through the objective lens and a beamsplitter, and is focused to illuminate a pinhole in front of the detector. The beam is defocused by moving the object out of the focal plane. Light at the pinhole is defocused and very little light gets back to the detector. Typically, with a large aperture lens, the 3 dB points of the response are of the order of 500 nm apart.

A second version of this microscope, illustrated in Fig. 1b, uses direct illumination through a pinhole and brings back the light through the same pinhole to the detector. In both cases, an image is formed by scanning the object or the beam and displaying the output as a video image.

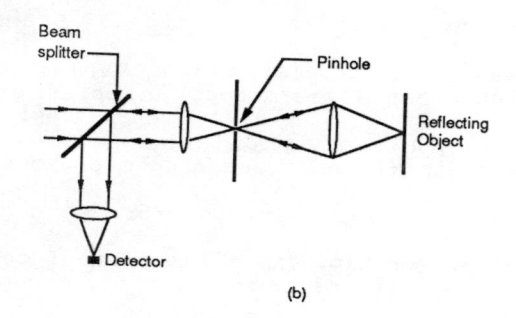

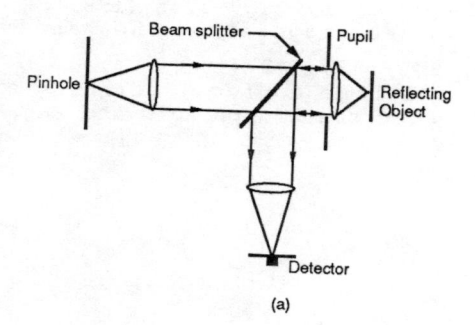

Fig. 1. (a) A CSOM excited by a collimated beam formed by passing the beam through a pinhole.
(b) A CSOM excited by a beam passing through a pinhole, with the beam received at the same pinhole.

Recently, several microscopes, which are derivatives of the original tandem scanning optical microscope (TSOM) of Petran and Hadravsky, have been developed to give a real-time rather than a video image, which takes several seconds to form.[3-7] These microscopes make use of a rotating Nipkow disk in which a large number of pinholes spaced of the order of ten pinhole diameters apart are formed in an interleaved spiral configuration on the disk. In the real-time scanning optical microscope (RSOM) made by the authors, there are 200,000 pinholes of 20 μm diameter. As shown in Fig. 2, several thousand of these pinholes are illuminated at once by an incident beam from a mercury vapor arc to form images on the object and are imaged on the object. Light returns through the same pinholes as that through which it entered. The image is observed through transfer lenses and an eyepiece. The disk is rotated at approximately 2,000 rpm to form an image with approximately 5000 lines at a frame rate of approximately 700 frames/s. Various measures are taken, such as tilting the disk, using polarized light, and using a stop to eliminate the reflected light from the disk. These techniques have been described in other papers.

It is apparent that in all of these confocal microscopes, the size of the pinhole is of critical importance. If the pinhole is too large, the transverse and range resolution are deteriorated. In the RSOM, if the pinholes are too small, the amount of light passing through the disk is decreased and the light budget becomes critical. It is therefore important to determine the optimum size of pinhole for good resolution while at the same time maintaining reasonable efficiency.

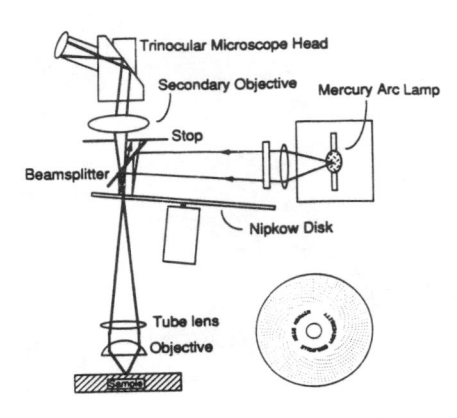

Fig. 2. A schematic of the real-time scanning optical microscope.

In this paper, we discuss the theory of the RSOM for a finite pinhole size and compare the result with those for a mechanically-scanned microscope using either a collimated input beam, a case that has been dealt with by Wilson et al, or an input beam which passes in and out of the same pinhole.[8] In the first case, the spot formed by the transmitted beam is always the same size. The definition of the received beam deteriorates as the pinhole size is increased. In the second case, as the pinhole size is increased, the illumination of the object lens becomes more nonuniform until the aperture of the lens is not fully filled. Similarly, the return beam is focused to a spot smaller than the pinhole size if the pinhole is large. Again, the definition of the image deteriorates. For the RSOM, the situation is somewhat different. As far as the range resolution is concerned, it is similar to the case of a mechanically-scanned microscope in which the light enters a pinhole and is received on the same pinhole. As far as the transverse resolution is concerned, we must take into the account the fact that the pinholes are rotating and determine the average field observed by an observer located at a point on the axis of the plane of the pinholes. This leads to a transverse resolution with infinitesimal size pinholes, which is that of a perfect confocal microscope, and to a transverse resolution with large pinholes, which is that of the standard microscope, i.e., approximately 1.4 times as large between 3 dB points as that for the perfect confocal microscope. For the mechanically-scanned version of the microscope, shown in Fig. 1b, as the pinhole diameter increases, the transverse resolution gets steadily worse since a large area detector is employed.

A second type of theory which we shall outline in this paper is concerned with the form of the images, and in particular, linescans obtained with confocal microscopes. We have set up an initial theory for the problem of a deep trench in metal or silicon. The theory determines the form of the linescan when the beam is focused on the surface of the silicon and the form of the line scan when the beam is focused into the trench. We have compared the results with both experiments on both acoustic and optical microscopes and have obtained fairly good agreement with the theory. The theory is variational in form and can be extrapolated as we develop it to deal with multiple dielectric layers and so on.

2. THEORY OF THE REAL-TIME SCANNING OPTICAL MICROSCOPE

Range Resolution

The theory used will be summarized, but will given in greater detail in other work by Kino et al.[7] We use scalar theory and suppose that the field of the wave incident on the pinhole is of value ψ_0 and that paraxial conditions are satisfied on the pinhole side but not on the sample side of the objective lens. We consider the system illustrated in Fig. 3. It follows from the Rayleigh-Sommerfeld diffraction theory, using the Fraunhofer approximation, that at a distance h_1 from the the pinhole, the field at the pupil plane D_1 an angle θ_1 and radius r_1 from the axis is:[9-10]

$$\psi_1(\theta_1) = 2\pi \frac{\psi_0}{j\lambda h_1} \frac{a^2 J_1(ka\sin\theta_1)}{ka\sin\theta_1} e^{-jk\left(h_1 + \frac{r_1^2}{2h_1}\right)} \tag{1}$$

where $k = 2\pi/\lambda$, and λ is the optical wavelength in free space.

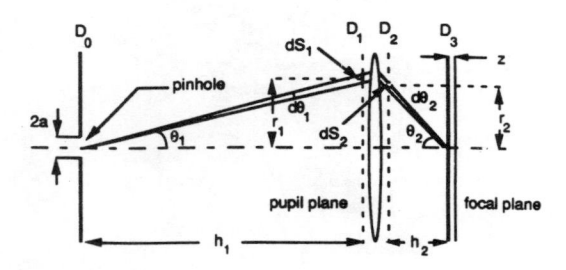

Fig. 3. A schematic of the system analyzed in the theory.

Using this formula, we may make a rough estimate of the optimum pinhole radius a(opt) by choosing the diameter d_P(3 dB) at the half power points of the beam diffracted by the pinhole to be equal to the pupil diameter 2b of the objective. This leads to the result:

$$a(\text{opt}) = \frac{0.25\lambda h_1}{b} \tag{2}$$

where h_1 is the spacing between the pupil and the objective (the tube length). Since $h_1 = 180$ mm, $\lambda = 546$ nm, and $b = 2$ mm, this leads to a pinhole radius a(opt) ≈ 11.6 μm, which may be compared to the pinhole radius of 10 μm used in the present RSOM of Xiao et al.[6] More exact results will be given below.

If the pinholes are too closely spaced, there is interference between their images on the object, and speckle effects will be apparent when a narrowband light source is employed. Furthermore, as the spacing between the pinholes is decreased, the area of the pinholes becomes closer to the total area illuminated. In this case the intensity of the image no longer falls off rapidly with the defocus distance z and part of the light reflected from the object can return through alternative pinholes to the eyepiece. For this latter reason, it is advisable to keep the area of the pinholes within the illuminating beam to be less than 1% of the total illuminated area.

Since it is important to obtain as good a definition as possible, which implies the use of large aperture lenses, it is desirable to develop a nonparaxial theory for the design criteria required. This has been done by use of the theory of stigmatic imaging, which leads to the sine condition:[9]

$$\frac{\sin\theta_2}{\sin\theta_1} = M \frac{n_1}{n_2} \tag{3}$$

where M is the magnification of the image, θ_1 and θ_2 are the entrance and exit angles, respectively, of a ray passing through the objective lens, and n_1 and n_2 are the refractive indices of the media on each side of the lens, respectively. We will assume from now on that $n_1 = 1$.

We have used a nonparaxial scalar theory to determine the range resolution and power efficiency of the microscope when the beam passing through the objective lens is reflected by an ideal plane reflector a distance z from the focus by considering the power $P(z)$ passing back through the pinhole. We define the ratio $I(z) = P(z)/P_0$ where P_0 is the power entering the pinhole from the light source, $P_0 = \pi a^2 I_0$, and I_0 is the intensity of the beam incident on the pinhole. It is assumed that the radius of the pinhole is large enough so that $ka \gg 1$.

We find the reflected field $\psi_0'(r_0)$ at the plane of the pinhole to be:

$$\psi_0'(r_0) = \psi_0 \frac{k_2 a}{M} \int_0^{\theta_0} J\left(\frac{k_2 a \sin\theta_2}{M}\right) J_0\left(\frac{k_2 r_0 \sin\theta_2}{M}\right) e^{-2jk_2 z \cos\theta_2} \sin\theta_2 \, d\theta_2 \tag{4}$$

where the relation between θ_2 and θ_1 is given by Eq. (3) and $\sin\theta_0 = NA$, the numerical aperture of the lens. The value of $I(z)$ may be found from the result of Eq. (4) by writing:

$$I(z) = \frac{2 \int_0^a |\psi_0'|^2 r_0 \, dr_0}{\pi a^2 \psi_0^2} \tag{5}$$

For an infinitesimal pinhole size, $\psi_0'(r_0)$ is uniform and we can write Eq. (4) for ψ_0' in normalized form:

$$V(z) = \frac{\psi_0'(z)}{\psi_0'(0)} = \frac{2 \int_0^{\theta_0} e^{-2jk_2 z \cos\theta_2} \sin\theta_2 \cos\theta_2 \, d\theta_2}{\sin^2\theta_0} \tag{6}$$

We note that if the system is paraxial, i.e., if θ_2 is not too large, we may put $\cos\theta_2 \approx 1$. Equation (6) then takes the simple form:

$$V(z) = \frac{\sin k_2 z (1 - \cos\theta_0)}{k_2 z (1 - \cos\theta_0)} e^{-jk_2 z (1 + \cos\theta_0)} \tag{7}$$

The magnitude of $|V(z)|^2$, or normalized intensity $I_{norm}(z)$, given by Eqs. (6) and (7) for a 0.95 aperture, has been determined. It is found that even with a 0.95 aperture, the difference between the two results for the 3 dB resolution $d_z(3\,dB)$ is small. For an infinitesimal pinhole size, $\lambda = 546$ nm and the numerical aperture is 0.95. The depth resolution given by Eq. (7) is $d_z(3\,dB) = 353$ nm, while a more exact calculation from Eq. (6) yields $d_z(3\,dB) = 375$ nm; thus, there is a 6% error in the use of Eq. (6), which reduces to a 3% error for $NA = 0.9$. Thus, the analytic form of Eq. (7) is a very convenient one.

We may derive from Eq. (7) a very useful formula for the spacing of the 3 dB points:

$$d_z(3\,dB) = \frac{.45\lambda}{n_2(1 - \cos\theta_0)} \tag{8}$$

Plots of the efficiency I(0) with respect to the parameter $n_2 a/\lambda M$ are given for different numerical apertures in Fig. 4. For a typical real-time scanning optical microscope with $n_2 = 1$, $M = 65X$, $\lambda = 546$ nm, $\sin \theta_0 = 0.9$, and $a = 10 \mu m$, $n_2 a/\lambda M = 0.28$ and the reflected power passing back through the pinhole is 22% of the incident power. Plots of the resolution $d_z(3\text{ dB})$ as a function of $n_2 a/\lambda M$ for different numerical apertures are given in Fig. 5. It will be observed that, as might be expected, the resolution becomes worse as the pinhole size is increased. It will be noted that the case treated by Wilson et al for the confocal microscope is that for a collimated beam illuminating the objective, which is equivalent to a transmitting pinhole of infinitesimal size, but a receiving pinhole of finite size.[8] Thus, their results are slightly different from ours.

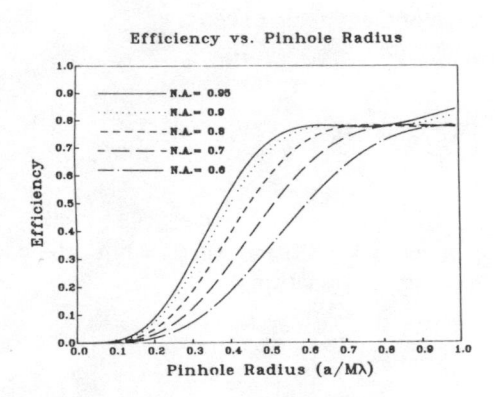

Fig. 4. A plot of the efficiency of the RSOM as a function of the parameter $n_2 a\lambda/M$ for different numerical apertures.

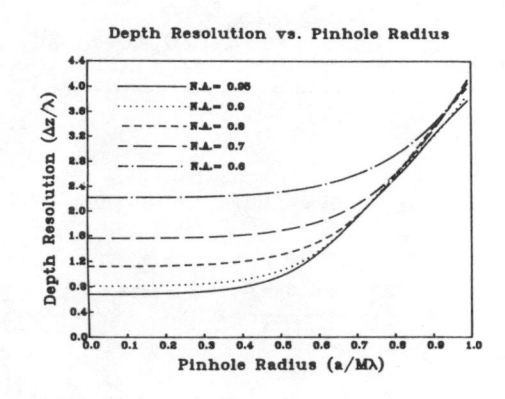

Fig. 5. Plots of the range resolution $d_z(3\text{ dB})$ as a function of $n_2 a\lambda/M$ for different numerical apertures.

A plot of the intensity response as a function of distance z is shown in Fig. 6 for our microscope with an N.A. of 0.9 and an objective lens magnification of 65X. It is seen that theory and experiment are in good agreement. The $d_z(3\text{ dB})$ width is theoretically 460 nm for a 20 μm pinhole diameter and $\lambda = 546$ nm. This compares to the value of 440 nm for an infinitesimal size pinhole. The measured value of $d_z(3\text{ dB})$ is 501 nm. Results taken with other lenses are in similarly fairly good agreement with theory; for instance, with N.A. = 0.8, the theoretical value of $d_z(3\text{ dB}) = 621$ nm while the experimental value is 651 nm. The sidelobes of the response do tend to be somewhat different than the theoretical levels. This is due to spherical aberration in the lens and can be accounted for.[11] More careful studies with mechanically-scanned confocal microscopes lead to the same conclusion.

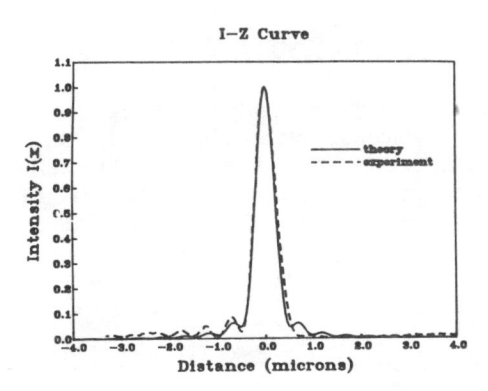

Fig. 6. A plot of the intensity variation as a function of distance z for a numerical aperture of NA = 0.9. This theoretical result is compared to an experimental result taken on the RSOM.

The Transverse Response I(x,0) of a Confocal Microscope

We have determined the transverse response of a confocal microscope without using the paraxial approximation. We first calculate the fields at the pupil plane of the objective. Then we determine the excitation at the pinhole by the waves reflected from a point reflector located at the focal plane a distance x from the axis. Since we need to determine the image seen by a stationary observer through a rotating set of pinholes, the analysis cannot be identical in form for either the case of a beam entering and leaving through a stationary pinhole, or that of a collimated transmitting beam and a stationary receiving pinhole of finite size.

We therefore need to define a *time averaged point spread function*. We assume that an observer looks at one point in space and determines the signal arriving at this point as the point reflecting object is moved a distance x from the axis. The point object is illuminated by a set of moving pinholes, and the image is observed through this same set of rotating pinholes. Therefore, we first determine how the illumination of the point object varies as the position of the center $R_0 \alpha$ of the illuminating pinhole varies. Then we determine the average signal received at a point of observation on the axis on the plane of the pinholes. This is the *time averaged point spread function*.

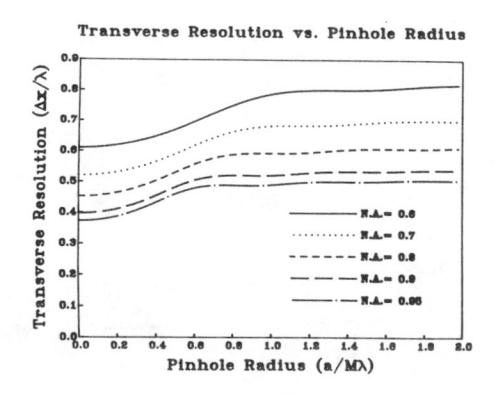

Fig. 7. Plots of the 3 dB transverse resolution as function of the parameter $n_2 a \lambda / M$ of the RSOM for different numerical apertures.

Plots of the 3 dB resolution $d_{RS}(3\ dB)$ as a function of the parameter $n_2 a/\lambda M$ for different numerical apertures are given in Fig. 7. It will be observed that the 3 dB resolution increases by a factor of approximately 1.4 as the pinhole size is increased from infinitesimal to very large values. When the pinhole size is large, the system becomes a standard microscope. When it is infinitesimal, the point spread response becomes that of a perfect confocal microscope.

3. IMAGING OF A TRENCH

In this paper, we have considered the definition of the microscope. However, that is only part of the problem. We must also determine the nature of the image when typical objects are observed with the microscope. For instance, if a simple step in a perfect reflector is observed with the microscope, the amplitude of the reflected wave should be the same on both sides of the step. However, the phase will change from one side of the step to the other. If the depth of the step is within the defocusing distance, and the reflected signal is V_0, then if the phase difference in passing over a step of height h is $2kh$, the signal received by the microscope from the lower part of the step is approximately $V_0 \exp -2jkh$. When the center of the beam is located at the step edge, the total signal received is of the form:

$$V = \frac{V_0}{2}\left[1 + e^{-2jkh}\right] = V_0 e^{-jkh} \cos kh \qquad (9)$$

This implies that the amplitude of the signal received has a dip when the center of the beam is located at the edge of the step.

Such simplifying assumptions have been shown to work well experimentally for measurement of phase and somewhat less well for measurement of amplitude. When the step sizes are small, the theories can be adapted to account for a change in reflectivity from one side of the step to the other, but they fail when the step size becomes large. Such theories also fail in observations of a channel. Even the amplitude variation given by such theories tends to not be very reliable when the beam is passed over a large step on, for instance, the edge of a piece of silicon.

Therefore, it is necessary to develop a more complete theory to work even with the simplest of cases, that of a channel in a perfect conductor or in silicon, an imperfect conductor. We have started, therefore, to carry out a variational theory for imaging of a simple trench. This is difficult since the beam from a scanning optical microscope is three-dimensional, although the trench is two-dimensional. However, the penetration of the fields into the trench depends on the way the Fourier components of the beam penetrate the trench, and this depends in turn on the variation of the field of the beam in both the x and y directions.

We have looked at the problem shown in Fig. 8 and have divided it into two parts. We call the incident field from the microscope ψ_{inc}, and the reflected field ψ_{refl} with the transmitted field into the trench ψ_{inside}. Ideally, we must satisfy the continuity conditions at the surface of the trench:

$$\psi_{inc} + \psi_{refl} = \psi_{inside} = \psi_{tot} \qquad (10)$$

and

$$\frac{\partial \psi_{inside}}{\partial z} = \frac{\partial \psi_{tot}}{\partial z} = \frac{\partial \psi_{inc}}{\partial z} + \frac{\partial \psi_{refl}}{\partial z} \qquad (11)$$

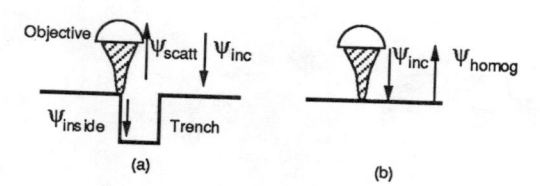

Fig. 8. (a) Schematic of a trench illuminated by a microscope. (b) A perfect reflecting mirror illuminated by a microscope.

In practice, it is not easy to work out a mathematical theory which satisfies these conditions exactly, especially in a three-dimensional case. We have therefore divided the problem into two parts. We call the microscope beam reflected by a perfect conductor ψ_{homog}, and the scattered field by the object ψ_{scatt} where:

$$\psi_{scatt} = \psi_{refl} - \psi_{homog} \tag{12}$$

$$\frac{\partial \psi_{scatt}}{\partial z} = \frac{\partial \psi_{refl}}{\partial z} - \frac{\partial \psi_{homog}}{\partial z} \tag{13}$$

We solve for ψ_{scatt} and $\partial \psi_{scatt}/\partial y$ rather than ψ_{refl}, for now we need only work with fields which are finite at the surface of the channel.

It can be shown that the basic relation for the response of the microscope follows from the reciprocity theorem:

$$V(x,y,z) = \int_{S_2} (\psi_a \nabla \psi_b - \psi_b \nabla \psi_a) \cdot \mathbf{n} \, dS \tag{14}$$

Where the integral is taken over the surface $z = 0$ of the object, one of the fields corresponds to the incident wave ψ_{inc} and the other to the reflected wave field, which we shall call ψ_{refl}, where:

$$\psi_{tot} = \psi_{inc} + \psi_{refl} \tag{14}$$

If $\psi_b = \psi_{inc}$ and if $\psi_a = \psi_{refl}$, it follows by direct substitution that we could put $\psi_a = \psi_{tot}$. Since the total field ψ_{tot} must obey the boundary condition $\psi_{tot} = 0$ on the surface $z = 0$ of a perfect conductor, but ψ_{tot} is finite at the surface of the slot, it is convenient to be able to state the above integral in terms of the of ψ_{tot} alone, and not be concerned with its gradient too. This is normally done by making use of the Rayleigh-Sommerfeld relation, and replacing the reference field or incident field (the equivalents of the Green's function) by a Green's function corresponding to that of the sum of the infinite boundary Green's function and its image in the plane of interest, i.e., the sum of the incident field and the field reflected from a perfect reflector. There may, in addition, be a homogeneous term related to the incident field. Following this procedure, we end up evaluating the following integral:

$$V_{scatt} = 2 \int_{S_2} (\psi_a \nabla \psi_b) \cdot \mathbf{n} \, dS = 2 \int_{S_2} (\psi_{tot} \nabla \psi_{inc}) \cdot \mathbf{n} \, dS \tag{15}$$

where, now, $\psi_b = 0$ and $\nabla \psi_b = 2 \nabla \psi_{inc}$, with $\psi_a = \psi_{tot}$, as before. The total value of V is:

$$V = V_{scatt} + V_{homog} \tag{16}$$

where V_{homog} is the value of V for a perfect reflector located at the interface and $V_{scatt} = 0$ for a perfect conductor. We do not know the value of ψ_{tot} at the interface; we only know ψ_{inc} at the interface. Therefore our whole aim is to evaluate the integrals required by using what knowledge we have of the boundary conditions and ψ_{inc}.

It is possible to use the basic boundary condition on the fields ψ, and either attempt to satisfy the second boundary condition or use some approximation to it, such as a finite series of functions which almost satisfy it. We have tried to do this by expanding the fields in the trench as a series of guide modes, and expanding the exciting field of the microscope as a two-dimensional Fourier series, and then solving a finite series with a matrix formalism. The computations are long and inaccurate. Instead, we have found it more convenient to use a variational condition in which we use a trial field for the potential ψ at the surface of the slot, i.e., a guess at the field at the surface of the slot. If the trial field is in error to first order, such variational methods give an error to second order in the parameter V_{scatt}. The variational solution is of the form:

$$V_{scatt} = \frac{2\left[\int_{-\infty}^{\infty}\int_{-\infty}^{\infty}\psi\frac{\partial\psi_{inc}}{\partial z}dx\,dy\right]^2}{\int_{-\infty}^{\infty}\int_{-\infty}^{\infty}\psi(x,y)\left[\frac{\partial\psi_{inside}}{\partial z} - \frac{\partial\psi_{scatt}}{\partial z}\right]dx\,dy} \quad (17)$$

where the terms $\partial\psi_{inside}/\partial z$ and $\partial\psi_{scatt}/\partial z$ are calculated from ψ by the use of an expansion in waveguide modes in the trench, and by the use of Green's functions in the region above the trench, respectively. We use, as a trial function, the potential derived from matching the incident field of the microscope beam with the forward wave fields in the trench, assuming that the field is zero outside the trench. A further modification is to multiply these fields by a tapering function calculated from a numerical solution for a plane wave incident normally on the trench.

We have used the scanning optical microscope to take images of trenches by focusing on the top surface and then on the lower surface. The trench is approximately 6 μm deep. Focusing still occurs when the beam is focused on the bottom of the trench because, although the beam is cut off by the trench, there is still a cylindrically-focused beam. The results obtained for different depths of trenches for where the maximum intensity occurs corresponding to the depth of the trench are in good agreement between the experimental results and the experimental results obtained by taking cross sections.

A line scan may be made from the images, as is shown in Fig. 9. It will be seen that there is a dip in intensity as the line scan is taken over the top surface and a maximum when the line scan is taken over the bottom surface. The theoretical calculations have been carried out by averaging the theoretical results in a series of scans over a variation in depth of 0.5 λ. The reason is that the amplitude varies radically over this small distance because of phase changes of the signals from the bottom of the trench. Experimentally, the same thing occurs with narrowband laser light. However, in the experiments with the RSOM, this effect does not occur because the bandwidth of the light is sufficient to eliminate such coherent effects. Thus, the theory has to be modified to take account of this effect in the way that we have described.

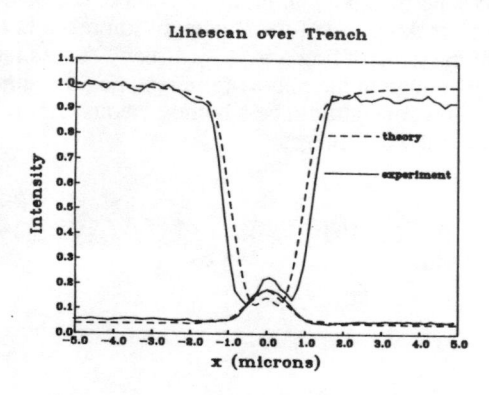

Fig. 9. A comparison of the theoretical and experimental line scans of a trench 2 μm wide and 6 μm deep..

The results are not perfect, but are a fairly good start. It should be added that without the tapering function, the errors are quite a bit larger, especially near the corners. The next stage will be to take account of dielectric materials using much the same formalism as is given here.

4. CONCLUSION

The theory of the real-time scanning optical microscope is now well understood and gives fairly good agreement between theory and experiment. We are developing a new theory for imaging complicated structures which can be sufficiently compact to be used on a personal computer or small work station. Results obtained are promising, although it is desirable to obtain better accuracy than has been obtained so far. It is not clear whether the inaccuracies are due to the nature of the experimental structure itself, which is not perfectly square at the top and bottom corners, as is assumed in the theory, or is due to problems with the theory.

5. ACKNOWLEDGMENT

This work was supported by the Joint Services Electronics Program under the Office of Naval Research on Contract No. N00014-84-K-0327, the Department of Energy under Contract No. DE-FGO3-87ER13797, and Semiconductor Research Corporation under Contract No. 87-MJ-120.

6. REFERENCES

1. T. Wilson and C. J. R. Sheppard, Scanning Optical Microscopy (Academic Press 1984).

2. Scanned Image Microscopy, Ed.: E. A. Ash (Academic Press 1980).

3. M. Petran, M. Hadravsky, M. D. Egger, and R. Galambos, "Tandem Scanning Reflected Light Microscope," J. Opt. Soc. America 58, 661-664 (1968).

4. M. Petran, M. Hadravsky, and A. Boyde, "The Tandem Scanning Reflected Light Microscope," Scanning 7, 97-108, (1985).

5. G. Q. Xiao and G. S. Kino, "A Real-Time Confocal Scanning Optical Microscope," Proc. SPIE, Vol. 809, Scanning Imaging Technology, T. Wilson & L. Balk, Eds. 107-113 (1987).

6. G. Q. Xiao, T. R. Corle, and G. S. Kino, "Real-Time Confocal Scanning Optical Microscope," Appl. Phys. Lett. 53 (8), 716-718 (1988).

7. G. S. Kino and G. Q. Xiao, "Real-Time Scanning Optical Microscopes," submitted to Scanning Optical Microscopes, Ed.: T. Wilson (Pergamon Press 1988).

8. T. Wilson and A. R. Cartini, "Size of the Detector in Confocal Imaging Systems," Opt. Lett. 12 (4), 227-229 (1987).

9. M. Born and E. Wolf, Principles of Optics (Pergamon Press 1975).

10. Gordon S. Kino, Acoustic Waves: Devices, Imaging, and Analog Signal Processing, Chapter 3: Wave Propagation with Finite Exciting Sources, 154-318, Prentice-Hall, New Jersey (1987).

11. T. R. Corle, C-H. Chou, and G. S. Kino, "Depth Response of Confocal Optical Microscopes," Opt. Lett. 11, 770-772 (1986).

Phase imaging in scanning optical microscopes

T. R. Corle and G. S. Kino

Edward L. Ginzton Laboratory, Stanford University
Stanford, California 94305 USA

1. INTRODUCTION

The confocal scanning optical microscope (CSOM) has the major advantages of a very short depth of focus, transverse definition, and image contrast that are better than with a standard microscope. The depth resolution of these microscopes is of the order of a wavelength. However, if it is necessary to carry out thickness measurements of films a small fraction of a wavelength thick, or profile steps with this order of height change, phase measurement techniques become a useful tool. Our aims have therefore been to incorporate within-CSOM phase contrast techniques and differential techniques for quantitative measurement of the position of edges, edge slope, and the thickness of thin films. In conventional microscopes this is most widely accomplished with Zernike phase contrast systems or by using Nomarski differential interference contrast (DIC).[1] In CSOMs, heterodyne interference techniques have been used.[2,3] This paper will describe two methods of phase imaging that we have been investigating, phase contrast and differential interference contrast. Most of the experiments were initially carried out on a single pinhole mechanically-scanned CSOM. Our recent goal has been to test our initial concepts on the real-time scanning optical microscope (RSOM). This microscope uses a Nipkow disk with 200,000 pinholes to dramatically increase the scan speed.[4]

2. PHASE CONTRAST TECHNIQUES

Phase contrast imaging is able to generate intensity variations in an image proportional to phase differences. This enables us to extend the useful range of CSOMs and, in many cases, to obtain quantitative measurements of the heights of surface features. We have been working on both an electro-optic and mechanical method of generating phase contrast images.

The operation of the electro-optic cell has been previously described in conference literature.[5] Physically it is composed of PLZT with transparent indium tin oxide electrodes deposited on it in the configuration of concentric parallel plate capacitors, as shown in Fig. 1. The cell is driven with an a.c. voltage so that it phase shifts either the central or outer portion of the incident illumination, causing a periodic defocus at the sample. If the a.c. signal is detected, a phase contrast image of the sample is generated.

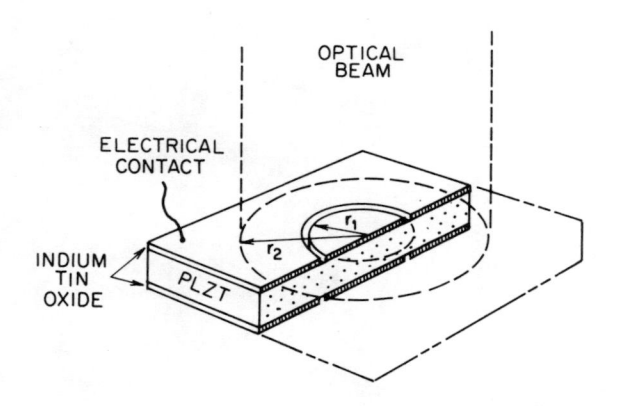

Fig. 1. Configuration of the PLZT electro-optic cell.

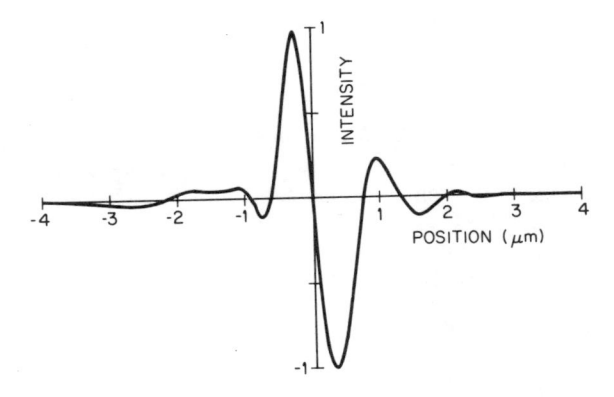

Fig. 2. Depth response of the microscope as a plane reflector is scanned axially through the focal plane.

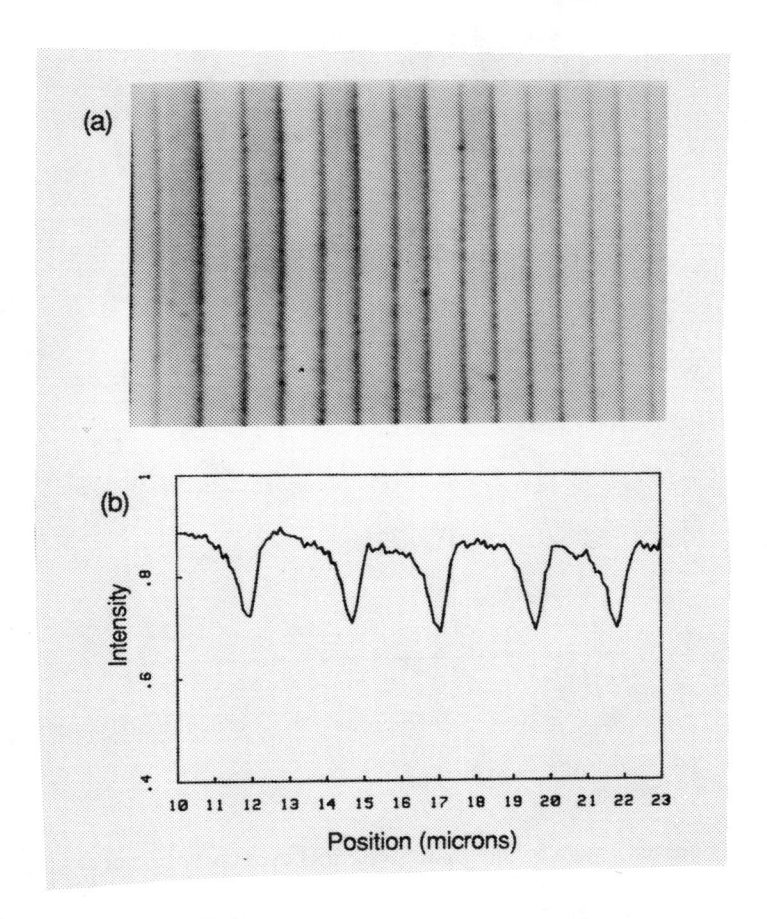

Fig. 3. (a) Mechanically-scanned CSOM image of an aluminum on aluminum grating. (b) A line scan through the image.

The cell was placed in the back focal plane of a single-pinhole CSOM and driven with 300 d.c. (P-P) a.c. at 60 kHz. Figure 2 is a plot of the output of a synchronous detector as a plane reflector is scanned axially through the focal plane. A N.A. 0.95 lens at $\lambda = 0.633$ µm was used. As the reflector is moved away from the focal plane, a signal is generated whose amplitude is proportional to the defocus distance. A synchronous detector is used at the output of the photodiode so that the output signal reverses in amplitude as the object passes through the focal point in Fig. 2. It should be noted that the phase contrast images in this system depend on the product of the signals passing through the center and outer region of the electro-optic cell rather than on the sum, as in a conventional Zernike phase contrast system. Therefore, the definition and contrast of the images are far better than in a conventional phase contrast system.

To illustrate the phase contrast capabilities of this microscope, we have taken pictures of a 102 nm tall 2.4 μm wide grating of aluminum on aluminum. Figure 3(a) is a conventional confocal image; Fig. 3(b) is a line scan through the image. Figure 4(a) is a phase contrast image of the same sample made with the electro-optic cell in a single pinhole microscope at $\lambda = 0.633$ μm and with a N.A. 0.95 lens; Fig. 4(b) is a line scan through the image. The grating pattern is clearly visible in the phase contrast image. By measuring the intensity variations in the line scan, and comparing this with the slope of the depth response, the height of the film was measured to be 101 nm.

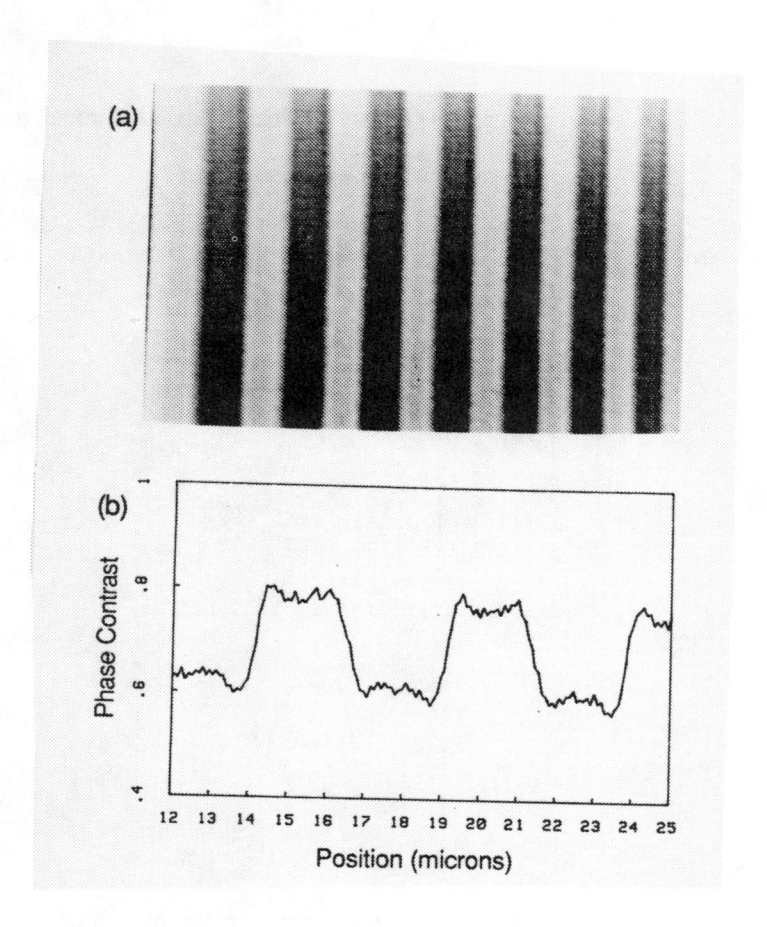

Fig. 4. (a) Phase contrast image of the grating in Fig. 3. (b) A line scan through the image.

In addition to electro-optics, we have been working on a mechanical method of generating phase contrast images.[6] This involves periodically defocusing the sample and detecting either the a.c. signal or, in the case of the RSOM, subtracting consecutive frames on an image processor. The operation of this device can be most easily understood by referring to Fig. 5. In Fig. 5 we show the depth response of a CSOM. To obtain a phase contrast image, we first slightly defocus the microscope and digitize a frame using a frame grabber or image processor. The microscope is then defocused in the opposite direction and a second digitized image is subtracted from the previously stored image. If the sample is in the focal plane, as in Fig. 5(a), the subtraction will result in zero output from the image processor. If the sample moves a small distance out of the focal plane, in this example 102 nm, a large signal will be generated. For small defocus and small displacements from the focal plane, the signal will be linearly proportional to the distance from perfect focus. A depth

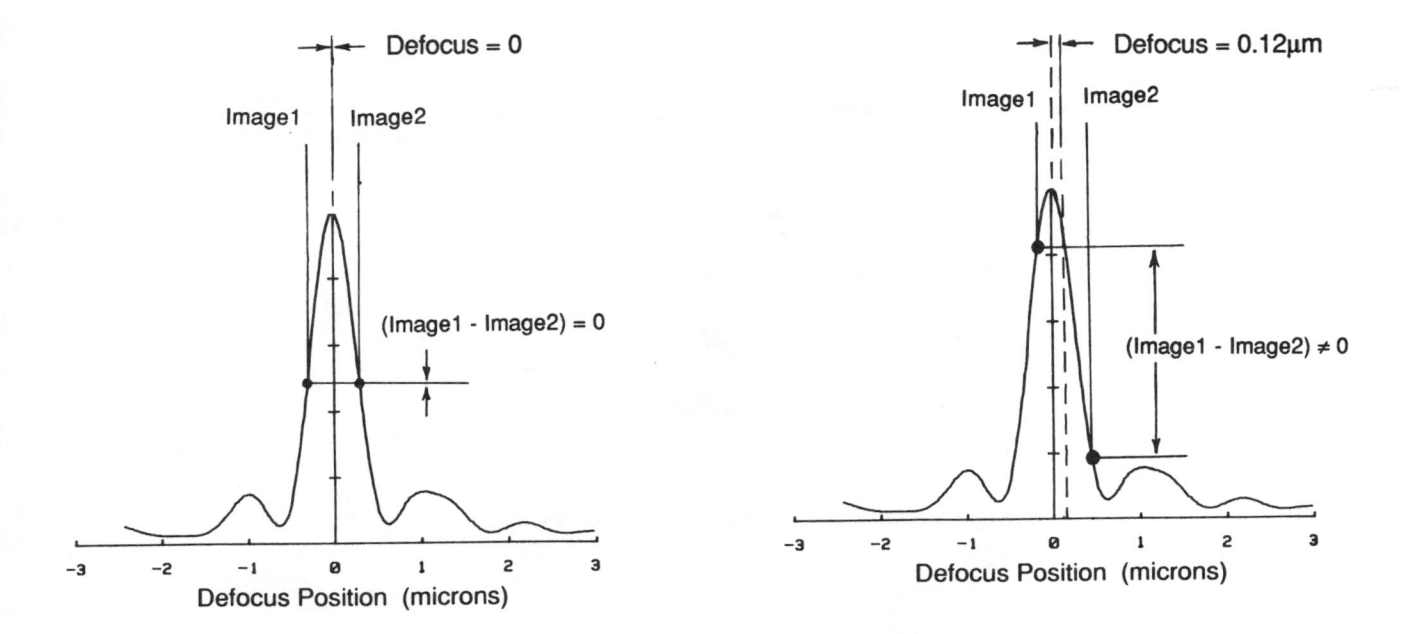

Fig. 5. Operation of the mechanical phase contrast microscope. (a) Zero defocus. (b) 0.12 μm defocus.

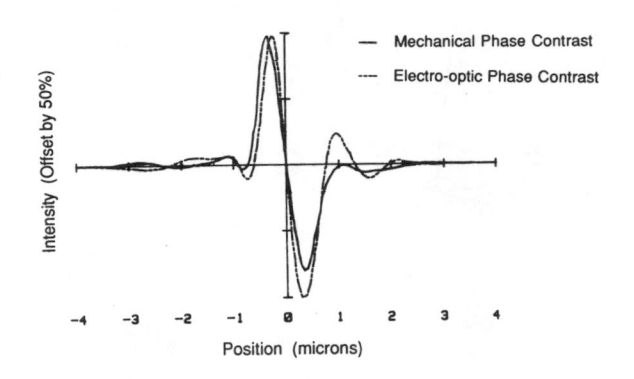

Fig. 6. Depth response curves comparing the mechanical and electro-optic phase contrast systems.

response curve for a N.A. 0.95 lens.at $\lambda = 0.633$ μm and defocus distance of $\lambda/5$ is shown in Fig. 6. For comparison, the electro-optic depth response is plotted in the same figure.

3. DIFFERENTIAL INTERFERENCE CONTRAST

We have also worked with a second phase imaging technique, Nomarski DIC.[7,8] With this technique a Wollaston prism is used to produce two slightly displaced spots on the sample, as illustrated in Fig. 7. The edges of the image are enhanced by the coherent interference of the two spots. An up step will be either brighter or dimmer than the surrounding region, depending on whether the two spots add in phase or out of phase; the down step will generally exhibit the opposite behavior. Superimposed on the edge-enhanced image will be a conventional reflectivity image of the sample. Because edges are enhanced, the distance between the top and bottom of the edge can readily be measured. This feature, coupled with the ability of CSOMs to measure height, enables quantitative measurements of edge slopes to be made.

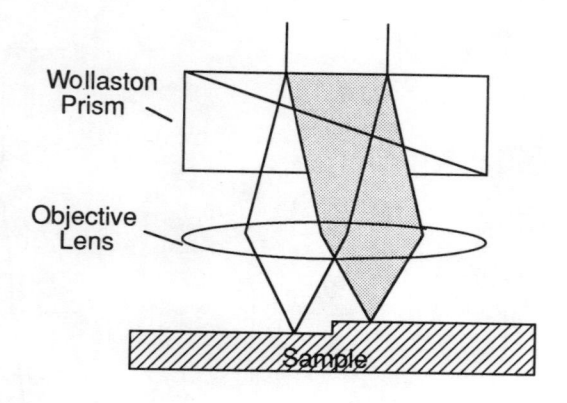

Fig. 7. Schematic of a Wollaston prism producing two spots on a sample.

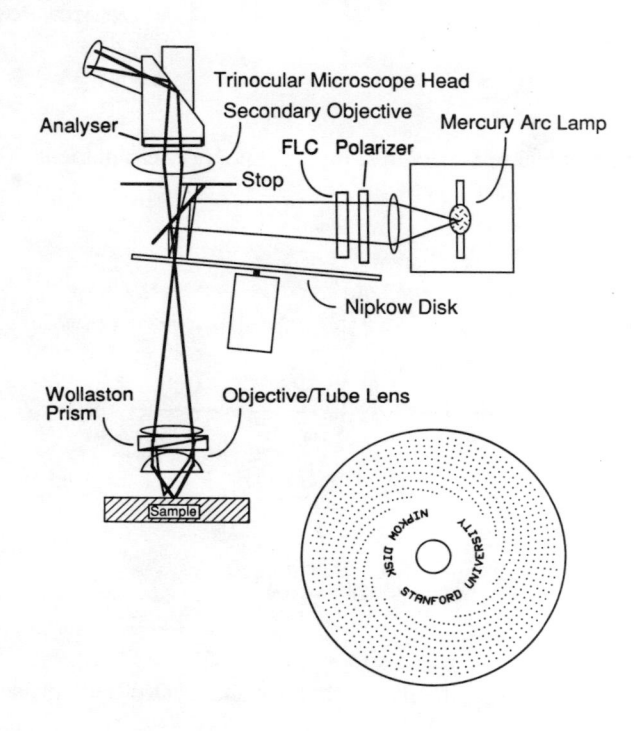

Fig. 8. Real-time scanning optical microscope with liquid crystal half-wave plate.

We can improve the contrast of our edge measurements by eliminating the reflectivity changes of the sample. We do this by reversing the relative phase of the two spots and subtracting successive frames using an image processor. We control the phase by using a ferroelectric liquid crystal (FLC) half-wave plate in the illumination path of the RSOM, as shown in Fig. 8.[9] The FLC has electronically-induced birefringence similar to a Pockel cell. The difference is that the liquid crystal runs at much lower voltages but at slower speeds.

To obtain a DIC image, the illumination is first polarized at 45 degrees to the optical axis of the Wollaston prism so that both spots are illuminated and the voltage on the FLC cell is set for constructive interference. An image is digitized and stored using a frame grabber. Next, the voltage on the FLC is reversed so that one of the spots receives a π phase shift.

The phase shift causes the two beams to interfere destructively at the detector. A second image is then digitized and subtracted from the first. The subtraction removes any changes in reflectivity from the image, thus increasing the edge contrast.

This technique has been outlined by M. Vaez Iravani and C.W. See using a Pockel cell on a single-pinhole scanning optical microscope.[10] Figure 9 shows the results taken with our RSOM. Figure 9(a) is a line scan through an image of a 2.4 μm bar pattern of aluminum on aluminum; Fig. 9(b) is a line scan through the sample with the microscope operating in DIC mode. Note that the reflectivity changes between the aluminum and glass are not present in the DIC image so that only the edges are visible. It will also be noted that only one edge is imaged. This is due to limitations imposed by the image processor. The visible edge (a down step) consists of constructive interference (edge brighter than the surroundings) minus destructive interference (edge darker than the surroundings). The opposite edge has destructive interference minus constructive interference; this yields negative values which the image processor sets to zero since standard video has no provisions for negative intensities.

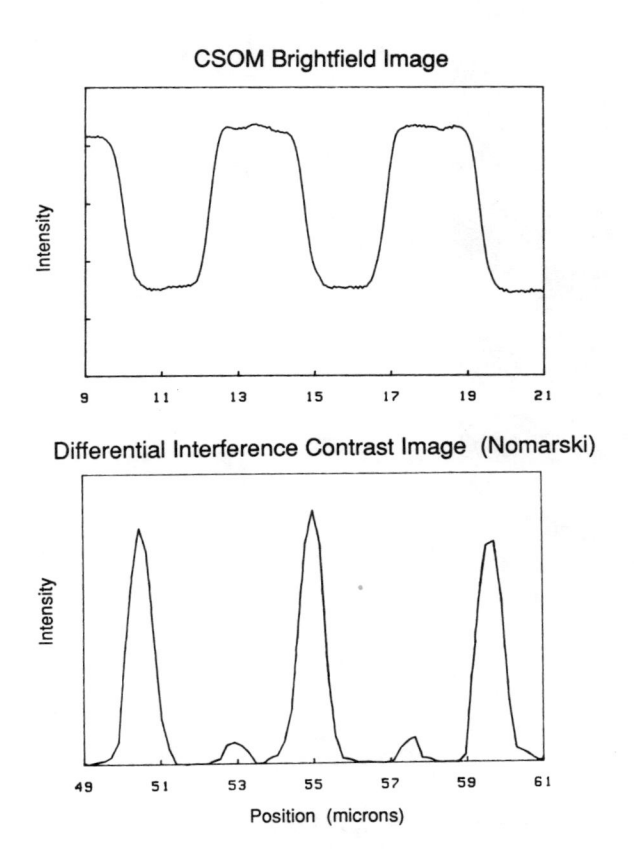

Fig. 9. (a) Line scan through an image of a 2.4 μm bar pattern of aluminum on aluminum made with the RSOM.
(b) Line scan through a DIC image of the same sample.

This technique works well for samples which are thinner than the depth of focus of the microscope. For taller samples, greater than approximately $\lambda/2$, the shallow depth response of a CSOM prevents interference of the two beams at the detector. Another stratagem must be adopted. To image these samples we rotate the analyzer and half-wave plate by 45 degrees so that they are aligned along the optical axis of the Wollaston prism. In this case only one spot is illuminated on the sample and a frame is digitized. Switching the voltage on the half-wave plate causes the other spot to be illuminated and a slightly displaced image of the sample is obtained. When these two are subtracted, an image of the edges is left, as shown in Fig. 10. Figure 10 shows a conventional image, made with the RSOM, and a DIC image of 2600 Å tall oxide isolation lines on silicon. The heights of the lines were measured using the RSOM and confirmed with a contact profilometer. The width of the edges can be measured from the line scan in Fig. 10. Given these two values, the edge slope was calculated to be 21°, about what is expected for silicon dioxide on silicon. It may be noted that the line scans show asymmetric edges on what should be a symmetric structure. The asymmetry is due to the fact that the FLC cell is not a true half-wave plate. Hence, both spots from the Wollaston prism are illuminated simultaneously, leading to constructive interference at one edge and destructive interference at the other.

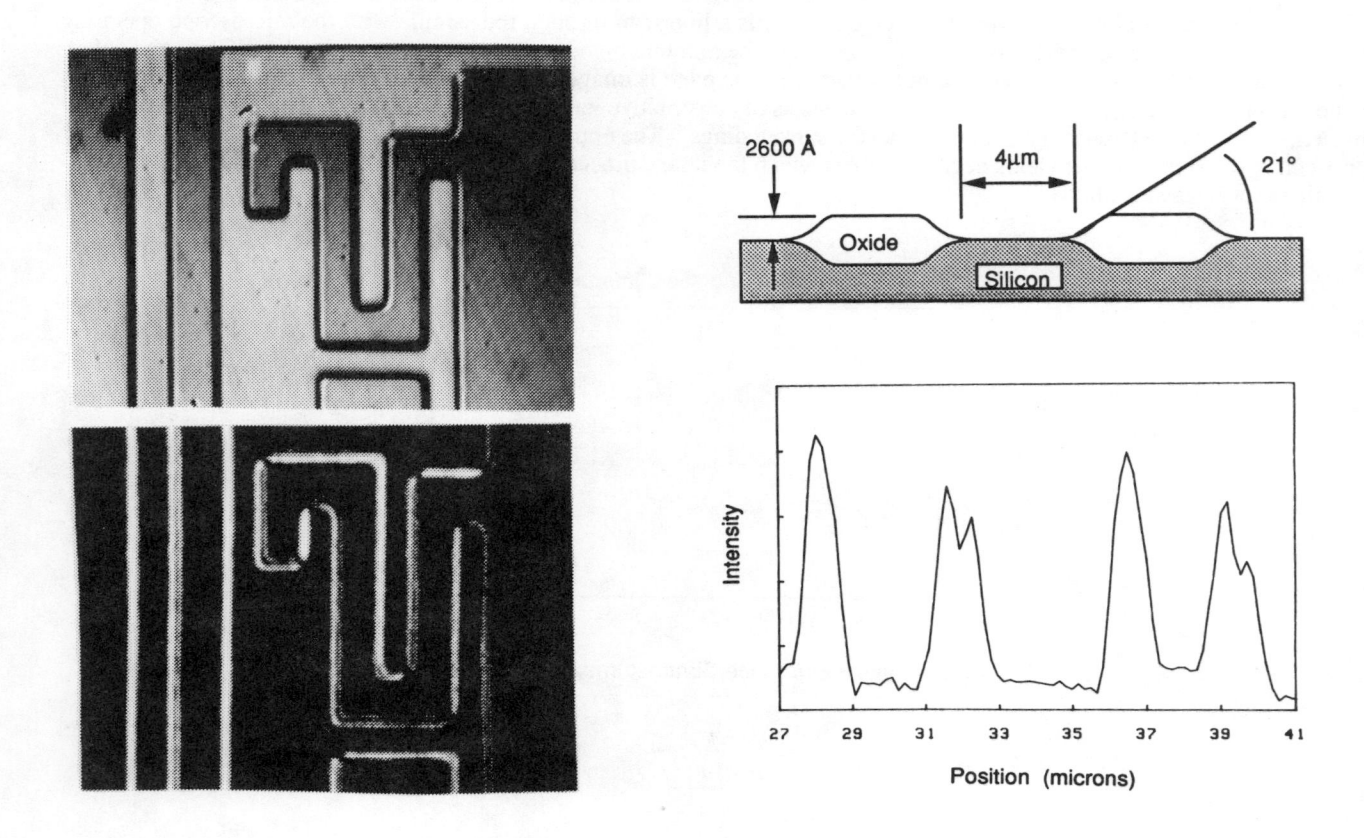

Fig. 10. Conventional and DIC images of silicon dioxide isolation lines on silicon.

4. CONCLUSION

We have examined a number of techniques for phase imaging in CSOMs. Through our electro-optic and mechanical phase contrast techniques we can extend the shallow depth response of CSOMs to enable them to quantitatively measure thinner structures. Differential interference contrast techniques make it possible to accurately determine the locations of edges and quantitatively determine edge slopes.

5. ACKNOWLEDGMENTS

We would like to thank Larry Pendergrass and Roger Jungerman of Hewlett Packard, Santa Rosa, as well as Ling Liauw of Intel, Santa Clara for providing us with samples. This work was supported by the National Science Foundation on Contract No. ECS-86-11638 and Semiconductor Research Corporation under Contract No. 87-MJ-120.

6. REFERENCES

1. M. Born and E. Wolf, Principles of Optics, 6th Ed., Pergamon Press (1975).

2. R. L. Jungerman, P. C. D. Hobbs, and G. S. Kino, "Phase Sensitive Scanning Optical Microscope," Appl. Phys. Lett. 45, 846-848 (October 1984).

3. H. K. Wickramasinghe, S. Ameri, and C. W. See, "Differential Phase Contrast Optical Microscope with 1 Å Depth Resolution," Elect. Lett. 18, 973-975 (1982).

4. G. Q. Xiao and G. S. Kino, "A Real-Time Confocal Scanning Optical Microscope," .Proc. SPIE, Vol. 809, Scanning Imaging Technology, T. Wilson and L. Balk, Eds., 107-113 (1987).

5. G. S. Kino, T. R. Corle, and G. Q. Xiao, "New Types of Scanning Optical Microscopes," Presented at SPIE's 1988 Santa Clara Symposium on Microlithography, Santa Clara, CA (28 Feb-4 Mar, 1988); Published in Proc. of SPIE, Integrated Circuit Metrology, Inspection, and Process Control II, Vol. 921, Ed.: Kevin Monahan (1988).

6. T. R. Corle, J. T. Fanton, and G. S. Kino, "Distance Measurements by Differential Confocal Optical Ranging," Appl. Opt. 26 (12), 2416-2420 (15 June 1987).

7. G. Nomarski, and A. R. Weill, "Application a la Metallographie des Methodes Interferentielles a Deux Ondes Polarisees," Rev. Metallurgie L11, 121-134 (1955).

8. W. Krug, J. Rienitz, and G. Schultz, Contributions to Interference Microscopy, London, Hilger and Watts (1951).

9. G. Q. Xiao, T. R. Corle, and G. S. Kino, "Real-Time Confocal Scanning Optical Microscope," Appl. Phys. Lett. 53 (8),716-718 (22 August 1988).

10. M. Vaez Iravani and C. W. See, "Linear and Differential Techniques in the Scanning Optical Microscope," to be published in J. Appl. Phys.

In Vivo Confocal Imaging of the Eye Using Tandem Scanning Confocal Microscopy (TSCM):

James V. Jester, Ph.D., H. Dwight Cavanagh, M.D., Ph.D., and Michael A. Lemp, M.D.

From the Center for Sight, Georgetown University, Washington, D.C.

ABSTRACT

The tandem scanning reflected light microscope (TSRLM) is a confocal light microscope which has the capability of looking into living tissue and obtaining high resolution, high magnification images of cellular structure. TSRLM can be used to study living tissue such as all layers of the corneal epithelium including basal epithelial cells, keratocytes, nerves, inflammatory cells, bacteria, and corneal endothelium. For the first time in vision research, real-time, in vivo, microscopic images of normal and pathologic tissues can be obtained from human or animal eyes using the TSRLM. Compared to other methods of vital microscopy, TSRLM has no present rival.

Specifically, TSRLM will: (1) Allow the hitopathologic analysis of living eyes, in vivo, over multiple observation periods without the need for tissue fixation and/or processing; (2) Assist in the acquisition and analysis of histopathologic images from human eyes, in vivo, in corneal disease; and (3) Greatly reduce the need for large numbers of animals in the histopathologic evaluation of experimental corneal disease and surgical procedures.

1. INTRODUCTION

1.1. Principles of confocal microscopy

The essential principle of confocal imaging is point illumination and detection which is achieved by both the illuminating and objective lenses having the same focus. In conventional microscopy, the specimen is broadly illuminated resulting in light reflected from above and below the plane of focus contributing to the final image. High resolution and contrast are achieved in conventional light microscopy by using thin (one cell thick) tissue sections. In thick tissues, reflections from above and below the plane of focus result in blurring of the image leading to the pink haze seen in living tissue. These limitations have prevented extension of light microscopy to whole living tissues except in special cirumstances such as specular microscopy. These problems are greatly lessened by using a confocal optical system design.

Confocal microscopy takes advantage of the principal of Lukosz,[2] which states that resolution may be improved at the expense of field of view. By limiting the field of view, ie point illumination, light is focused on only a very small area within a tissue. The objective, which has the same focus as the illuminating lens, collects light from the same area within the tissue. A point detector at the same focal distance as the point illuminator effectively eliminates scattered light from above and below the plane of focus. As the size of the field of view approaches the width of the focal plane (1-2 microns), the amount of scattered light contributing to the final image decreases, thereby increasing resolution and contrast while enhancing depth discrimination.[1] A larger field of view if built-up by scanning the image which can be achieved by either moving the tissue or moving the point illuminator and detector in synchronization.

1.1.1. TSRLM design.
The Tandem Scanning Reflected Light Microscope was developed to permit the examination of internal structure in living whole tissues by Professors Petran and Hadravsky at Charles University in Pilzen, Czechoslovakia in 1967.[3] In 1983 the microscope was used to study bone and teeth in the West.[4] Since that time, nine microscopes have been manufactured and seven are located outside Czechoslovakia. The only TSRLM dedicated to Vision research was obtained in the fall of 1986 by the Center for Sight at Georgetown University.[5]

Tandem scanning is achieved by using a miniaturized Nipkow disc consisting of a thin ceramic based, copper disc containing 14,000 (40-80 micron) holes arranged as 40 Archimedean spirals.[3] The arrangement of the holes give the disc a central symmetry such that each hole has an exact conjugate pair on the same diameter and at the same radial distance from the center of rotation, but on the opposite side. These holes constitute the apertures which limit the illumination and detection of the sharply defined spots in the plane of focus necessary to achieve confocal imaging of the specimen.

The optical transmission of the Nipkow disc is approximately one percent. The holes allow white light to come through openings in the disc. The objective lens forms images of these spots originating from the illuminating side of the disc at the focused-on plane within the specimen. The remaining portion of the TSRLM is constructed such that light reflected from the illuminated spots in the object plane is directed and focused on the conjugate aperture holes on the opposite side of the disc. A very high portion of the light reflected from the focused-on plane passes to the observation side of the disc to form an image. Light scattered from planes above or below the plane of focus will not be in focus on the opposite side of the disc and will therefore be intercepted by solid portions of the Nipkow disc.

As the Nipkow disc turns, each single spot scans a single line. The arrangement of holes in the disc is such that when the disc is rotated, the spots of light cover in succession the whole field of illumination and the whole field of view. The TSRLM therefore allows for wide field imaging without loss of resolution and contrast in thick tissue.

Light passing through the illuminating side of the disc is directed toward the objective by a series of mirrors. Reflected light returning from the specimen is deflected by the beam splitter and a final mirror to arrive at the observation side of the disc. The eyepiece in the microscope is of the Ramsdem type, focused on the Nipkow disc. Alternatively, a low light level video imaging system can be positioned at this location.

The microscope is designed to accept all objective lenses having a standard RMS thread and 160 mm working distance. The particular choice of lens is determined by the specific application. Immersion lenses are essential for optical sectioning of fresh biologic tissue in order to reduce the effects of strong reflections at the surface. When using water immersion lenses, Goniosol or Dacriose can be used as an immersion medium for the eye. The main determinant of the depth to which optical sectioning can be achieved is the working distance of the objective lens. Using a high power, high numerical aperture lens, it is possible to section down to 200 microns below the surface since no cover slip is used, before objective lens and specimen come into contact. Lower numerical aperture lenses with longer working distances permit sectioning to greater depths. For example, the 16x and 25x objectives routinely permit full thickness views of the corena to a depth of .5 to 1.0 mm.

2. RESULTS

Research areas where the TSRLM has already proven successful have been previously summarized.[4] The TSRLM developed for ophthalmic use provides high resolution and contrast images of living ocular tissue. Currently, the technique is most successful forthe cornea since some water immersion objective lenses of suitable working distance are available. The corneal epithelium, stroma, and endothelium have been studied in vivo and photographed in situ.

Compared to conventional microscopic techniques, the TSRLM provides greater detail and more information about the number and shape of surface epithelial cells (Figure 1, 2). Confocal scanning images of the superficial epithelium, similar to those obtained by scanning electron microscopy, show both light and dark surface epithelial cells (Figure 2). Unlike scanning electron microscopy and in vivo biomicroscopic techniques, the TSRLM also provides serial images of deeper layers of the corneal epithelium. Immediately below the surface, superficial epithelial cell nuclei can be seen (Figure 3). Below the superficial epithelium, the cell borders of the wing cells appear as an irregular honeycomb. These images are seen as punctate and/or linear reflections which perhaps originate from desmosomal junctions of intercellular spaces. At the deepest layer of the epithelium, basal epithelial cell nuclei can be seen. In some eyes, reflections can be imaged which appear to originate from the epithelial basal lamina/Bowman's membrane (Figure 4). Marked infolding of the basal lamina zone appears to be present which may increase the surface area available for basal epithelial cell attachment.

Unlike any conventional biomicroscopic technique, the TSRLM is capable of providing information about the cellular structure of the corneal stroma. Corneal nerves, which can only be appreciated by special histopathologic stains are seen coursing through the anterior corneal stroma in the normal living rabbit eye (Figure 5). Keratocytes, generally thought to be sparsely distributed, are seen densely packed with branching processes and distinct nuclei (Figure 6). Confocal imaging can also be used to detect inflammatory cells within the cornea following corneal abrasions (Figure 7). Inflammatory cells are highly reflective and appear to stream into the site of injury. High resolution, wide field views of the corneal endothelium can also be seen (Figure 8).

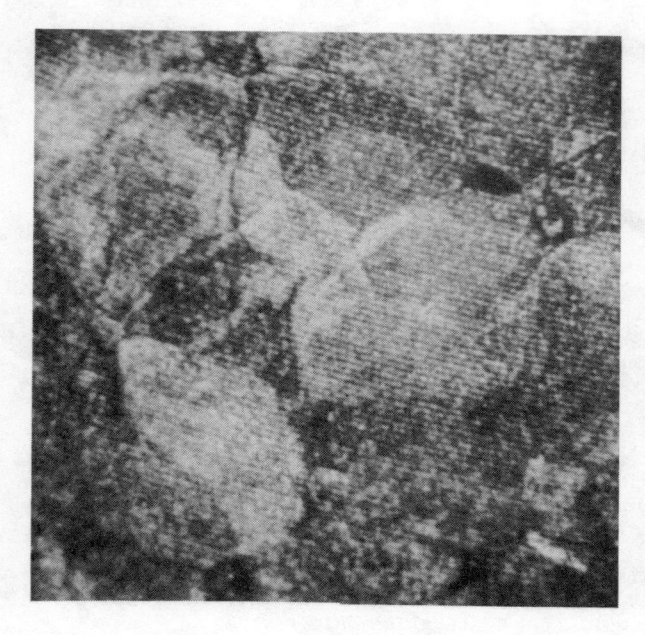

Figure 1. Surface structure of the superficial epithelium.

Figure 2. Below the surface of the superficial epithelium show epithelial cell nuclei and cell borders.

Figure 3. Basal epithelial cells.

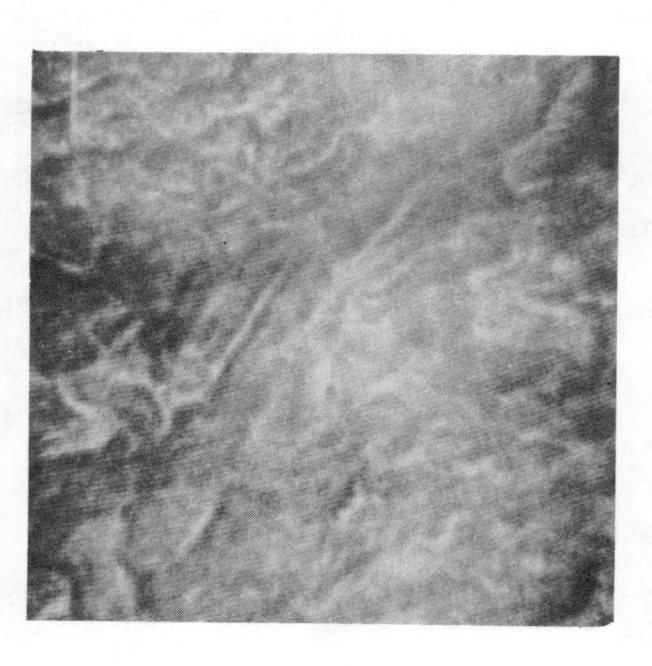

Figure 4. Basal lamina/Bowman's membrane with marked infolding.

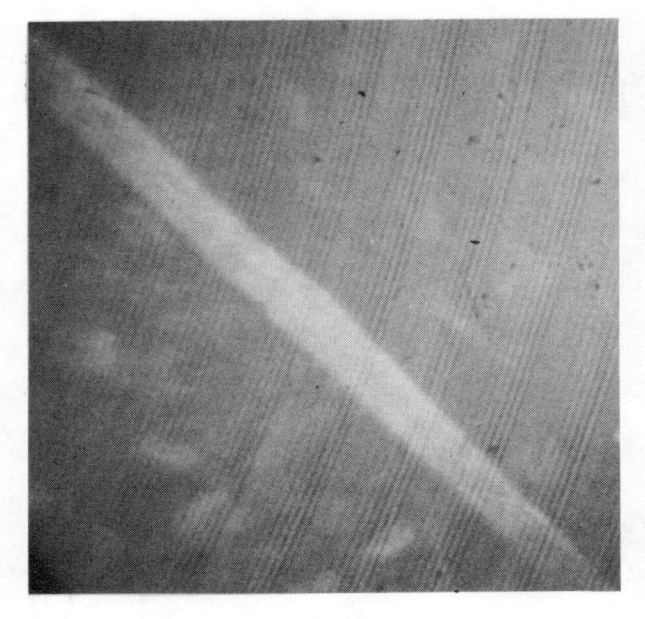

Figure 5. Stromal nerves.

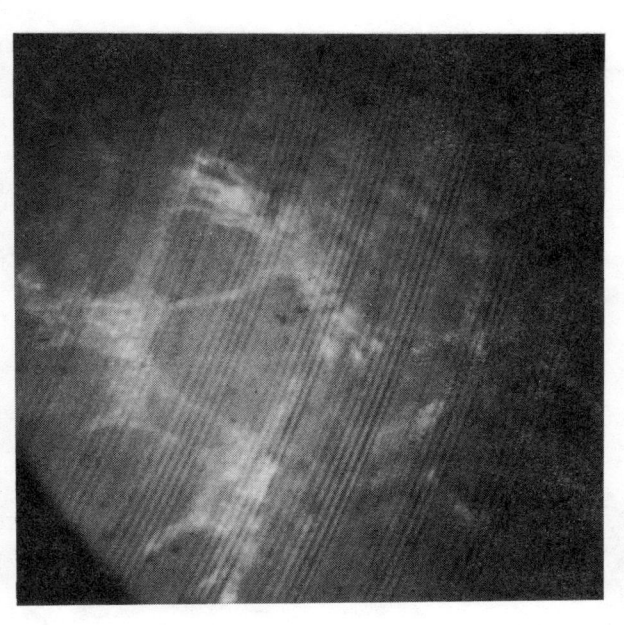

Figure 6. Stromal keratocytes with interconnecting cell processes.

Figure 7. Acute inflammatory cells.

Figure 8. Corneal endothelium.

3. REFERENCES

1. J. Wilson, C.J.R. Scheppard, Theory and practice of scanning optical microscopy, Academic Press, London, 1984.
2. W. Lukosz, Optical systems with resolving powers exceeding the classical limit, J. Opt. Soc. Am. 57:1190, 1966.
3. M. Petran, M. Hadravsky, J. Benes, A. Boyde, In vivo microscopy using the tandem scanning microscope, Ann. N.Y. Acad. Sci. 483:440, 1986.
4. A. Boyde, Applications of tandem scanning reflected light microscopy and three dimensional imaging, Ann. N.Y. Acad. Sci. 483:428, 1986.
5. J.V. Jester, H.D. Cavanagh, and M.A. Lemp, Confocal Microscopic Imaging of the Eye and the Tandem Scanning Confocal Microscope (TSCM): in Non-Invasive Diagnostic Techniques in Ophthalmology, ed. B.R. Masters, Springer-Verlag, New York (1988, in press).

Confocal Laser Scanning Microscopy for Ophthalmology

G. Zinser, R.W. Wijnaendts-van-Resandt, C. Ihrig

HEIDELBERG INSTRUMENTS GMBH, Im Neuenheimer Feld 518, 6900 Heidelberg, FRG

1. ABSTRACT

Many problems in ophthalmology deal with the three-dimensional structure of various parts of the living human eye. A complete confocal laser scanning microscope for eye diagnosis and first clinical results are described. Geometrical measurements of the cornea and topographical measurements of the retinal substructures are presented.

2. INTRODUCTION

Among many other applications confocal laser scanning systems can also be used for ophthalmological imaging [1, 2, 3, 4]. Up to now morphometrical measurements of the corneal curvature are realised with imaging processes [5, 6, 7]. The disadvantages of these indirect processes are that they cannot show the topography in a specific region and that the accuracy is very poor in the middle of the cornea. Measurements of the corneal thickness require contact with the eye [8, 9]. The biomorphometry of the posterior segment of the eye has been investigated with the fundus camera or with the laser scanning ophthalmoscope [10, 11, 12] without any possibility of depth measurements.

By using the modern confocal laser scanning and detection method a system has been developed which produces three-dimensional images of both the anterior and posterior eye segment. The principle of confocal microscopy has been discussed in detail [13, 14] and will not be repeated here. It was explained that, by scanning a laser beam in the appropriate way, optical section images with high spatial resolution are produced either parallel (transverse x-y-images) or perpendicular to the structure's surface (longitudinal x-z-images).

A series of transverse or longitudinal images results in a three-dimensional image of the structure.

3. THE LASER TOMOGRAPHIC SCANNER

The Laser Tomographic Scanner (LTS) is subdivided into two physical units. The operator unit and the patient unit [Fig. 1].

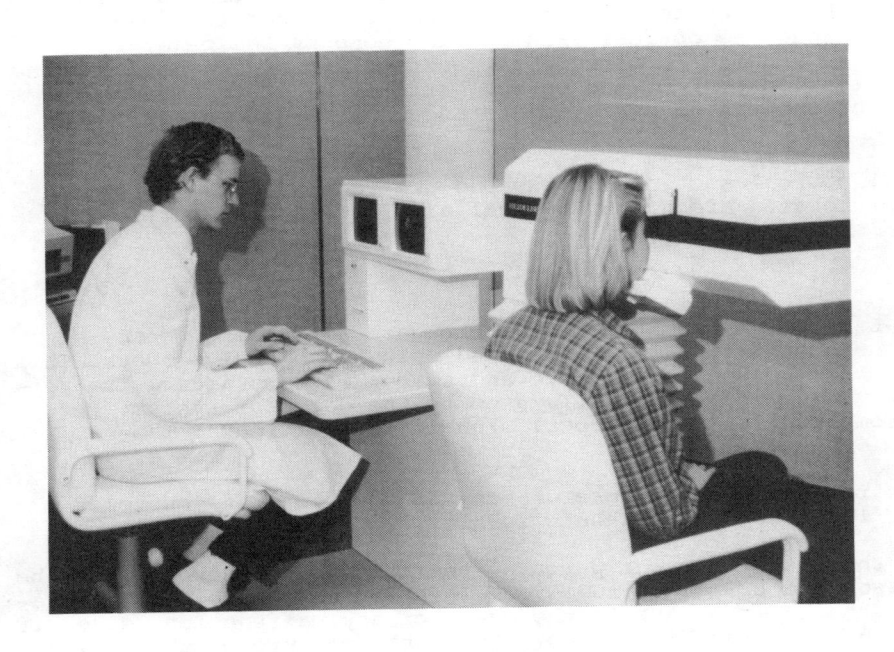

Fig. 1:
The HEIDELBERG INSTRUMENTS Laser Tomographic Scanner

The Operator Unit.

The operator unit integrates the operator desk with keyboard and monitor to display the data and the results. A second monitor shows the examined eye. The operator unit also houses the computer and the complete electronical system.

The Patient Unit.

The patient unit includes all optical and mechanical parts of the scanner and the patient chair as well as the adjustment of the system.

3.1 The system adjustment

The patient is sitting in the chair with his head in the head rest. The scanner housing is lifted to the altitude of the patient's eyes; the head rest is adjusted to the position of chin and forehead. The patient looks into the system with both eyes. The scanner optics is positioned to the center of the eye under examination. By adjusting the pupil distance, the second eye is positioned in front of an eye piece through which a fixation light is presented.

All movements are done with stepper motors under automatic control and are initiated from the keyboard.

3.2 The Fixation

Through the eye piece the patient looks with his second eye on a fixation target [Fig. 3, Fig. 6]. The target consists of an array of 40 x 40 light emitting diodes. As both eyes do not move independently the fixation of a luminous point keeps the examined eye quiet, too. By changing the diode the position of the eye can be varied and different locations of the eye be brought into the scanning field. The angular range is 30 degrees. A diffractive error between - 10 and + 4 dpt can be corrected by the eye piece.

3.3 The Reference Image System

In order to have a permanent control of the patient's fixation and of the measuring position, a video image of the eye under examination is recorded continuously. For that reason the iris is illuminated by an infra-red semiconductor laser through wave guides. The IR-light scattered by the iris is collected by the measuring optics, deflected out via a hotlight mirror and imaged to the video-camera.

3.4 The Scanning Unit

The comparison between various confocal laser scanning systems is a comparison between the various technical realisations of three-dimensional recording of the object. With a diagnostic system like the LTS this is achieved by scanning the stationary object (the eye) by a focussed laser beam. The scanning along the optical axis is also done by the beam. The transversal scanning is done with galvanometric mirrors. A resonant mirror scanner does the fast deflection. The movement of the focus along the optical axis is realised differently for the cornea measurements and for the retina measurements and will be discussed separately.

Eight transversal or longitudinal optical sections per second are recorded. The direction of the longitudinal images can be rotated around the optical axis by rotating parts of the scanning system.

3.5 The Image Formation System

The light source is a 7 mW HeNe-Laser. The beam is guided through a polarising, beamsplitter, expanded, scanned by the scanning unit and directed onto an objective tubus. This tubus contains the last beam forming in front of the eye and also a wave retarder. The reflected light is transferred back via the same optical system to the detector system. In front of the high sensitive photomultiplier the confocal pinhole is positioned. The dynamic range is 12 bit.

All optical, mechanical and electronic components are of very high quality and customer designed. The loss of the signal light is 50 % without pinhole.

As the laserbeam touches one of the most important parts of the human body there have to be safety precautions. For that reason the LTS contains several systems to control and regulate the laser power. Shut down of the laser occurs in the event of any malfunction, e. g. if one of the scanners does not reach a minimum elongation. The light exposure of the retina is below the maximum permissible energy as defined in the ANSI/IEC standard [15].

3.6 The Control and Operating System

The operating system is the interface between the user and the ophthalmologic system. It consists of several hardware and software subsystems.

An alphanumercial keyboard permits the input of text and of absolute coordinates for the mechanical movement. Function keys are provided to move all mechanical and optical subsystems, to initiate measurements and to change between measuring modes. A high resolution color monitor displays the laser scan images and measurement results. Image data can be stored permanently using an optical disk drive. One removable disk cartridge has a capacity of up to 256 examinations.

The electronic hardware [Fig. 2] is connected to a VME-bus. The entire system is controlled by a Motorola 68000 CPU with 2 MByte RAM and an additional 2 MByte image buffer. Some special hardware is used for acquisition and processing of image data. Image acquisition is done with a special dual ported data buffer. Data acquisition speed is 2400 lines/sec with 256 pixels per line and 256 lines per image.

The modular structured software package includes the system management, the drivers for all system components and the image and data presentation. Most important is the image data processing, such as the reconstruction of the cornea curvature or the computing of the cornea thickness. Three-dimensional image processing enables to display the topography of the retina and to compute the volume of the optic nerve cup.

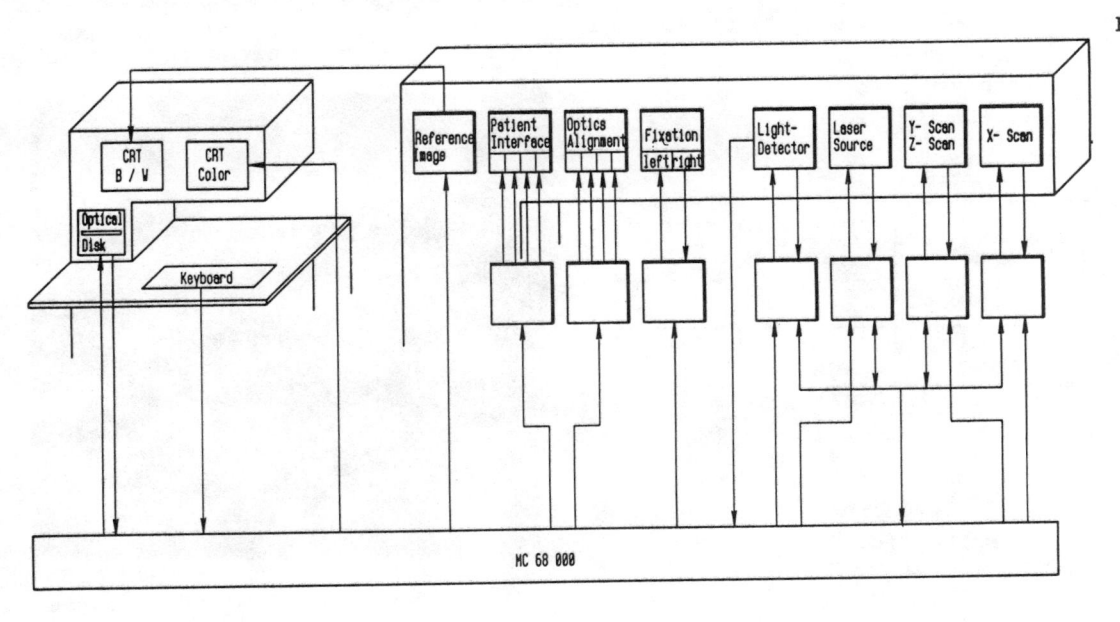

Fig. 2: LTS Hardware Structure

4. IMAGE FORMATION AND RESULTS

4.1 LTS Mode for Anterior Segment Measurements

To scan the anterior segment of the eye the laser is directly focussed onto the eye [Fig. 3]. The focussing microscope lens with a NA 0.4 and a working distance of 11 mm is moved along the optical axis. By simultaneous scanning along x- and z-direction, optical section images (perpendicular to the surface) of the cornea are produced. Such images reflect the topography of both the anterior and posterior corneal surface. They serve to measure the curvature of the surface [Fig. 4], the thickness of the cornea and the depth of incisions in the cornea.

The transverse field has a maximum length of 2.5 mm. The depth of the longitudinal field is 1.5 mm for corneal measurements. With a maximal longitudinal field of 10 mm cornea, anterior chamber and lens can be imaged simultaneously [Fig. 5]. When determining the geometrical distances along the z-axis, the various refractive indices of air and the different parts of the eye have to be considered.

The optical resolution is 1 um transversal and 4 um longitudinal. The accuracy for the measurement of the radius of curvature is 0.1 mm, corresponding to 0.5 dpt for the refractive power. The accuracy of thickness measurements of the cornea is 5 um.

4.1 continued: LTS Mode for Anterior Segment Measurements.

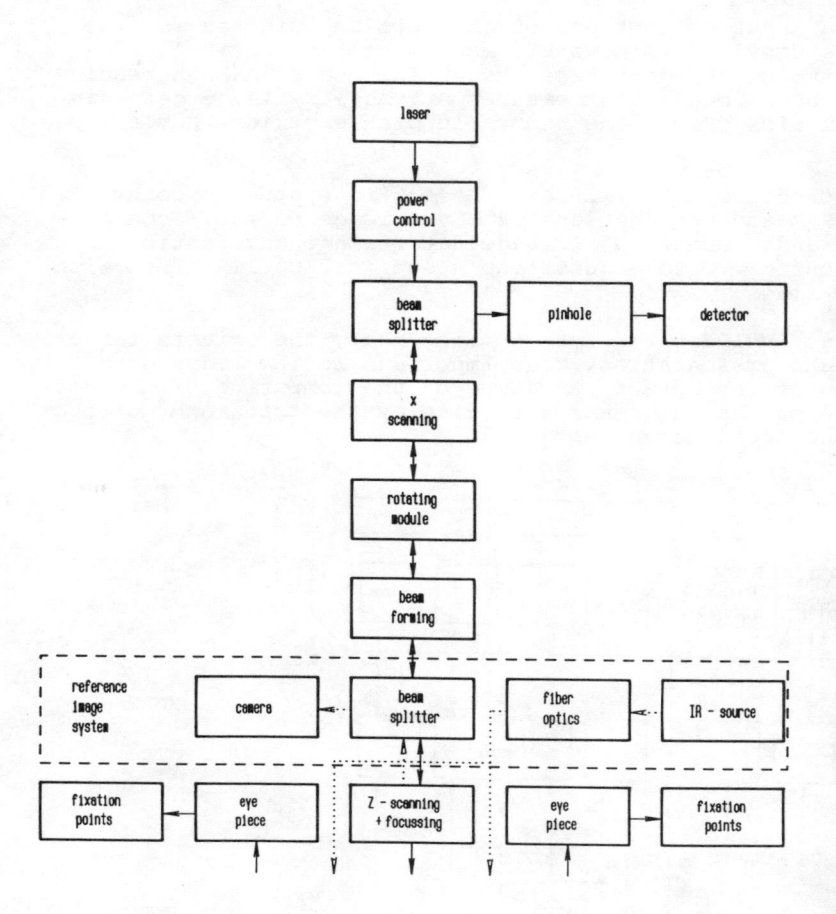

Fig. 3: Image formation system for anterior segment measurements.

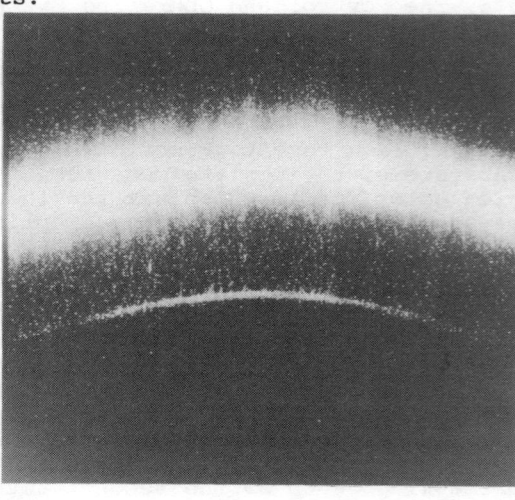

Fig. 4: The anterior and posterior surface of the cornea. Dimensions: 2.5 mm along the x-axis, 1.5 mm along the z-axis.

Fig. 5: The cornea, the iris and the eye lens.

4.2 LTS Mode for Posterior Segment Measurements

For focussing the laser beam onto the retina, cornea and lens are part of the optical system [Fig. 6]. The scanning beam leaves the LTS as a parallel bundle. Variation of this parallelism produced by scanning of the last lens moves the focal plane.

The parallel beam has a diameter of 3 mm so that for imaging of the retina no dilation of the patient's pupil is necessary.

A series of 32 x-y-images at different focal planes with a z-distance of 20 um [Fig. 7, Fig. 8, Fig. 9] is normally recorded to measure the topography of retinal structures. The field of view has a maximum angle of 20 x 20 degrees. Parallel to the optical axis the series covers typically a region of 1.5 um (maximum 6 mm).

The optical resolution is limited by the eye lens itself. Transversal resolution is 7 um, longitudinal resolution is 150 um.

The poor optical resolution along the optical axis can be improved by processing of the data. Since the signal along the z-axis is mainly due to the reflex of a surface, the position of that surface can be determined with significantly higher accuracy. The accuracy of a single light measurement is below 50 um.

4.2 continued: LTS Mode for Posterior Segment Measurements.

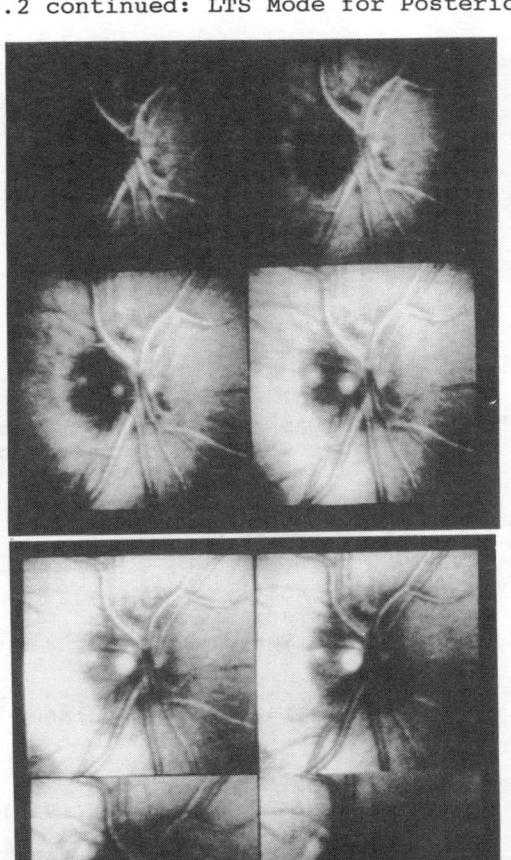

Fig. 7: The optic nerve head.
Every 4th plane of a series.

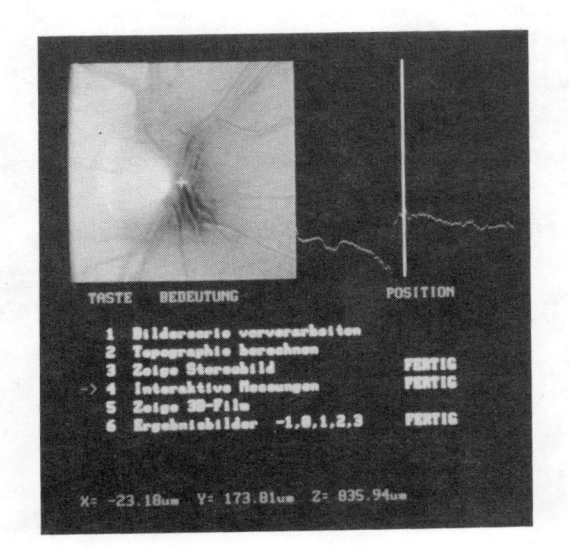

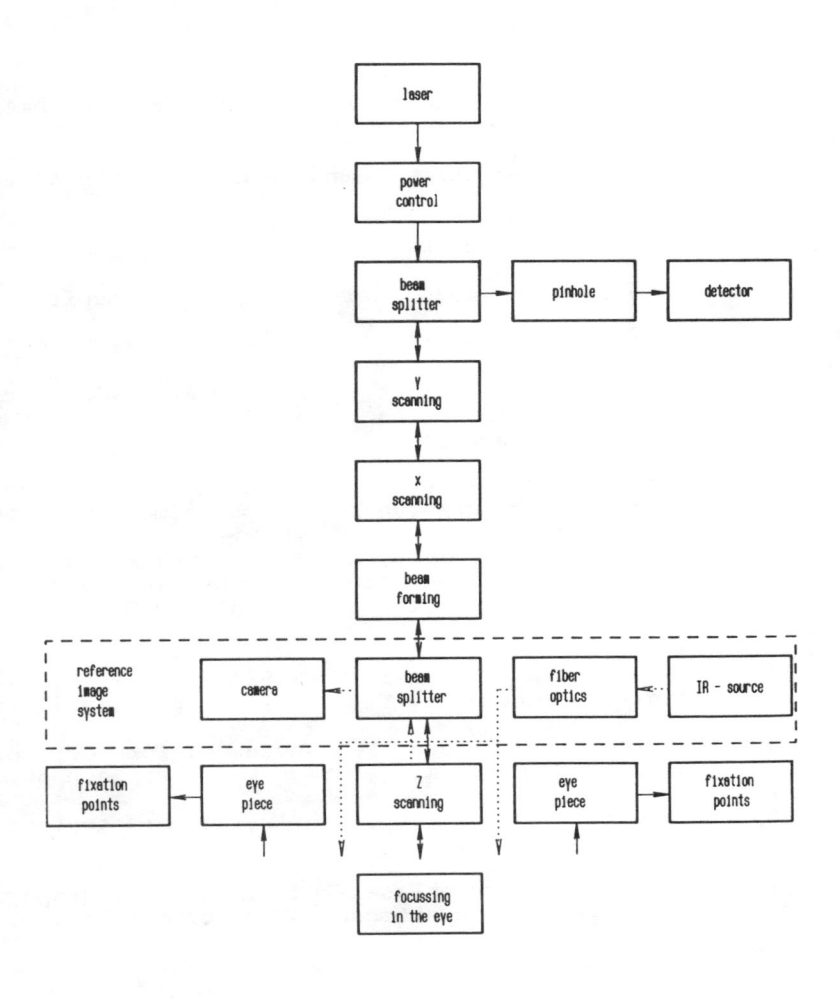

Fig. 6: Image formation system for posterior segment measurements.

Fig. 8: The topography of the retina. The color in the right part signifies a determined height.
Left: A vertical section through the retina. The curve means the height profile one line along. The bright line corresponds to the cross in the right image.

5. CONCLUSIONS

The LTS
- produces high resolution optical section images, both perpendicular or parallel.
- measures precisely the curvature and thickness of the cornea, even in the central region.
- produces three-dimensional and topographical images of the retina and its components.
- operates contact free.
- does not require dilation of the pupil.

6. REFERENCES

[1] F.E. Kruse, R.O.W. Burk, H.E. Völcker, G. Zinser, U. Harbarth: Laser Tomographic Scanning of the cornea topography. Proc. of the second international workshop on laser corneal surgery, Boston (1988)

[2] F.E. Kruse, R.O.W. Burk, H.E. Völcker, G. Zinser, U. Harbarth: Laser Tomographic Scanning of the optic nerve head. Ophthalmology 95 (9): 165 (1988)

[3] G. Zinser, U. Harbarth, C. Ihrig, F.E. Kruse, R.O.W. Burk, H.E. Völcker: Konfokales Laser Tomographie Scanning. Physikalische Grundlagen und apparativer Aufbau. ABC - Erstes int. Symposium für die digitale Bildverarbeitung und -analyse in der Ophthalmologie (1988)

[4] J.W. Bille, A.W. Dreher, W.F. Sittig, S.I. Brown: 3D-corneal imaging using the laser tomographic scanner (LTS). Invest. Ophthalmol. Vis. Sci. 28: 223 (1987)

[5] F. Berg: Vergleichende Messungen der Form der vorderen Hornhautflaeche mit dem Ophthalmometer und mit photographischer Methode. Acta Ophthalmol. 7: 386 - 423 (1929)

[6] T. Kuwara: Corneal topography using Mare contour fringes. Appl. Opt. 18: 3675 - 3678 (1979)

[7] S.D. Klyce: Computer-assisted corneal topography. High resolution graphic presentation and analysis of keratoscopy. Invest. Ophthalmol. Vis. Sci. 25: 1426 - 1430 (1984)

[8] T. Chan, S. Payor, B.A. Holden: Corneal thickness profiles using an ultrasono-graphic pachometer. Invest. Ophthalmol. Vis. Sci. 24: 1408 - 1410 (1983)

[9] R.D. Lepper, H.G. Trier: Computer-aided ultrasonic measurements of the corneal thickness. Ophthalmic Res. (1986)

[10] Th. Behrendt, K.E. Doyle: Reliability of images size measurements in the new Zeiss fundus camera. Am. J. Ophthalmol. 59: 896 - 899 (1965)

[11] R.H. Webb, G.W. Hughes: Scanning laser ophthalmoscope. IEEE Trans. Biomed. Eng. BME-28: 488 - 492 (1981)

[12] U. Klingbeil, J. Caprioli, M.L. Sears: Quantitative Analysis of optic disc topography and pallor. Invest. Ophthalmol. Vis. Sci. 29: 122 (1985)

[13] T. Wilson, C. Sheppard: Theory and Practice of Scanning Optical Microscopy, Academic Press, London (1984)

[14] R.W. Wijnaendts-van-Resandt, C. Ihrig: Application of confocal beam scanning microscopy to the measurement of submicron structures, SPIE Vol. 809: 101 - 106 (1987)

[15] International Electrotechnical Commission IEC 825: Radiation safety of laser products, equipment classification, requirements and user's guide (1984)

Scanning microscope for optically sectioning the living cornea

Barry R. Masters

Emory University School of Medicine, Department of Ophthalmology
1327 Clifton Road, N.E., Atlanta, Georgia 30322

ABSTRACT

A prototype of a clinical optically sectioning, two-dimensional redox fluorescence imaging microscope is described. Ultraviolet light from an arc lamp is conducted to the optical system with a quartz optical fiber. A variable slit projects the light onto the cornea after passing an excitation interference filter. The dipping cone of the microscope applanates the cornea and focuses the light onto the endothelial cell layer which is 6 microns thick. The fluorescence emission from the mitochondrial reduced pyridine nucleotides (NADH + NADPH) is detected by a microchannel plate-gated intensifier attached to a Newvicon Video Camera and a digital image processor. The intrinsic natural cellular fluorescence is imaged and is indicative of the cellular state of oxidative metabolism. In addition, an optically sectioning microscope was developed with an optical spectrum analyzer to characterize the fluorescence spectra from thin layers in the cornea and ocular lens. Its unique feature is that the scanning objective is attached to a piezoelectric driver and scans the eye from the tear film to the aqueous humor. The depth resolution is 6 microns with an 100 x objective and 18 microns with a 50 power objective (100 micron slits). The applications include fluorescence measurements on biological layered structures. The present study involves the noninvasive measurement of oxidative metabolism of the component layers of the in vivo cornea. Finally, the utility of confocal microscopy in ophthalmology is demonstrated as a series of confocal images of the rabbit cornea with a depth resolution less than one micron.

1. INTRODUCTION

The rapid development of confocal microscopy[1,2] together with recent advances in low light imaging detectors promises a revolution in ophthalmic diagnostic imaging. The ability to image ocular tissue in situ at the cellular and subcellular levels permits morphological studies to be performed which heretofore could only be made with fixed, sectioned, and stained materials. The quality of the images obtained rivals those made with scanning electron microscopy; but without the morphological and structural artifacts usually associated with the fixation process. The two key features are better rejection of out-of-focus light and higher resolution than with conventional light microscopy[3,4]. Clinical and research investigations with confocal imaging microscopes would complement the use of wide field specular microscopes[5,6].

This paper describes three instrumental approaches to scanning microscopy of ocular tissue. The first instrument is a microscope that is confocal in one dimension[7]. This scanning optically sectioning microscope is used as a research instrument to obtain mitochondrial fluorescence signals which serve as a cellular diagnostic index of oxidative metabolism[8,9,10,11]. The objective lens is scanned along the optic axis to obtain a profile of the fluorescence intensity throughout the thickness of the cornea. This instrument is not capable of making a two-dimensional image. It has been applied to the living rabbit to study oxidative metabolism of the limiting layers of the cornea[12]. An optical spectrum analyzer is connected to the microscope to permit spectral analysis of the fluorescence emission. The characterization of the source of the fluorescence emission is performed with the spectrum analyzer at each morphological region of the cornea or ocular lens. The depth resolution can approach 5 microns.

The second prototype instrument is a modification of the Keeler Wide Field Specular Microscope. The instrument can record a two-dimensional redox fluorescence image of the corneal epithelium or the corneal endothelium[13]. The low level light emission from the mitochondrial reduced pyridine nucleotides required the use of a high sensitivity, low noise microchannel plate intensifier coupled to a Newvicon Video Camera, and a digital image processor. While the scanning optically sectioning microscope previously described has a depth resolution of about 5 microns, it is not an imaging device; it detects photons in planes along the optic axis of the eye. However, the two-dimensional imaging microscope presents an image of the fluorescence intensity in thin planes perpendicular to the optic axis of the eye. Its limitation is the large depth of focus. This instrument can measure the redox state of human endothelial cells, thus, providing a noninvasive diagnostic measure of oxidative cellular metabolism.

Finally, to emphasize the utility and the image quality of the laser scanning confocal microscope, morphological investigations were made using the Lasersharp MRC-500 Confocal Fluorescence Imaging System[14,15]. In the reflectance mode, studies were performed on both human eye bank eyes and enucleated rabbit eyes. High resolution images of the cornea, the ocular lens, and the retinal layers were obtained. The advantage of the laser scanning confocal microscope are the submicron depth of field and the high contrast images obtained of almost transparent material. Structures such as nerve fibers, endothelial microvilli, structural fibers, and details of keratocyte cell processes are clearly imaged. Previously, it was only possible to observe these images with scanning electron microscopy and transmission electron microscopy of fixed material. The laser confocal microscope in the reflectance mode forms the image from differences in refractive index in the ocular tissues. The ability to resolve focal planes of less than one micron thickness provides the ability to study the cellular metabolism of the basal epithelium, the wing cells of the epithelium, and the surface epithelial cells in a variety of normal and pathological cases. In addition, the cellular oxidative metabolism of the individual endothelial and basal epithelial cells can be investigated with the use of redox fluorometry.

A clinical instrument for use in diagnostic ophthalmology would combine the features of the three instruments. A narrow depth of field would permit measurements of the various cell types which comprise the epithelium of the cornea. The oxidative metabolism state of the individual cells of the corneal basal epithelium and endothelium could be studied. The use of an optical spectrum analyzer would characterize the source of the fluorescence. Digital image processing together with software for three-dimensional reconstruction of the cornea from confocal serial sections is within current technology.

INSTRUMENTATION AND RESULTS

The details of the optically sectioning ocular fluorometer microscope using two 100 micron conjugate slits has been previously described. It is used with living rabbits and perfused rabbit cornea. This instrument was used in an investigation of the effect of various contact lens materials on the oxygen concentration in the tear film and on the induced cellular hypoxia. This instrument has been modified by replacing the slits with a set of 50 micron apertures to improve the depth resolution. The photomultiplier detection system is connected to a dual channel photon counting system for improved signal to noise.

Alternatively, a thin quartz fiber optic connects the optically sectioning microscope with a Princeton Instruments, Inc., Optical Spectrum Analyzer. This instrument contains a monochromator and a double microchannel plate intensifier with a 1024 linear diode array. The spectrum analyzer permits the spectral analysis of the fluorescence emission from any focal plane in the cornea. It is used to identify and characterize the source of the fluorescence, as well as for fluorescence indicators of pH and calcium levels in the various layers of the cornea. The entire instrument is computer controlled.

The schematic of the revised instrument is shown in Fig. 1. The optical principle of the optically sectioning microscope is shown in Fig. 2. This instrument has a scanning objective lens and uses a drop of physiological saline as the optical coupling fluid. The objective lens does not make physical contact with the surface of the cornea. An example of the depth resolution for both reflected and fluorescence light is shown in Fig. 3.

The Keeler Wide Field Specular Microscope is used to obtain human specular photomicrographs of the corneal endothelial cell layer. It is used clinically to document endothelial morphology; however there is no direct link between endothelial morphology and cellular function. Our goal is to investigate the correlation between the endothelial morphology and its cellular oxidative metabolism. This can be accomplished by the use of noninvasive redox fluorometry and confocal microscopy. In the specular mode (halogen lamp illumination) a reflected light specular image of the corneal endothelial cell pattern is obtained. The a mirror focuses the 366 nm excitation light onto the corneal endothelium and a fluorescence image of the same field is obtained. Fig. 4 shows the schematic diagram for two-dimensional fluorescence imaging. The digital image processor is used to correlate and analyze the two images after digital filtering, and image registration is performed. The modified wide field fluorescence imaging instrument is shown in Fig. 5. The present limitation of the instrument is the depth of field. While it is possible to image the intrinsic cellular fluorescence from the reduced pyridine nucleotides, the rejection of out of plane light is minimal. It is not possible to image the individual cell layers that comprise the corneal epithelium; for example the single cell layer forming the basal epithelium.

Since the intensity of the fluorescence from the reduced pyridine nucleotides is very low, a highly sensitive detection is required. The system consists of a microchannel plate intensifier coupled to a low noise Newvicon Video Camera. A MegaVision, Inc. Digital Image Processor is used to average several frames and to analyze the fluorescence image. Fig. 6. shows an image of the pyridine nucleotide fluorescence of rabbit corneal endothelial cells. This is the negative of the video image for increased resolution. Dark pixels correspond to high fluorescence intensity, and light pixels correspond to low fluorescence intensity. The corneal endothelium is about 6 microns thick. Although this is less than the optical sectioning thickness of the instrument, it is possible to obtain the image because the planes adjacent to the endothelium are devoid of fluorescence at the wavelengths used.

To illustrate the high rejection of out of plane light and the enhanced lateral resolution of the scanning laser confocal microscope the structure of the cornea was investigated with the Lasersharp MRC-500 unit. The light source was a 25 mW argon ion laser with the combined 488nm and 514 nm lines. The objective was a 50X, N.A. 1.00 on a Nikon Microscope. Kalman filtering was used to improve the signal to noise ratio of the images, and about 16 frames were averaged. Our investigation used both human eye bank eyes and rabbit corneas. Confocal imaging of the sections through the cornea, as well as the ocular lens and the layers of the retina were made. This paper will show several images from the morphological structures in the rabbit cornea which were made from enucleated rabbit eyes. These images are made in the reflected light mode with the fluorescence emission filter of the scanning laser confocal microscope removed. Figs. 7-11 illustrate the depth resolution and image quality that can be obtained on living tissue. <u>The coupling of confocal microscopy and noninvasive redox fluorescence provides a new tool to investigate cellular oxidative metabolism in basic and clinical ophthalmology</u>.

2. SUMMARY AND CONCLUSIONS

The application of low light imaging detectors coupled to a confocal microscope presents a new technology for diagnostic imaging in ophthalmology. The ability to image any plane in the living, unstained, unfixed cornea in near real-time represents an advance in ocular pathology. The use of confocal microscopy together with redox fluorescence imaging permits the diagnostic evaluation of cellular oxidative metabolism at a resolution of individual cells. The confocal microscope can obtain redox fluorescence images of individual endothelial cells in the cornea. This will permit investigation of heterogeneities of cellular oxidative metabolism on a cell-to-cell basis. In addition, the depth resolution available from the confocal microscope allows the various layers of the corneal epithelium to be optically sectioned. <u>Thus, for the first time it is possible to investigate the differences of oxidative metabolism not only for the various cell layers which comprise the epithelial layer, but also to study differences of cellular function on a cell-to-cell basis.</u>

3. ACKNOWLEDGEMENTS

This work was supported by a grant from N.I.H. EY-06958, and a departmental grant from Research to Prevent Blindness, Inc. Dr. Steve Paddock of the Integrated Microscopy Resource for Biomedical Research, University of Wisconsin, Madison, Wisconsin, collaborated in obtaining the laser confocal images.

The authors also wish to thank Mrs. Gussie Damon and Ms. Mary Lynn Covington for their assistance in the preparation of this manuscript.

4. REFERENCES

1. M. Petran, M. Hadravsky, M.D. Egger and R. Galambos, "Tandem-scanning reflected-light microscope," J. Opt. Soc. Am. 58, 661-664 (1968).
2. M.A. Lemp, P.N. Dilly and A. Boyde, "Tandem-scanning (confocal) microscopy of the full-thickness cornea," Cornea 4, 205-209 (1986).
3. T. Wilson and C.J.R. Sheppard, <u>Theory and Practice of Scanning Optical Microscopy</u>, Academic Press, London (1984).
4. C.J.R. Sheppard, "Scanning optical microscopy," in <u>Advances in Optical and Electron Microscopy</u>, R. Barer, V.E. Cosslett, eds., Vol. 10, pp. 1-98, Academic Press, London (1987).
5. D.M. Maurice, "A scanning slit optical microscope," Invest. Ophthalmol. 13, 1033-1037 (1974).
6. C.J. Koester, "Scanning mirror microscope with optical sectioning characteristics: applications in ophthalmology," Appl. Opt. 19, 1749-1757 (1980).
7. B.R. Masters, "Optically sectioning ocular fluorometer microscope: applications to the cornea," in <u>Time-Resolved Laser Spectroscopy in Biochemistry</u>, J.R. Lakowicz, ed., Proc. SPIE 909,

343-348, (1988).

8. B. Chance, B. Schoener, R. Oshino, F. Itshak and Y. Nakase, "Oxidation-reduction ratio studies of mitochondria in freeze-trapped samples. NADH and flavoprotein fluorescence signals," J. Biol. Chem. 254, 4764-4771 (1979).

9. B. Chance, "Pyridine nucleotide as an indicator of the oxygen requirements for energy-linked functions of mitochondria," Circ. Res. 38(5 Suppl. I), 31-38 (1976).

10. B.R. Masters, "Noninvasive corneal redox fluorometry," Curr. Eye Res. 4, 139-200 (1984).

11. B.R. Masters, "Noninvasive redox fluorometry: how light can be used to monitor alterations of corneal mitochondrial function," Curr. Eye Res. 3, 23-26 (1984).

12. B.R. Masters, "Effects of contact lenses on the oxygen concentration and epithelial mitochondrial redox state of rabbit cornea measured noninvasively with an optically sectioning redox fluorometer microscope," in *The Cornea: Transactions of the World Congress of the Cornea III*, H.D. Cavanagh, ed., pp. 281-286, Raven Press, Ltd, New York (1988).

13. B.R. Masters, "Two-dimensional fluorescent redox imaging of rabbit corneal endothelium," Invest. Ophthalmol. Vis. Sci. 29(Suppl.), 285 (1988).

14. W.B. Amos, J.G. White and M. Fordham, "Use of confocal imaging in the study of biological structures," Appl. Opt. 26, 3239-3243 (1987).

15. J.G. White, W.B. Amos and M. Fordham, "An evaluation of confocal versus conventional imaging of biological structures of fluorescence light microscopy," J. Cell Biol. 105, 41-48 (1987).

5. FIGURE LEGENDS

Fig. 1. Schematic diagram of the scanning optical sectioning microscope showing a simplified ray path. The source is either a laser or a mercury arc lamp connected to the aperture by a quartz fiber optic. F1, and F2 are narrow band interference filters to isolate the excitation wavelengths. F3 is a narrow band interference filter to isolate the emission light. The aperture is labelled A. M1, M2 and M3 are front surface mirrors, and B.S. is a quartz beam splitter. L3 is the scanning objective 50X, N.A. 1.00. The objective is scanned along the optic axis of the eye.

Fig. 2. The optical principle of the optical sectioning microscope. A simplified ray path is shown. Saline is used as the optical coupling fluid between the surface of the cornea and the scanning objective of the microscope. The position of the objective determines the volume of the object that is optically sectioned. There are two conjugate apertures, one adjacent to the source and one at the detector. The light rays from the upper surface of the cornea (epithelium) are detected after passing through the detector aperture; the rays from the lower surface (endothelium) impinge upon the conjugate detector aperture and are not detected.

Fig. 3. An optical section through the rabbit cornea illustrating the resolution for the reflected light (solid line) and the 460 nm fluorescence emission. The reflected intensity is 10 times the fluorescence intensity. The ordinate is relative intensity and the abscissa is distance into the cornea. The tear film is on the right side and the aqueous humor is on the left side of the figure. The fluorescence from the region between the endothelial and the epithelial intensity peaks (stromal region) is non-specific fluorescence.

Fig. 4. Schematic diagram of the wide-field specular microscope modified for two-dimensional fluorescence imaging. The source of light is a 100 W mercury arc lamp with an I.R. filter and a narrow band interference filter to isolate the 366 nm line. A fiber optic cable connects the source to the instrument. F1 is the excitation filter and F2 is the emission filter to isolate the fluorescence. The applanating cone lens has a thin layer of saline or an index matching gel as the coupling fluid. The optical sectioning is obtained by the scanning lens used for focusing. M1, M2 and M3 are mirrors. The image is detected with an intensified video detector. The light rays focus on the corneal endothelium.

Fig. 5. Photograph of the modified Keeler Wide Field Specular Microscope showing the patient in the headrest, the fiber optic connecting the source of ultraviolet light and the microscope, and the intensified video detector. For standard widefield reflected specular microscopy, a halogen lamp is the source and the filters are removed. The instrument is readily switched between the two modes.

Fig. 6. Two-dimensional redox fluorescence image of rabbit corneal endothelial cells. <u>This image is the negative of the fluorescence image; dark pixels correspond to high fluorescence intensity, and light pixels refer to low fluorescence intensity. The cell borders show low fluorescence intensity</u>. The excitation was the 366 nm line of the mercury arc lamp, and the fluorescence is imaged in the range of 430-470 nm. The Keeler Widefield Specular Microscope was modified for fluorescence imaging. The detector was a Newvicon Video Camera coupled to a microchannel plate light intensifier. The intensified video output was averaged on a digital image processor. The figure is a magnified image of the original digitized image.

Fig. 7. Confocal reflected light image of rabbit basal epithelial cells focused in the center of the cells. The plane of focus is close to the bottom of the cells. These cells are about 20 microns thick and about 10 microns in diameter. They are the only cells that undergo mitosis.

Fig. 8. Confocal reflected light image of rabbit basement membrane showing folds. This membrane is at the base of the basal epithelium and separates it from the stromal region. The image width is about 75 microns.

Fig. 9. Confocal reflected light image of a submicron optical section in the anterior stroma of the rabbit cornea. The focus is centered on the nerve fiber, with some loss of focus on the keratocyte cell bodies in adjacent focal planes. It shows the stromal keratocytes and their long processes. The thin horizontal line in the image is a nerve fiber with a diameter in the micron range.

Fig. 10. Confocal reflected light image of the endothelial cells of the rabbit cornea. The image shows the junctions between adjacent cells. The focus is within the plane of the endothelial cells. These cells are 15-20 microns in diameter.

Fig. 11. Confocal reflected light image of endothelial cells of rabbit at higher magnification. The focus is within the plane of the endothelial cells. These cells are 15-20 microns in diameter.

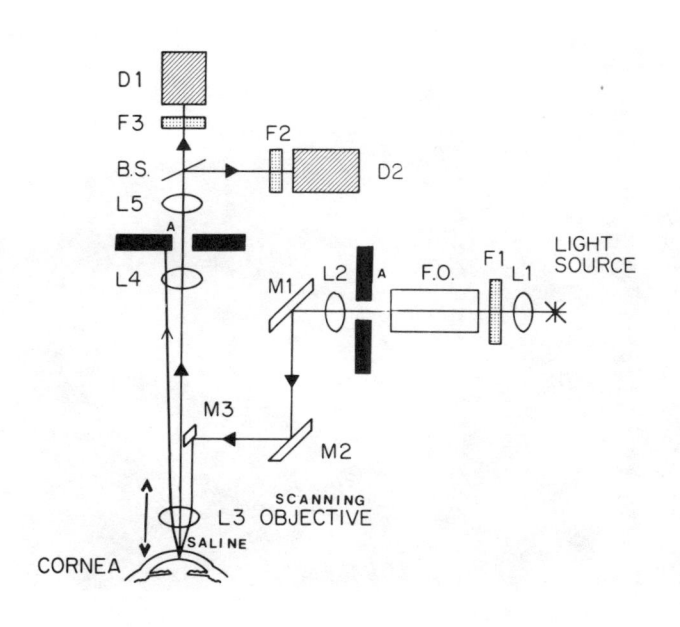

FIG. 1

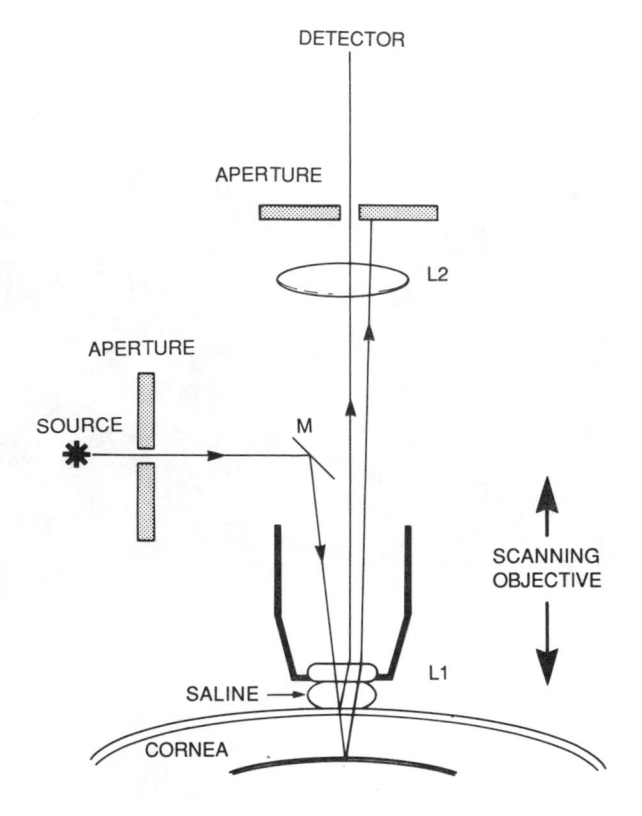

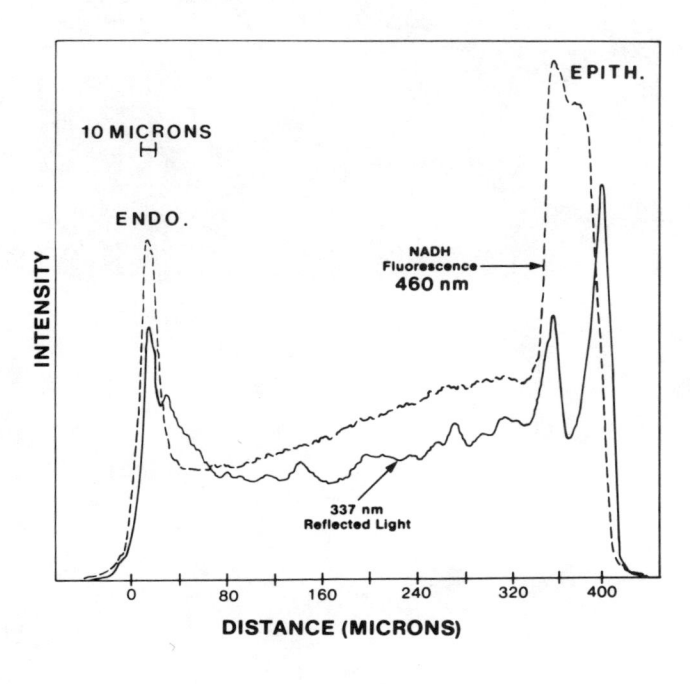

FIG. 2

FIG. 3

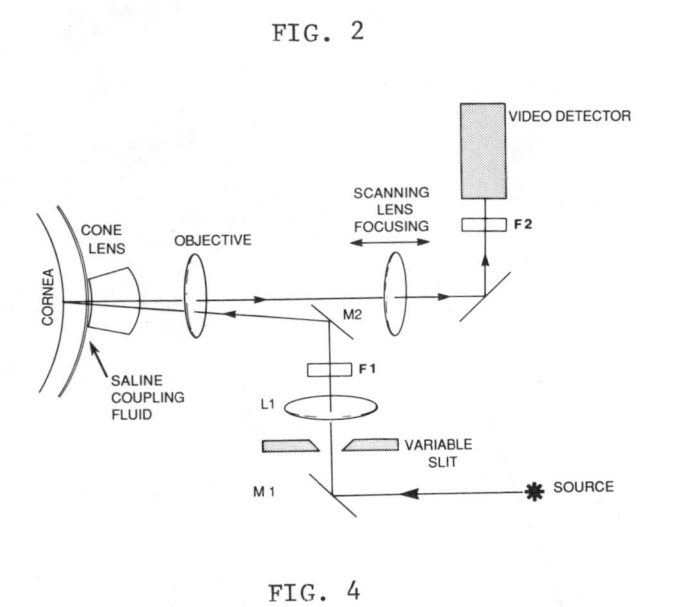

FIG. 4

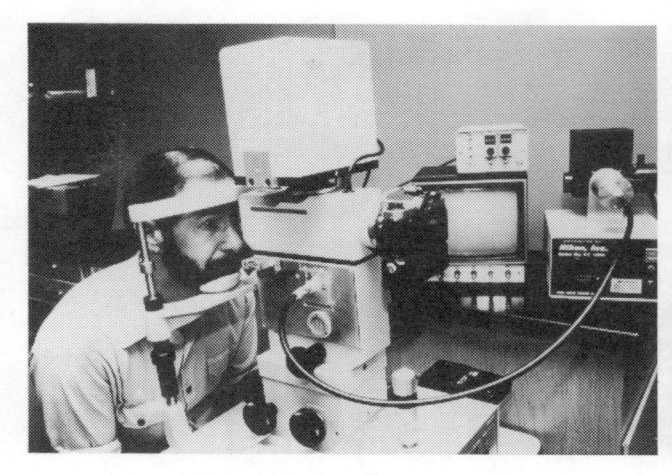

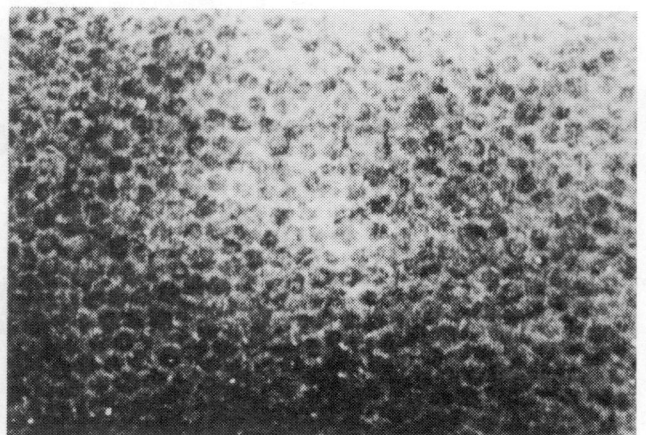

FIG. 5

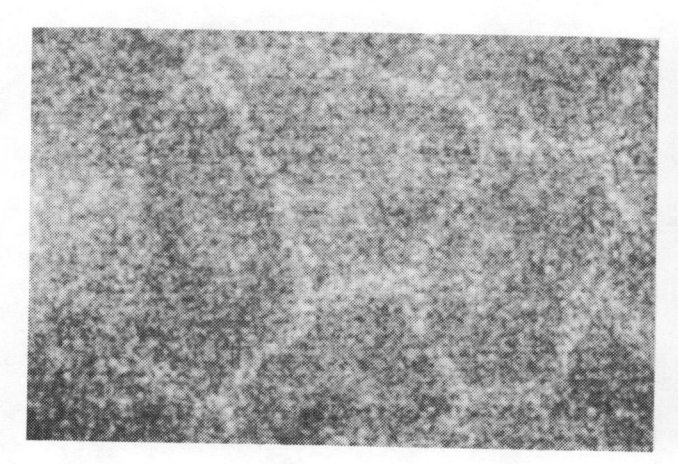

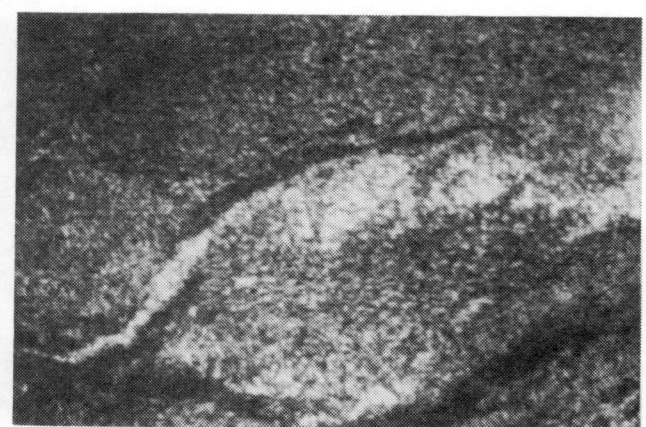

FIG. 6

FIG. 7

FIG. 8

142 / SPIE Vol. 1028 Scanning Imaging (1988)

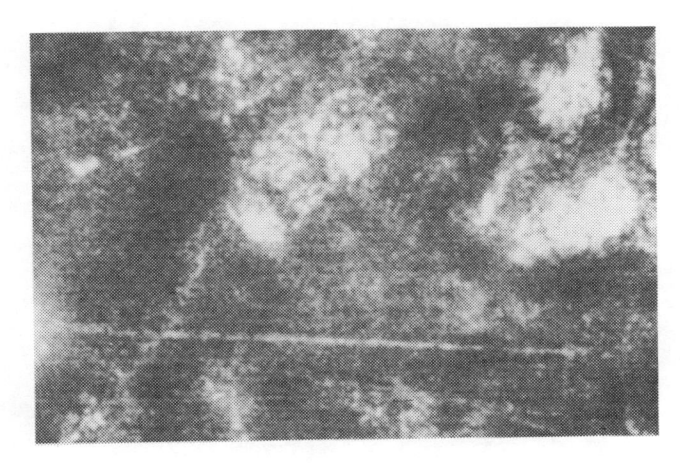

FIG. 9

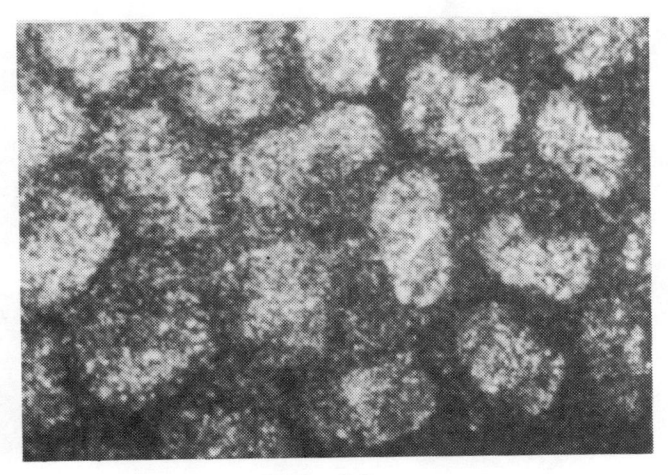

FIG. 10

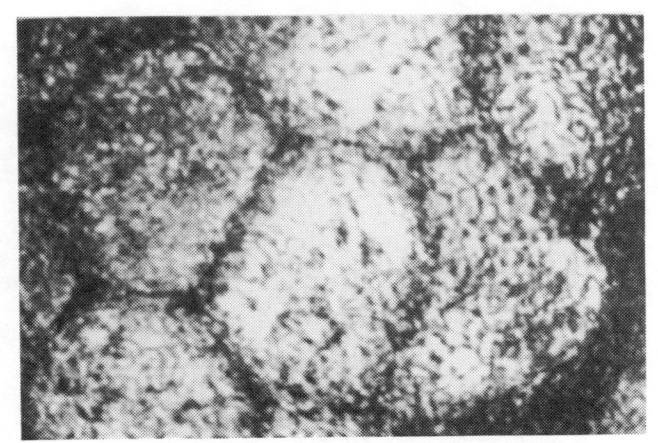

FIG. 11

ECO1
SCANNING IMAGING

Volume 1028

SESSION 4

Biological and Materials Microscopy

Chair
S. Siegel
Carl Zeiss GmbH (FRG)

Confocal Fluorescence Microscopes for Biological Research

Ernst H.K. Stelzer, Reiner Stricker, Reinhard Pick, Clemens Storz and Pekka Hänninen

European Molecular Biology Laboratory (EMBL), Confocal Light Microscopy Group (Programme of Phys. Instr.)
Meyerhofstraße 1, Postfach 10.2209, D-6900 Heidelberg, West Germany (FRG)

ABSTRACT

A confocal beam scanning laser microscope (CBSLM) for confocal fluorescence and confocal reflection microscopy has been built at the European Molecular Biology Laboratory (EMBL) in Heidelberg. The instrument has now been used for almost one year with a large number of different biological specimens, experimental protocols and fluorophores. The instrument is stable, has a high detection efficiency and is easy to use.

SCANNING PRINCIPLE

The central component of the CBSLM is a scanning mirror that is turned in two axes. This is accomplished by mounting a galvanometer (General Scanning, GF120DT) on a turnable motordriven device. The galvanometer provides the movement of the light spot in the object plane along the x-axis and turning the galvanometer provides the same movement along the y-axis. This arrangement reduces the number of optical components considerably as no relaying optics for two scan mirrors are necessary. The movement of the light spot along the z-axis is accomplished by moving the micropscope objective which is mounted on a linear motor. The configurations allows images to be recorded in an x/y- or an x/z-plane.

IMAGE SIZE

The standard image has a size of 512 x 512 pixels but may be as large as 4096 lines by 1024 columns. Each pixel is recorded with an accuracy of 8bit but accumulations are always performed with an accuracy of 16bit. The field in an x/y-plane has a maximum size of 170µm x 170µm and may be as small as 20µm x 20µm. The field along the z-axis can vary between 80µm and 10µm. The amplitudes along all axes are selected with 12bit DA converters. This precision is sufficient to set any desired amplitude with an accuracy of one micron. Any combination of x/z-field sizes is allowed. The aspect ratio, i.e. the number of lines along the z-axis over the number of pixels per line along the x-axis is usually one. If desired, it may be varied between one and five.

SCANNING SPEED

The scanning frequency in the x-axis is fixed to 250Hz. The scanning movement along the x-axis drives both the frequencies along the y- and the z-axis. The maximum frequency along the y-axis is 0.4Hz and along the z-axis is 2Hz. A whole frame in an x/y-plane of size 512 x 512 pixels is recorded in 2.8sec (including retrace and software overhead), while a frame of size 512 x 170 pixels in an x/z-plane is recorded in 0.8sec. Averages consume exact multiples of these times. Scanning is performed in forward direction only.

CONTRASTS

The CBSLM provides a number of different contrasts such as confocal and nonconfocal fluorescence, confocal double fluorescence, confocal reflection and nonconfocal transmission. However, only one contrast is available at any time.

PERFORMANCE

The lateral resolution is less than 0.2µm while the 20% to 80% value of the discrimination along the z-axis is 0.6µm (see figure 5). These values have been measured with an excitation wavelength of λ_{exc} = 514nm and an emission wavelength λ_{em} = 550nm with a lens of numerical aperture 1.32. The detection efficiency (the product of all transmission and reflection factors) is better than 50% and depends mainly on the quality of the filters. The repeatability of the voxel access in the object plane is probably better than 50nm. This value is important, since it determines the minimum field size that may be used to record averaged images.

LIGHT SOURCE

The light sources are an Argon-ion (Spectra-Physics 2020-05) and a He-Ne laser. The He-Ne laser is used only to set up the light path. The Argon-ion laser provides several lines between 454.5nm and 514.5nm from which two, 476nm and 514.5nm, are used to excite dyes such as Lucipher Yellow, NBD, Fluorescein, Rhodamine or Texas Red. The principal layout of the separation is described in figure 1. To avoid any photodamage on the sample a slow shutter prevents any light illuminating the sample between the recording of two images while a fast piezo-driven shutter discriminates against 90% of the light during the retrace of the scanner. The source pinhole is fixed in position and in size.

DETECTOR

The emitted or reflected light is detected with photomultipliers (R1463-01, Hamamatsu). The photomultipliers are operated in analog mode and located directly behind pinholes of a fixed size. Three detector units (photomultiplier, preamplifier and amplifier) are permanently installed: confocal fluorescence, confocal reflection and non-confocal transmission. Any one of these may be selected as the input into the intensity A/D-converter.

MICROSCOPE

The microscope is designed as one of inverted type. A turnable mirror allows one to switch between a conventional transmission microscope and a confocal fluorescence or confocal reflection microscope. The optical components (commercial grade or standard equipment) are mounted on a vibration isolated table mostly with Spindler & Hover "Mikrobank" parts. The whole microscope is therefore relatively small, compact and still adaptable to new experiments. An even smaller design using a conventional microscope and specially fabricated parts would be possible especially since the beam is parallel within the scan and detector units. A detailed description of the optical layout is found in figure 1.

SAMPLE STAGE

The microscope stage is based on a Leitz product to which DC motors and LVDTs have been added to allow an automatic access to any x/y-position. The coarse movement along the z-axis is performed manually. The coarse positions are recorded along the x and y axes and are reproducible to better than 10μm. This arrangement is important since the CBSLM lacks a conventional fluorescence microscope that is necessary to get an overview over the sample and to select "good" (from the biologists' point of view) areas. The arrangement of the LVDT (the positioning device) has therefore been duplicated on a conventional fluorescence microscope that is located next to the CBSLM.

ELECTRONICS

Most of the electronic equipment has been built at the EMBL. Based on the Europe card standard it consumes three 19" racks. All units are programmable thus giving access to most of the features of the CBSLM ("digital microscope"). The basic idea was to use as much analog electronics as possible, controlling only via D/A-converters and retrieving any information via A/D-converters. The feedback loops (e.g. in the galvanometer driver and the motor control) are analog. The A/D-converter for the light intensity has three inputs for three detector units. Changing from fluorescence to reflection is therefore accomplished by changing the input. A very important part is a system of two line buffers which are alternately accessed by the intensity A/D-converter and by the computer that reads, averages and stores the stream of data bytes.

COMPUTERS

A VME bus based MC68010 computer controls every part of the microscope. A special board contains the frame grabber and gives access to the necessary electronic equipment that is accessible only through this computer. A second computer is used to deal with the Ethernet available at the EMBL.

CONTROL PROGRAMS

The control program is used to gather all the information concerning the status of the microscope. This information together with some comments on the specimen, the user's name, the current date and time are stored with every image that is transferred via the Ethernet link to the local area VAX cluster. The software controlling the microscope has been designed to do whatever is necessary to record images. These images or series of images as well as stereo pairs are then sent to a VAX and stored on the mass storage devices in the VAX cluster at the EMBL. Any further processing is performed on the VAXes which provide a number of workstations, plotters, printers, displays, image processing software etc. The program controlling the microscope provides, among many others, the following features: single image acquisition, acquisition and calculation of averaged images (line, frame, rolling), acquisition of a series of images with a varying position along the optical axis (z-series), calculation of stereo pairs from previously recorded z-series, summations and the display of series or anglyphs. Commands to the control program are either typed or selected from a menu. The program accepting the data on the VAX automatically calculates a histogram and generates individual and session protocols that are sent to a printer.

USAGE

During the past year the CBSLM has been used for approximately fifteen applications producing around 30000 images. Since most of this work has been done in collaboration with the Cell Biology Programme at the EMBL the research was concentrated on immunocytochemical investigations of mammalian cells: e.g. endo- and exocytosis, colocalization of different organelles and distribution of cytoskeletal elements.

REFERENCES

1. H.J.B. Marsman, G.J. Brakenhoff, P. Blom, R. Stricker and R.W. Wijnaendts-van-Resandt, "Mechanical scan system for microscopic applications," Rev. Sci. Instrum. 54, 1047-1052 (1983).
2. R.W. Wijnaendts-van-Resandt, H.J.B. Marsmann, R. Kaplan, J. Davoust, E.H.K. Stelzer and R. Stricker, "Optical fluorescence microscopy in three dimensions: microtomoscopy,"J. Microsc. 138, 29-34 (1985).
3. E.H.K. Stelzer and R.W. Wijnaendts-van-Resandt, "Applications of fluorescence microscopy in three dimensions: Microtomoscopy,"SPIE Proceedings 602, 63-70 (1986).
4. E.H.K. Stelzer, H.J.B. Marsman and R.W. Wijnaendts-van-Resandt, "A setup for a confocal scanning laser interference microscope,"Optik 73, 30-33 (1986).
5. E.H.K. Stelzer and R.W. Wijnaendts-van-Resandt, "Nondestructive sectioning of fixed and living specimens using a confocal scanning laser fluorescence microscope: Microtomoscopy,"SPIE Proceedings 809, 130-137 (1987).
6. G. van Meer, E.H.K. Stelzer, R.W. Wijnaendts-van-Resandt and K. Simons, "Sorting of sphingolipids in epithelial (MDCK) cells,"J. Cell Biol. 105, 1623-1635 (1987).
7. E.H.K. Stelzer, R. Stricker, R. Pick, C. Storz and R.W. Wijnaendts-van-Resandt, "Serial sectioning of cells in three dimensions with confocal scanning laser fluorescence microscopy (Fl-CSLM): Microtomoscopy," (to appear) SPIE Proceedings 909, (1988).
8. E.H.K. Stelzer and R. Bacallao, "Confocal Fluorescence Microscopy of Epithelial Cells," SPIE Proceedings, this volume (1989).
9. R. Bacallao and E.H.K. Stelzer, "Preservation of Biological specimens for observation in a confocal fluorescence microscope and operational principles of confocal fluorescence microscopy," in Methods in Cell Biology, A.M. Tartakoff, ed., (to appear), Academic Press, Orlando, Florida, USA (1989).
10. E.H.K. Stelzer, R. Stricker, R. Pick, C. Storz and R.W. Wijnaendts-van-Resandt, "A confocal single-mirror beam scanning laser fluorescence microscope," (to appear).

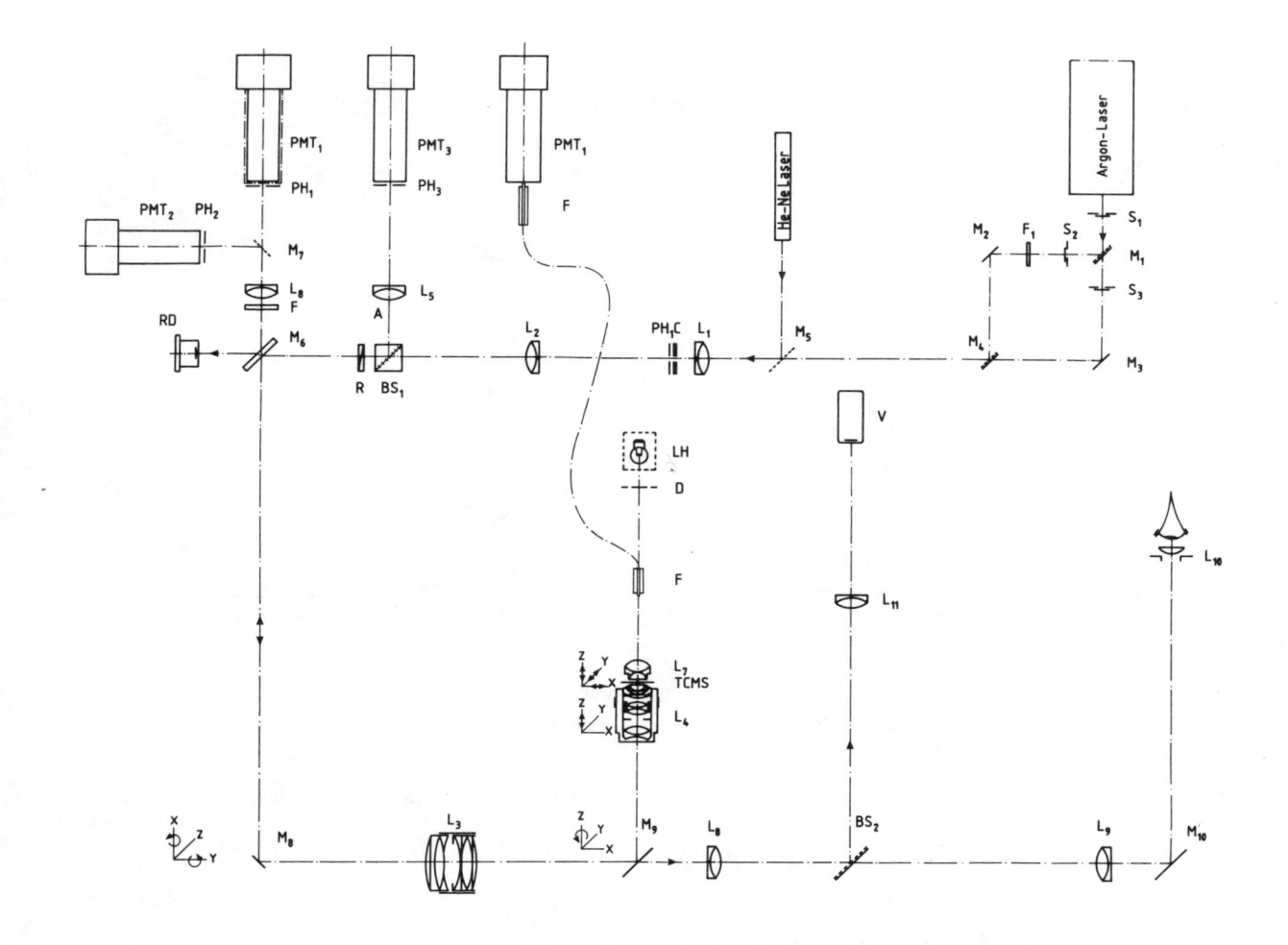

Figure 1. The optical layout of the CBSLM. The principal light source is an Argon ion laser. A set of two dichroic mirrors (M_1 and M_4) and a filter (F_1) separate two lines (488nm and 514.5nm). While one shutter (S_1) switches off all light, two other shutters (S_2 and S_3) may be used to select either of the two excitation beams. The expanded and spatially filtered beam (Lens L_1, pinhole PH_1 and lens L_2) passes a polarizing beam splitter (BS_1), is reflected by a dichroic mirror (M_6) and scanned with a twice mounted mirror (M_8). The scanned beam falls through a camera lens (L_3) into a microscope objective (L_4) and forms a light point moving across the object (TCMS). In other words, the lens L_3 forms an image of the scan mirror M_8 in the entrance aperture of the microscope objective L_4. The fluorescent light emitted in the object is collected with the same lenses. The lens (L_8) behind the dichroic mirror (M_6) focuses the light into a pinhole in front of a photomultiplier. For the conventional setup the mirror (M_9) below the microscope objective must be turned so that a transmission image may be viewed directly through the eyepiece or via the monitor as it is recorded with a video camera. The lambda/4 plate (R) polarizes the light circularly so that light reflected in the sample is vertically polarized when it enters the beam splitter and hence reflected in the direction of the photomultiplier PMT_2. Photomultiplier PMT_1 is supplied through a fiber optical cable. The diagram is complete in so far as it contains all the optical elements that are present in the CBSLM.

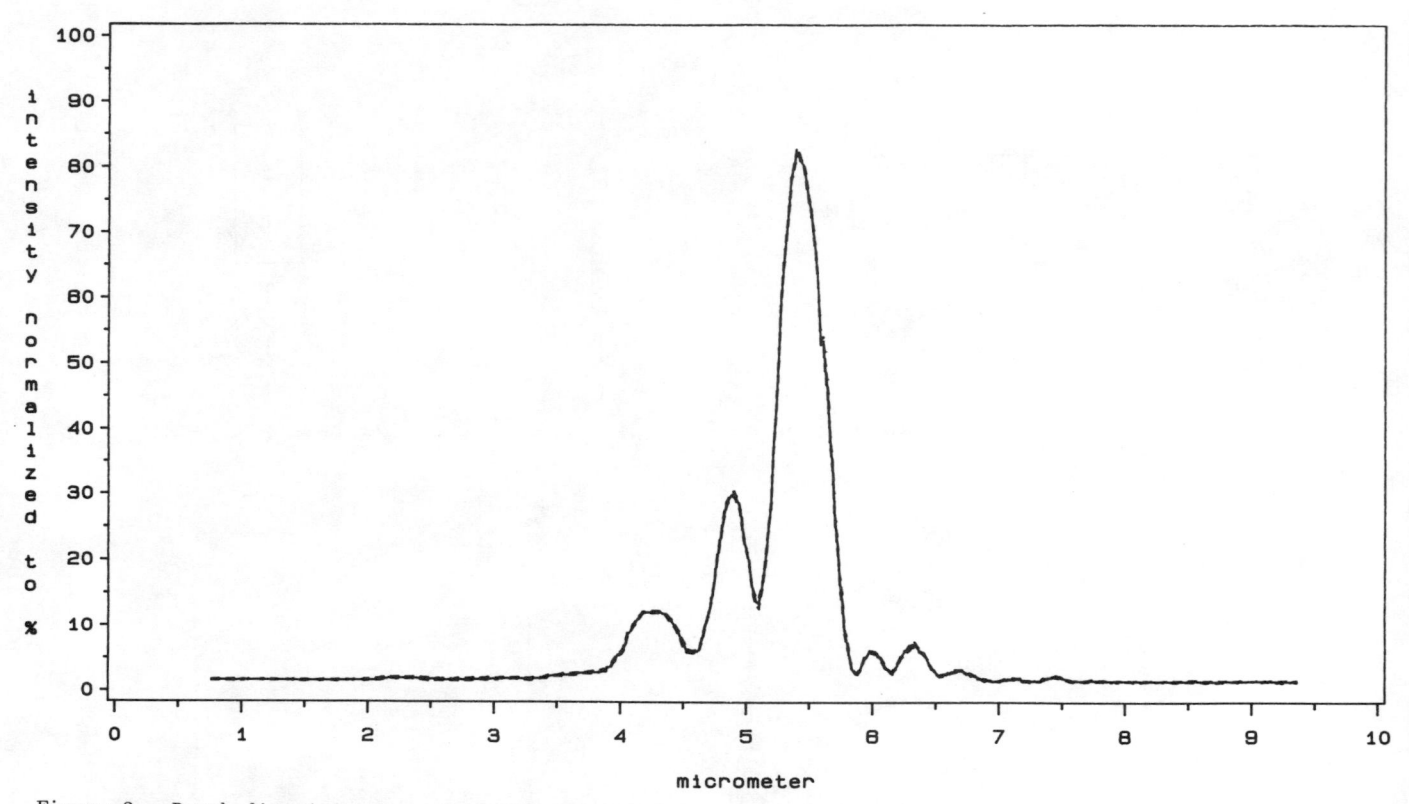

Figure 2. Depth discrimination along the optical axis in a confocal reflection microscope. Shown is the intensity as a function of the position of the beam along the optical (z-) axis. A polished surface on an integrated circuit has been used as a mirror. A scan in an x/z-plane (see figure 3) was analyzed to produce these curves. The main curve has a half width of 0.41µm (λ = 514nm, field size along x 30µm, along z 10µm, aspect ratio 3, recording time 2.8sec).

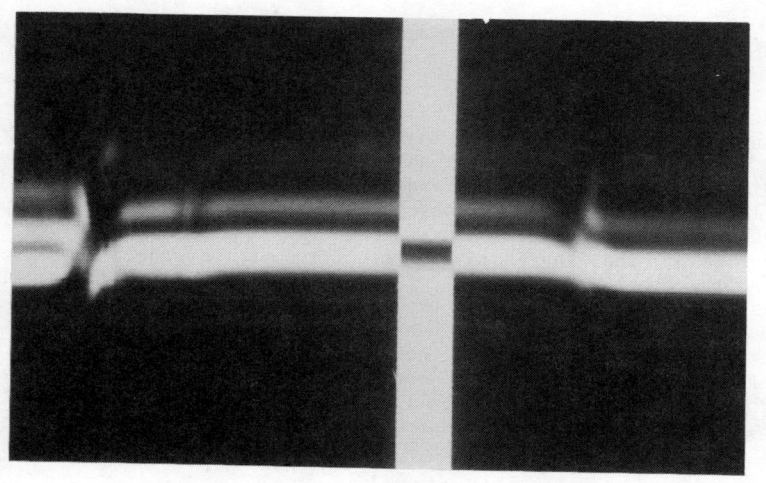

Figure 3. Depth discrimination along the optical axis in a confocal reflection microscope. Shown is one x/z-image. A polished surface on an integrated circuit has been used as a mirror. A scan in an x/z-plane (reversed area) was analyzed to produce the curve in figure 2 (λ = 514nm, field size along x 30µm, along z 10µm, aspect ratio 3, recording time 2.8sec).

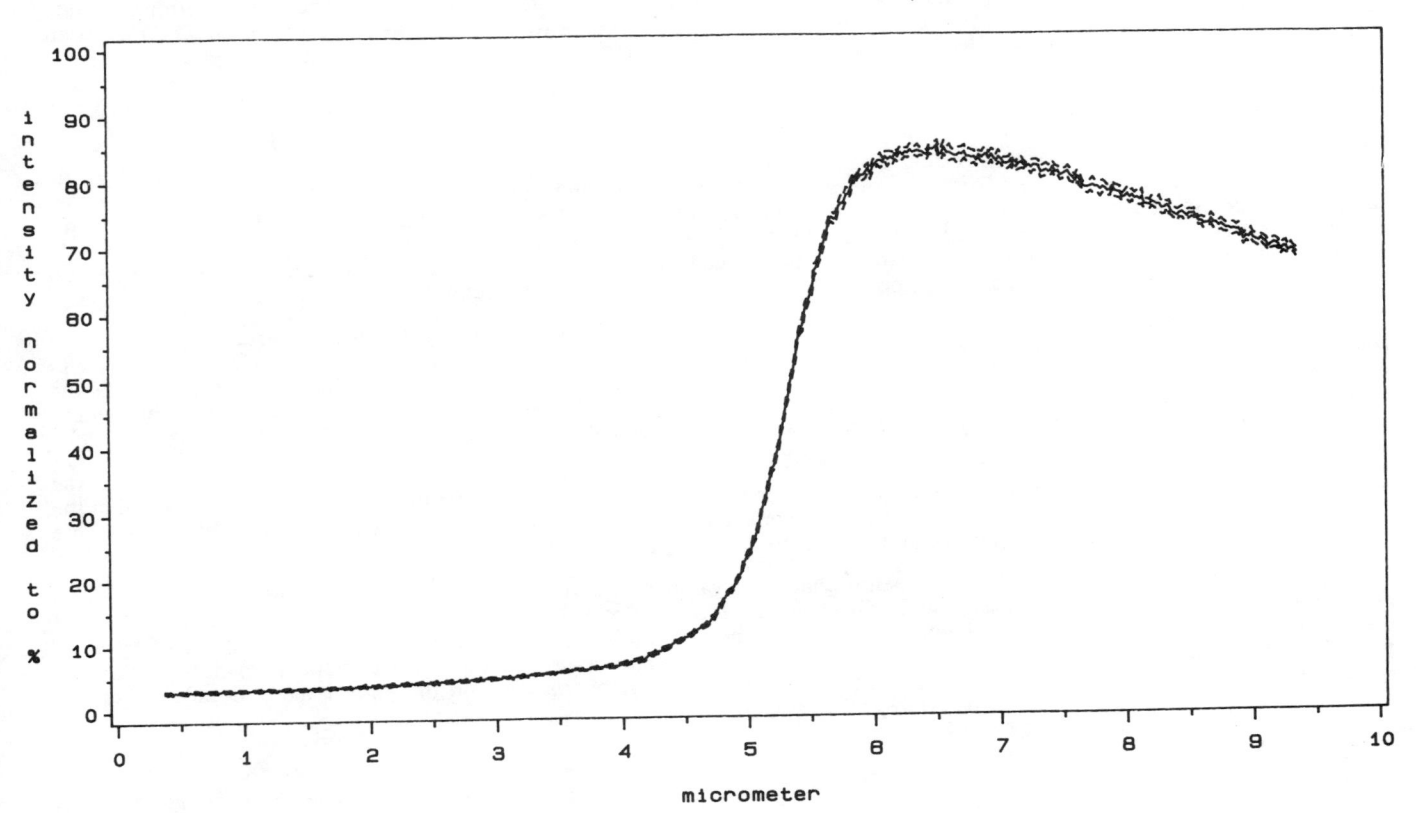

Figure 4. Depth discrimination along the optical axis in a confocal fluorescence microscope. The image shows an x/z-scan through a layer of Rhodamine 6G dissolved in oil. Visible is the transition from the cover glass to the Rhodamine 6G. The reversed area is plotted in figure 5 (λ_{exc} = 514nm, λ_{em} above 550nm, field size along x 30µm, field size along z 10µm, aspect ratio 3, recording time 2.8sec).

Figure 5. Depth discrimination along the optical axis in a confocal fluorescence microscope. Plotted is the intensity of the fluorescence emission as a function of the position of the beam along the optical axis. This can be analyzed and returns a value for the depth discrimination. Visible is the transition from the cover glass to the Rhodamine 6G (λ_{exc} = 514nm, λ_{em} above 550nm, field size along x 30µm, field size along z 10µm, aspect ratio 3, recording time 2.8sec).

An apparatus for laser scanning microscopy and dynamic testing of muscle cells

Ian Hunter, Serge Lafontaine, Poul Nielsen and Peter Hunter

Biomedical Engineering, McGill University, 3775 University Street, Montreal, Quebec, Canada H3A 2B4

Abstract

A Type 2 (confocal) transmission laser scanning microscope containing dynamic mechanical testing facilities for the purpose of isolated muscle fiber studies is under development in our laboratory. An embryonic skeletal muscle cell may be kept alive and grown in a temperature-controlled chamber which is confined by two microscope objectives. The muscle cell can be electrically stimulated or manipulated using two motor clusters in 3 dimensions for imaging purposes while independent mechanical experiments are performed. The microscope consists of an interferometer which works either in a Mach-Zehnder configuration to provide cross-correlation functions from which magnitude, phase, and polarization information is obtained, or in a Michelson arrangement to provide auto-correlation functions (eventually for Fourier transform Raman spectroscopy). The microscope is presently controlled by a MicroVAX-II/GPX system while a dedicated parallel computer (multiple-instruction multiple-data) is being developed to cope with the demanding analysis and real-time control requirements of the apparatus. The approach taken is to avoid acquiring 3-D images and instead to fit structural models to optical data acquired from the original object as needed by the model building algorithms.

Introduction

Rationale

Muscle is a linear stepping motor with remarkable mechanical characteristics. The mechanical behavior of a single living active (stimulated) muscle fiber (single cell) may be studied by applying a length perturbation (input) to it and recording the resulting force (output). The dynamic input-output relation (mechanical impedance spectroscopy) is found to be dynamically nonlinear and except over short durations must also be considered to be time varying. Nonlinear system identification techniques[1] have been used with considerable success to formulate dynamic nonlinear models directly from stochastic muscle fiber length perturbation experiments. Whereas such "black-box" or lumped models are able to make accurate force predictions they are of little scientific interest because a single muscle cell is not a homogeneous mechanical system. Rather it is a distributed system consisting of a large number of force generating elements, or sarcomeres, organized in series and in parallel. The sarcomeres interact with the result that the overall mechanics of a single muscle fiber is determined both by distributed interactions among sarcomeres in addition to intrinsic sarcomere mechanics. In order to go beyond the "black-box" characterization of muscle, two tools, one technological the other mathematical, must be developed.

Technological Tools

An understanding of sarcomere interactions requires measurements of sarcomere dimensions throughout a muscle cell while mechanical experiments are performed. This implies an apparatus capable of performing mechanical experiments while optical measurements are made. Sarcomeres together with other cell entities form a 3-D structure which must be measured with as much precision as possible. Electron microscopy cannot be used since the muscle cells must be kept alive. Regular optical microscopy lacks the required precision and furthermore yields rather limited optical information such as the spatial variation in intensity or phase. In contrast, laser microscopy provides increased precision and potentially can yield more optical information.

Our initial requirements are that optical measurements (samples) be made at least every 10nm over the volume of the cell. A small muscle fiber dissected from a whole muscle might have a length of 5mm and a diameter of 100μm. A spatial sampling increment of 10nm throughout the whole cell would result in a 3-D image having a total of about 10^{13} samples. This is clearly too many samples to store in existing fast access memory. A cultured muscle cell is smaller and may be imaged with fewer samples. For example a two-day old cultured muscle cell of 100μm length and 20μm diameter would result in 10^{10} samples. However, this is still too many for practical manipulation using current technology.

An alternative method is to avoid storing the 3-D images in the first place. In this approach a structural model (e.g. finite-element mesh) is fitted using samples taken from the original muscle cell rather than from a 3-D image stored in memory. Apart from circumventing the image storage and retrieval issue, this approach enables adaptive structural modeling in which local sampling increments (or sample integration times) are selected as required by the structure-fitting algorithm. Optimal (non-uniform) sampling strategy issues become relevant and, more particularly, feasible.

There are two major consequences of treating the original object as unsampled random access memory. One is that the motors controlling the 3-D positioning have to permit random access optical sampling from arbitrary 3-D locations within the object. The other is that the same computer used to implement the adaptive model-building techniques must also control the apparatus used to gather optical samples and perform the mechanical experiments.

The approach we have taken has evolved somewhat since we started 2 years ago. Earlier versions of the apparatus were rather limited in capability. Our present (and still evolving) apparatus is able to move objects in 3-D for both mechanical testing and optical sampling. The associated computational and control facility is not operational yet. A 6-degree of freedom force reflecting human interface has been constructed to permit human control of the 3-D positioners for purposes of cell manipulation and cell surgery. Work on 2-degree of freedom micro-grippers to go on the end of the 3-D positioners, for added flexibility in cell manipulation, has only just begun.

Mathematical Tools

The mathematical tools required are those of continuum modeling and system identification. Continuum modeling techniques are required to characterize the muscle cell as a 3-D mechanical structure. System identification methodology must be used to discover the dynamic and probably nonlinear constitutive equations of the mechanics used by the continuum model. We are developing two varieties of continuum model. The first variety utilizes the finite element method[2] and the second involves the theory of series and parallel coupled systems. The finite element method is to be used to model the mechanics of the muscle cell wall which for the present is considered as a continuous membrane. It is not clear yet what experimental method will yield the membrane constitutive equations. The coupled systems theory is to be used to model the interactions among the sarcomeres. Each sarcomere is considered as a separate system which is locally coupled in series to a sarcomere on either side of it. A series of sarcomeres extending (usually) from one end of the muscle cell to the other is termed a myofibril. A whole muscle cell consists of a parallel bundle of myofibrils. The problem is to measure the dynamic interactions among series coupled sarcomeres within the parallel myofibril organization, and then to determine from this the dynamics of the isolated sarcomere (under the assumption that all sarcomeres have similar dynamics).

Two issues arise in connection with these modeling approaches. The first is how to fit structures to 3-D optical measurements. The second is what type of optical information should be acquired. In a sense the second issue is the most important. Fitting structures to inappropriate optical measures is difficult. One approach is to acquire a large amount of optical information (e.g. magnitude, phase, polarization) at each point and then to attempt to find an appropriate transform to enhance the structures of interest. Another approach is to reveal the structures of interest according to their molecular content via, for example, their chemical spectra. Nuclear magnetic resonance imaging of specific isotopes is an example of this. The method we are attempting involves a combination of both of these approaches and involves the use of temporal correlation functions estimated using interferometry.

Correlation function approach

Temporal auto- and cross-correlation functions of light may be estimated using Michelson and Mach-Zehnder interferometers respectively. Specific lags in the correlation functions correspond to particular optical delays in the interferometers. If, $C_{ii}(\tau)$ denotes the auto-correlation function of the light entering (input) an object (determined using a Michelson configuration), and $C_{io}(\tau)$ denotes the cross-correlation function between the light entering (input) an object and the light emerging (output) from an object (determined using a Mach-Zehnder configuration), then $C_{ii}(\tau)$ and $C_{io}(\tau)$ are related by the convolution,

$$C_{io}(\tau) = \int_0^\infty h(\sigma) \, C_{ii}(\tau - \sigma) \, d\sigma$$

For sampled auto- and cross-correlation functions, a sampled version of $h(\tau)$ may be estimated using Toeplitz matrix inversion. If single frequency light is used then (neglecting nonlinearities such as Raman effects) $h(\tau)$ is simply an impulse whose height and delay can be transformed to a give the change in magnitude and phase caused by the object at that particular frequency. For optically non-isotropic objects (such as muscle), $h(\tau)$ is dependent on polarization angle. In this case it is convenient (though not general) to determine correlation functions at two orthogonal incident polarization angles (denoted by ↖ and ↗) (or left and right circular polarizations) and at two orthogonal output polarization components (denoted by ↖ and ↗). This yields 4 correlation functions which in the case of cross-correlation may be denoted

$$C_{io}^{\nwarrow\nwarrow}(\tau) \qquad C_{io}^{\nwarrow\nearrow}(\tau) \qquad C_{io}^{\nearrow\nwarrow}(\tau) \qquad C_{io}^{\nearrow\nearrow}(\tau)$$

and in the case of output auto-correlation,

$$C_{oo}^{\nwarrow\nwarrow}(\tau) \qquad C_{oo}^{\nwarrow\nearrow}(\tau) \qquad C_{oo}^{\nearrow\nwarrow}(\tau) \qquad C_{oo}^{\nearrow\nearrow}(\tau) \; .$$

These 8 correlation functions embody many of the linear and nonlinear optical properties of the object at the particular location (pixel) (of course constrained in the linear case by the input auto-correlation function bandwidth). Fourier transform of the output auto-correlation functions yields polarization-specific Raman spectra.

Construction

The basic mechanical configuration of an earlier version of our apparatus is given in Hunter et al.[3] However this earlier version suffered from a number of deficiencies which has resulted in major changes in our design. We will therefore present the new design in detail.

3-D Positioning System

A typical 2-D scanner moves rapidly in one axis during a line scan. The motor controlling this axis must be fast. The motor associated with the second axis controls movement from one line to the next and so need not be very fast. For example a 2-D scanner constructed by Ichioka et al.[4] uses a voice-coil type motor (fast scan axis) mounted on a stepping motor (slow scan axis). 3-D scans usually involve successive 2-D scans and thus the motor controlling the third axis movements can be quite slow. A 3-D scanner can therefore be configured with 3 similar linear motors mounted in series (i.e. one motor is mounted on another which is in turn mounted on a third). The top most motor controls the first (line scan) axis because it is free to move rapidly. The second motor (second axis) is somewhat limited in speed because it has to overcome the inertia of the first motor. The third motor "sees" the inertias of the first two motors and is consequently much slower. A 3-D scanner having this series configuration could, for example, be constructed by appropriately stacking 3 piezo-electric actuators or 3 optical table motor-driven translation stages.

As mentioned previously one of our objectives is to avoid recording images but rather to treat the original object (muscle cell) as random access memory from which data are retrieved as needed by the model building software. This means that the usual assumption that images may be built up line by line and slice by slice does not apply. Rather we require a 3-D positioning system which is able to move equally rapidly in all axes. Series configurations could be used to achieve this if each motor had an appropriate (but different) power. Indeed almost all commercial multi-axis robots utilize such a configuration. However, given the rather poor power-to-mass ratios of current motors this approach is not practical for high performance positioning. Another completely different motor configuration involves a parallel arrangement of similar motors. In a parallel drive arrangement each motor has its own connection to the point whose position is being controlled. Innovative parallel drive 2-D scanner designs have been developed by a number of groups. [5] [6] [7]

Our apparatus incorporates a pair of 3-D parallel drive micro-positioners. Each micro-positioner consists of 3 actuators connected in parallel and orthogonal to each other. Each actuator actually consists of two motors in series: an electromagnetic linear (voice-coil) motor (Bruel & Kjaer, model 4810) provides large displacements (10nm to 1mm) and a piezo-electric slab mounted on it produces small displacements (1nm to 1μm). In addition to the 3 linear axes provided by the 3 electromagnetic and 3 piezo-electric motors, each micro-positioner incorporates a rotary motor to give a fourth rotary axis (unused here). A quartz glass tube (1mm diameter, 80mm long) extends out from each actuator. The 3 orthogonal tubes meet at a point (apex of a pyramid) (see Figure 1). Simultaneous operation of the 3 actuators results in movement of this point in a 3-D workspace. The coordinate system defined by the 3 actuators is rotated (about 2 axes) with respect to the coordinate system convenient for mechanical testing and optical measurements. Thus even a simple translation in the latter coordinate system involves control of all three actuators.

An arbitrary transformation between two 3-D coordinate systems involves a 4 by 4 matrix. If the coordinate system local to the 3-D positioners has axes denoted by a, b and c, and if the coordinate system of the apparatus has axes denoted by x, y and z, then the transformation from a point in x,y,z space to a point in a,b,c space is given by,

$$\begin{bmatrix} a \\ b \\ c \\ 1 \end{bmatrix} = \begin{bmatrix} 3^{-0.5} & 0 & 2(6^{-0.5}) & 0 \\ 3^{-0.5} & -2^{-0.5} & -6^{-0.5} & 0 \\ 3^{-0.5} & 2^{-0.5} & -6^{-0.5} & 0 \\ 0 & 0 & 0 & 1 \end{bmatrix} \begin{bmatrix} x \\ y \\ z \\ 1 \end{bmatrix} \quad \text{right micro-positioner}$$

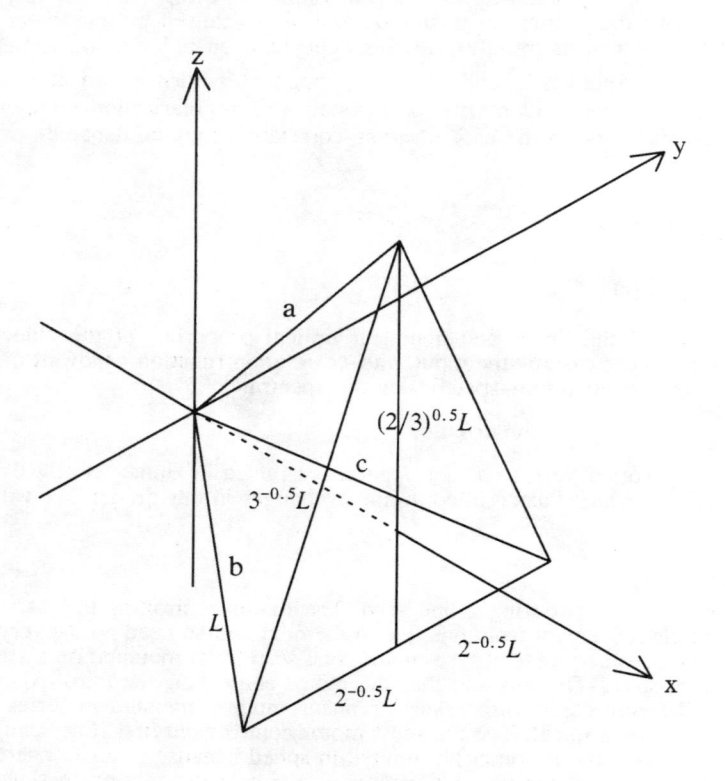

Figure 1. Coordinate systems of the SLM (x,y,z) and the right 3-axis micro-positioner (a,b,c).

$$\begin{bmatrix} a \\ b \\ c \\ 1 \end{bmatrix} = \begin{bmatrix} -3^{-0.5} & 0 & 2(6^{-0.5}) & 0 \\ -3^{-0.5} & -2^{-0.5} & -6^{-0.5} & 0 \\ -3^{-0.5} & 2^{-0.5} & -6^{-0.5} & 0 \\ 0 & 0 & 0 & 1 \end{bmatrix} \begin{bmatrix} x \\ y \\ z \\ 1 \end{bmatrix} \qquad \text{left micro-positioner}$$

These particular transformation matrices are associated with micro-positioner orientations which minimize interference with the optical paths (see below). It should be noted that more complicated transformations are required if bending of the quartz tubes must be taken into account. At present the effect of bending is small but eventually will be accounted for when the parallel computation and control unit is operational.

The apparatus is mounted on an optical table (Newport model M-GS-34-ST) to provide some passive low-pass filtering of building vibration. Unfortunately we have found that the resulting isolation is unsatisfactory at frequencies below 15Hz. An active high-pass system involving custom built low frequency accelerometers and servo-controlled mass has been constructed to alleviate the problem but requires further development. A lead-felt-copper-aluminum housing may be lowered onto the table to completely enclose the apparatus for purposes of sound, electrostatic and electromagnetic attenuation (all uncalibrated). Air may be pumped through the enclosure via dust and moisture filters. The temperature and humidity of the enclosure and surround is actively controlled (Liebert model Challenger 2) to within 1°C and 1% respectively. However we have found that the temperature control is inadequate (despite a thermal mass design approach) for determining the correlation functions to the required accuracy and must be improved by at least an order of magnitude. All of the electronics associated with the apparatus is custom built and in order to get low noise levels is run entirely from storage batteries (6mΩ impedance, deep cycle Lead-acid).

Optical Configuration

The optical configuration is shown in Figure 2. A water cooled Argon ion laser (Lexel, model 75) provides the illumination. The laser is wavelength tunable (7 lines from 457.9nm to 528.7nm) and may be amplitude modulated (2 to 300 mW, 0 to 1kHz). A spatial low-pass filter (18mm focal length objective, 10μm pin hole, 177mm focal length lens) expands (to 10mm diameter) the vertically polarized beam. The beam then enters a polarization angle rotator which can flip the polarization angle by plus or minus 45°. The beam is then amplitude split. One beam (object beam) traverses the sample via two similar objectives (Olympus, model IC100, 1.25 N.A.). The other (reference) beam traverses two more similar objectives followed by a vertically oriented polarizer, then, in the Mach-Zehnder mode (see below), combines with the object beam. In the Michelson configuration (see below) the reference beam is blocked and the object beam combines with itself. The combined beam then traverses a focusing lens (177 mm focal length) after which it is split by a polarizing beam splitter into two orthogonal

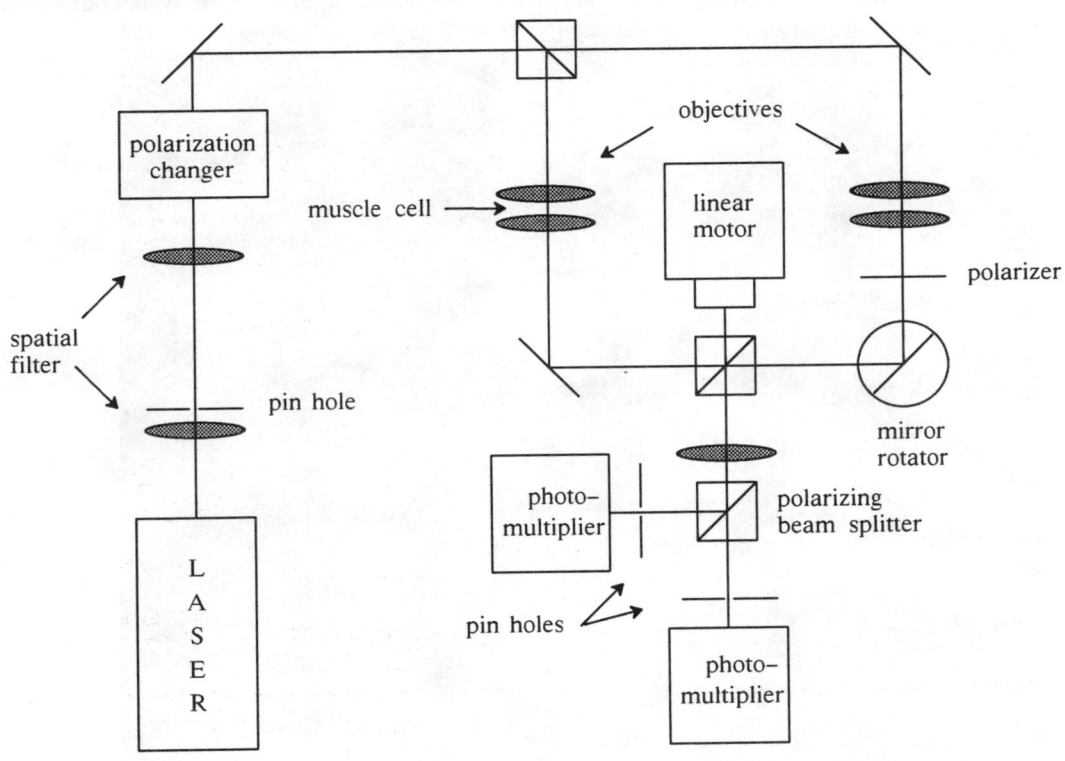

Figure 2. Optical configuration of the scanning laser microscope.

components. Each component beam then proceeds to a photomultiplier tube (Hamamatsu model 1P21) preceded by a pinhole (10μm diameter). The optical delay between object and reference beams (or object beam with itself) may be manipulated plus and minus 6mm for purposes of measuring the temporal auto- (Michelson) or cross- (Mach-Zehnder) correlation functions (see above). The actuator used to vary the delay is similar to those (i.e. piezo-electric disc mounted on linear electromagnetic motor) used in the 3-D positioners. When the 488nm wavelength is used a band-reject filter (Omega Optical, Raman notch model 488RB) centered at 488nm may be inserted to attenuate (10^6) the laser line which dominates the autocorrelation function.

Cell growth facilities

The muscle cells are grown, using cell-culture techniques, from muscle tissue dissected from frog (*Xenopus Laevis*) embryos. The cells begin as a 30μm diameter sphere which over a period of a few days grows into a muscle fiber (e.g. 150μm long, 20 μm wide). The cells grow in a clear solution containing a blend of various amino acids, vitamins, and inorganic salts. The cells will attach to thin layers of collagen (tendon). Our current objective is to get single cells to grow into the apparatus by first having the cells attach to 2 collagen-coated probes (one on each side of the cell) controlled by the 3-D positioners, and then servo-controlling the positioners to pull on each end of the cell with a constant tension. The "cell chamber" is formed by the 740μm gap between the 2 objectives (see Figure 2) and is filled with the cell-growth solution. A diamond slab projecting into the "cell chamber" is attached to a temperature controller.

Computation and control

A cluster of dedicated MicroVAX-II computers (0.1 million floating point operations per second (Mflops)) are used at present to control the apparatus. The combined computational power of these computers is 10^3 to 10^4 too small for effective control of the apparatus during model building. An attempt to use loosely coupled 32-bit floating point digital signal processing (DSP) chips (using AT&Ts 8 Mflops DSP-32) to control the apparatus was abandoned in favor of a tightly coupled (multiple-data multiple-instruction) parallel configuration (using Texas Instruments 33 Mflops TMS320C30) capable of both substantial computation and control. The parallel computational and control unit 8 is nearing completion and will when fully populated (32 processors) provide the required performance.

Analysis

As mentioned above our orientation is to avoid image acquisition where possible. The analysis approach centers on the use of continuum modeling (principally the finite element method) coupled with system identification techniques. In the usual use of the finite element method the constitutive equations are known or assumed. In our case these equations are not known and must be discovered via system identification techniques. We have developed the software for both finite element modeling (nonlinear, time-varying, 3-D) and system identification (nonlinear, time-varying).[9] A tensor-on-manifold data representation implemented in Common-LISP has been developed to promote a common data structure and syntax for continuum

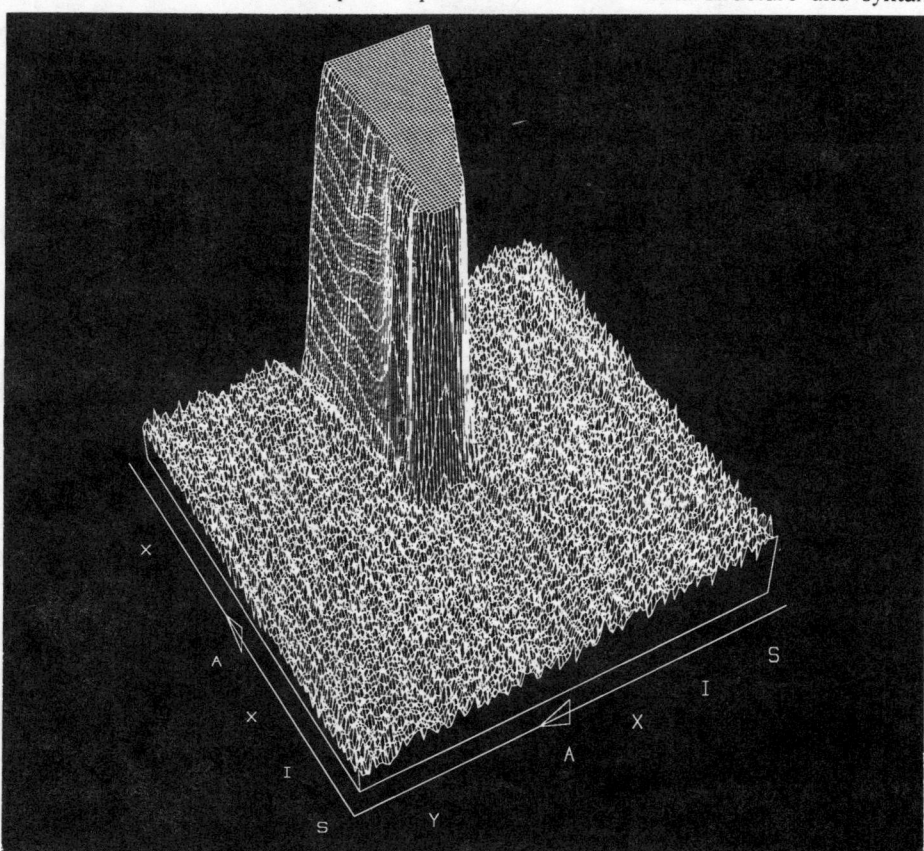

Figure 3. A 64μm by 64μm scan in the x-y plane of the end of a fine steel probe.

modeling, system identification and image processing. However the marriage of these methodologies for use in our muscle cell studies is not yet at a usable stage.

Human Interface

A 6-degrees of freedom force-reflecting human interface has been constructed to enable human control over the pair of 3-D micro-positioners. This interface consists of a pair of 3-D macro-positioners configured in the same way as the 3-D micro-positioners. Each hand controls one of the 3-D micro-positioners via the computer from one of the 3-D macro-positioners. Displacements of the macro-positioners are scaled down as much as 10^8. Forces experienced by the micro-positioners (e.g. drag forces, cell wall resistance) may be reflected back to the motors in the macro-positioners to provide the human with a scaled (up to 10^8) force sensation. This is principally to provide "tactile" feedback during cell manipulation.

Evaluation

Figure 3 shows a $\pm 32\mu m$ x-y axis scan (cross-section) of the end of a fine stainless-steel probe. Intensity (vertical polarization component) is plotted vertically and increases towards the bottom of the plot. Figure 4 shows the tip of the same probe at a 16 times higher magnification ($\pm 2\mu m$ x-y axis scan). In this figure the x- and y-axes have been rotated by 180°. The multiplicative nature of the optical noise is evident in both Figures 3 and 4 which show raw data.

Figure 5 shows intensity step responses obtained by making $4\mu m$ x-axis scans about the middle of the object shown in Figure 4. The left and right halves of Figure 4 correspond to scans made in opposing directions. Each curve in the figure was obtained at a different z-axis (optical axis) position (vertical axis on plot) about the "in focus" position. The rapid roll-off in spatial frequency bandwidth either side of the "in focus" position is clear from this figure.

Comparison of measured and theoretical intensity step response functions

For a coherent confocal transmission microscope the detected light intensity may be written for even point-spread functions as

$$I(x_s, y_s) = |h_1 h_2 \otimes t|^2 \qquad (1)$$

where h_1 and h_2 are the point-spread functions of the two objectives, 5 and t is the object transmittance. For a lens with a circular aperture the point-spread function is given by

$$h(v) = -jN\exp(-jkf)\ \exp\left(\frac{-jv^2}{4N}\right)\left(\frac{2J_1(v)}{v}\right) \qquad (2)$$

where $v = kr\sin(\alpha)$ (where $\sin(\alpha)$ is the numerical aperture of the lens)
$k = 2\pi/\lambda$ (where λ is the laser wavelength)
$N = \pi a^2/\lambda f$
f is the focal length of the lens
a is the lens radius
r is the distance from the optical axis
J_1 is a Bessel function of the first kind of order one.

For line structures where the object transmittance only varies along the x-axis the intensity can be expressed as

$$I(x_s) = \left| \int_{-\infty}^{\infty} g(x_0)\ t(x_s - x_0)\ dx_0 \right|^2 \qquad (3)$$

where $g(x_0)$ is the integral along the y-axis of the point-spread function and assuming identical objectives is given by

$$g(x_0) = \int_{-\infty}^{\infty} h^2(x_0, y_0)\ dy_0 \qquad (4)$$

In Equation 2 the term $-j\exp(-jkf)$ represents a constant phase and can be ignored for intensity calculations. Furthermore given that the Fresnel number of the objectives used here is large, the phase variations are negligible for a large range of v. Therefore the line-spread function can be simplified to

$$g(x_0) = 4N^2 \int_{-\infty}^{\infty} \frac{J_1^2[k\sin\alpha\ (x_0^2 + y_0^2)^{0.5}]}{k^2 \sin^2\alpha\ (x_0^2 + y_0^2)}\ dy_0 \qquad (5)$$

For a step change in intensity Equation 3 simplifies to

$$I(x_s) = \left| \int_{-\infty}^{x_s} g(x_0)\ dx_0 \right|^2 \qquad (6)$$

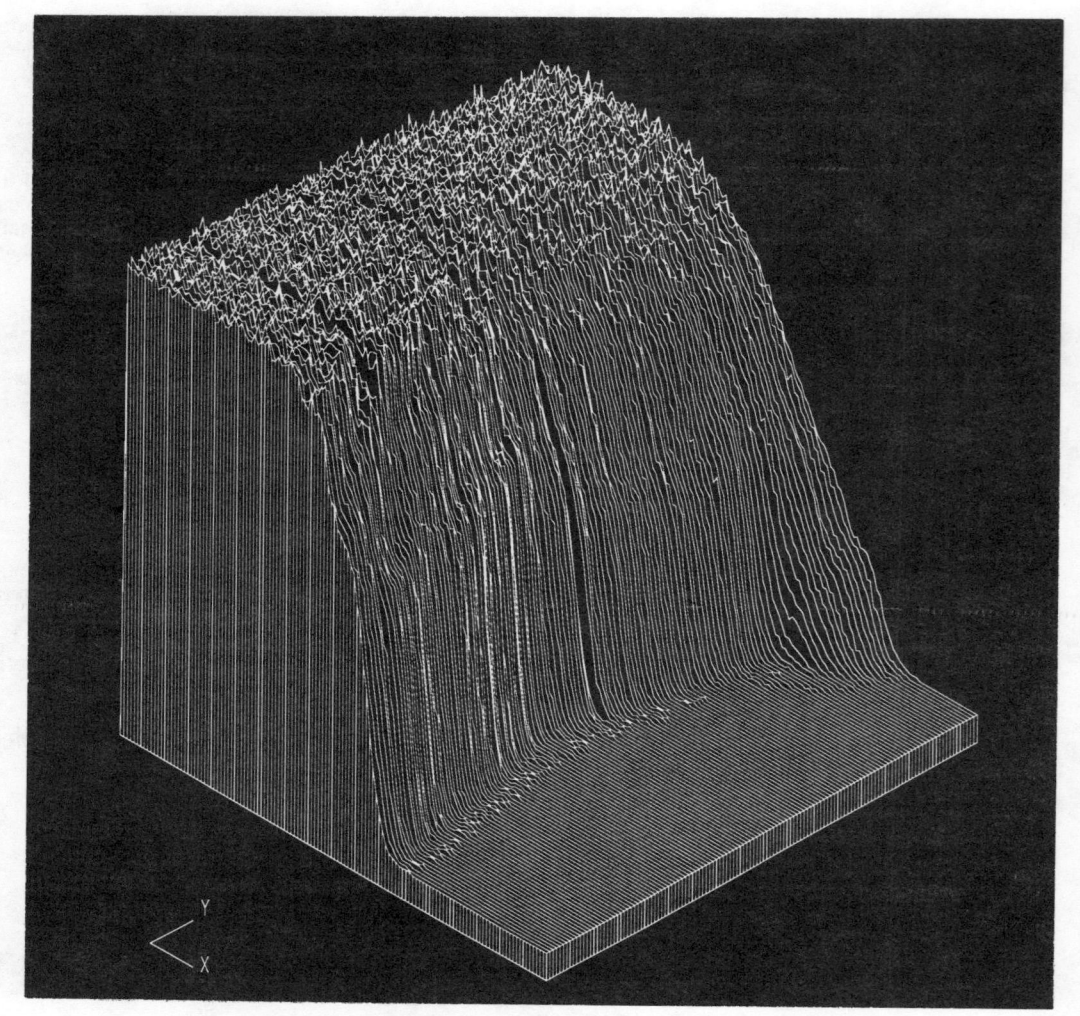

Figure 4. A 4μm by 4μm scan in the x-y plane of the tip of the probe shown in Fig 3.

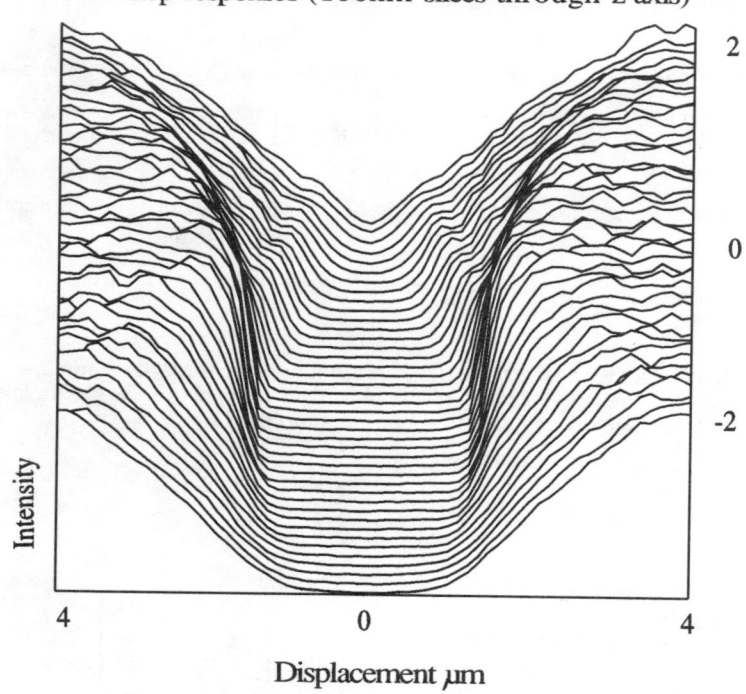

SLM step responses (100nm slices through z-axis)

Figure 5. 4μm scans in the x-axis at different z-axis positions (100nm steps) about the "in focus" position (z=0nm).

Intensity step response (488nm 1.25 NA)

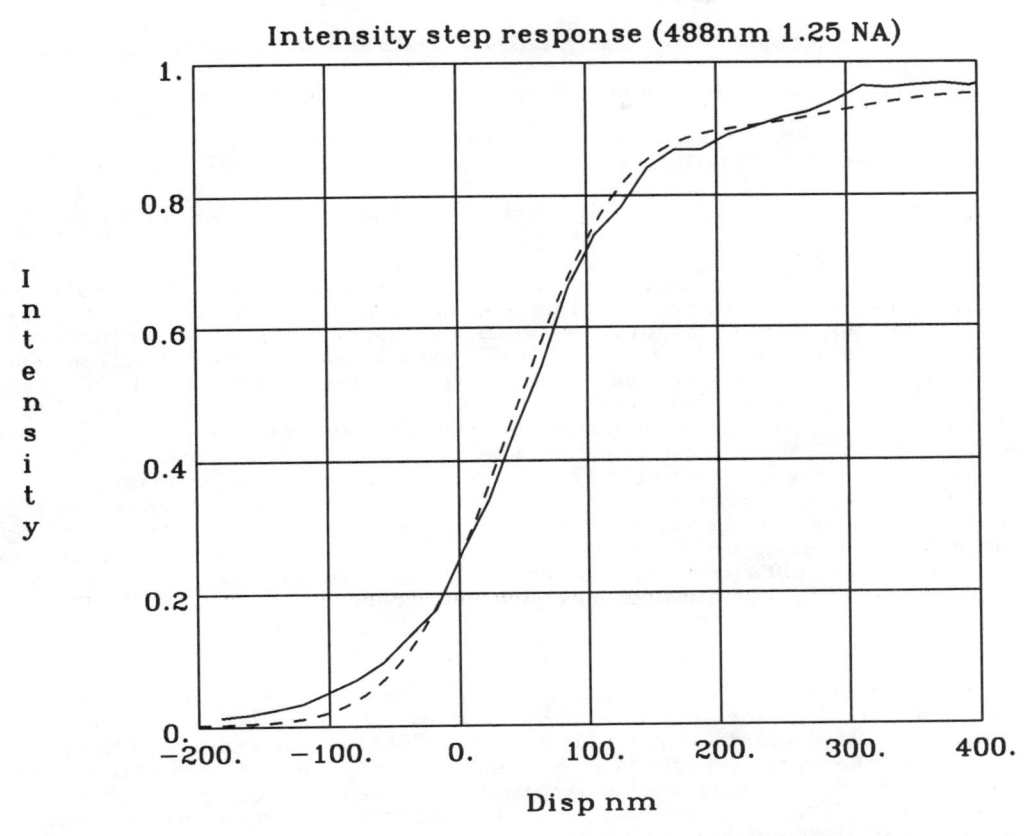

Figure 6. Comparison of corrected measured (solid line) and theoretical (dashed line) intensity step responses.

Equations 5 and 6 were evaluated numerically. The resulting intensity step response (given the laser wavelength and numerical aperture of the objectives) is plotted in Figure 6 (dashed line). Figure 6 also shows a measured "in focus" intensity step response (raw data taken from Figure 5) corrected (displacement * 0.67) for the non-step (ramp) object used.

Acknowledgments

We wish to thank the Medical Research Council (MRC) of Canada and the Natural Sciences and Engineering Research Council (NSERC) of Canada for supporting this research with operating and equipment grants and with personal support to IWH as an MRC Scholar and an NSERC International Scientific Exchange Award to PJH. We would also like to thank the Canadian Institute for Advanced Research (CIAR) for supporting IWH as a CIAR Scholar. The TMS320C30 software donation to us by Texas Instruments is gratefully acknowledged.

References

1. Hunter, I.W. and Korenberg, M.J. The identification of nonlinear biological systems: Wiener and Hammerstein cascade models. Biological Cybernetics, 55, 135-144 (1986).
2. Davies, A.J. The finite element method: a first approach. Clarendon Press, Oxford (1980).
3. Hunter, I.W., Lafontaine, S. and Hunter, P. A scanning laser microscope for muscle fiber studies. Proc. SPIE Scanning Imaging Technology, 809, 144-150 (1987).
4. Ichioka, Y., Kobayashi, T., Kitagawa, H., and Suzuki, T. Digital scanning laser microscope. Applied Optics, 24, 691-696 (1985).
5. Wilson, T. and Sheppard, C.J.R. Theory and practice of scanning optical microscopy. Academic Press, London (1984).
6. van der Voort, H.T.M., Brakenhoff, G.J., Valkenburg, J.A.C. and Nanninga, N. Design and use of a computer controlled confocal microscope for biological applications. Scanning, 7, 66-78 (1985).
7. Dixon, A.J., Doe, N. and Pang, T-M. Industrial applications of confocal scanning optical microscopy. Proc. SPIE Scanning Imaging Technology, 809, 36-43 (1987).
8. Martel, S., Hunter, I.W., Nielsen, P., Lafontaine, S. and Kearney, R.E. A parallel supercomputer for biomedical analysis and control. Proceedings of the 14th Canadian Medical and Biological Engineering Conference, 14, 43-44 (1988).
9. Hunter, I.W. and Kearney, R.E. NEXUS: a computer language for physiological systems and signal analysis. Computers in Biology and Medicine, 14, 385-401 (1984).

Laser scanning microscopy to study molecular transport in single cells

Manfred Scholz, Heinrich Sauer, Hans-Peter Rihs, and Reiner Peters

Max-Planck-Institut für Biophysik,
Kennedyallee 70, 6000 Frankfurt 70, Federal Republic of Germany

ABSTRACT

Laser scanning microscopy was used to study dynamic processes in single cells. The laser scanning microscope of Heidelberg Instruments was complemented with a 1W krypton laser and a microinjection set-up. Simple algorithms were worked out which permit determination of local and integrated intensities in fluorescence scans. In one application the lateral diffusion of macromolecules was studied. The krypton laser was used to irreversibly photolyse fluorescently labeled dextrans in small volumes of a thin fluid layer; equilibration of the local fluorescence inhomogeneity by lateral diffusion was followed by repetitive scanning. In a second application the permeability of single red blood cell membranes which had been exposed to the complement cascade was studied. In a further application artificial nuclear proteins, constructed by molecular genetic methods, were injected into the cytoplasm of hepatoma cells. The kinetics of protein transport from cytoplasm to nucleus were derived from fluorescence scans. In all applications a good agreement between results obtained by laser scanning microscopy and those obtained independently by other methods and instruments was observed.

1. INTRODUCTION

Laser scanning microscopy has inaugurated a new era in light microscopy. High resolution in both the focal plane and the direction of the optical axis, high sensitivity by photometric registration, and the digital format of the reconstructed image are just a few of many outstanding characteristics (for review, see Wilson and Sheppard[1]). So far, the method has been mainly used to image non-biological or fixed biological specimen. Our group is primarily interested, however, in the spatio-temporal dynamics of macromolecules in living cells and for that purpose has previously introduced[2] an analytical laser microscopic method, fluorescence microphotolysis (for review, see Peters[3]). Laser scanning microscopy and fluorescence microphotolysis have, at least from the technical point of view, many analogies. This prompted us to explore possibilities of combining the two methods and, more generally, of using laser scanning microscopy in quantitative studies of dynamic processes in single living cells.

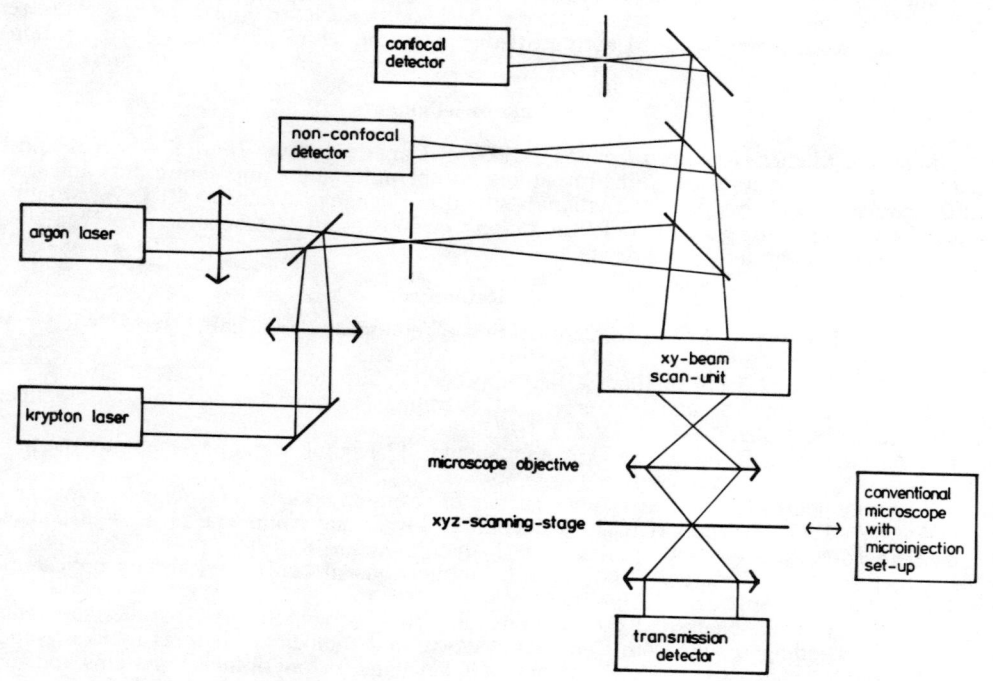

Figure 1. Scheme of the laser scanning microscope.

2. METHODS

The laser scanning microscope (LSM) of Heidelberg Instruments (Heidelberg, Federal Republic of Germany) was chosen because the system has an open configuration in which both hardware and software are readily accessible to modifications. The LSM was complemented with a 1W krypton laser as indicated in Figure 1. Furthermore, a microinjection set-up was added. This consisted of a conventional microscope equipped with a micropipette system for pressure injection of single cells. The conventional microscope was mounted on the optical table of the LSM. A motorized x-y-z-stage was employed to automatically transfer the specimen between the objective lenses of conventional microscope and LSM (accuracy of positioning better than 5 micrometers). Living cells were injected and imaged in Petri dishes using a water-immersion objective lens (numerical apperture 1.0). LSM scans of macromolecular solutions or erythrocyte ghosts were made with a 40-fold or a 100-fold oil-immersion objective lens.

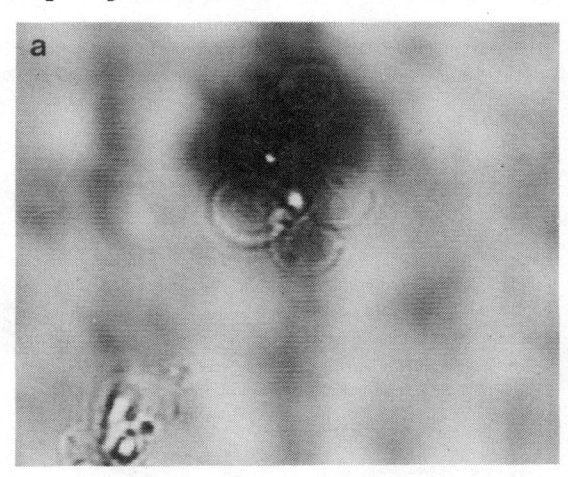

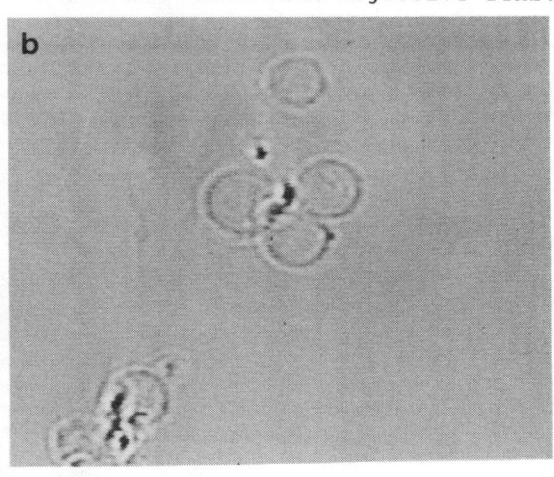

Figure 2. Through-light LSM scan of erythrocyte ghosts. a) Raw image. b) DFC-image (see text).

Some specimen such as erythrocyte ghosts were found to exhibit low contrast in through-light LSM scans. The image of these cells was severely blurred by background noise. Image quality could be largely improved, however, by the differential focal contrast (DFC) method illustrated in Figure 2. The DFC-method involves two scans of the same specimen - one exactly focussed, the other slightly defocussed. In the defocussed scan the difference of local background intensity from the global average is determined. This difference is then substraced from the in-focus scan.

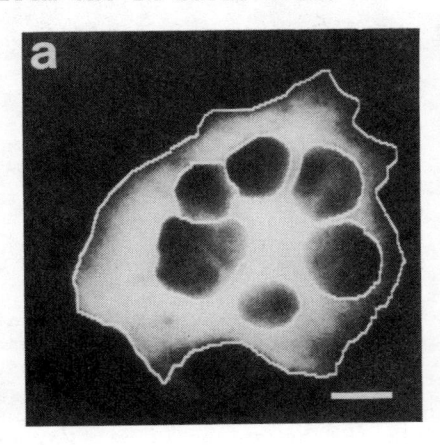

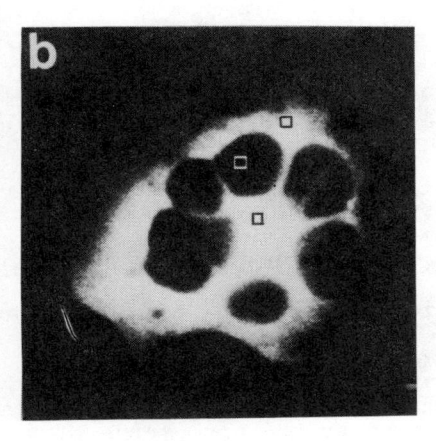

Figure 3. Quantitation of fluorescence in LSM scans. The fluorescence x-y-scan of a living HTC cell is shown which was injected with a fluorescently labeled 70-kD-dextran. a) Determination of integral fluorescence of cytoplasm and nuclei. b) Determination of local fluorescence.

For quantitative purposes an algorithm was developed which integrates over selectable areas of LSM scans. In Figure 3a, for instance, the boundaries of a cell and of its nuclei were delineated. The algorithm determines the relative integrated fluorescence within these boundaries and thus yields an approximate value of the total amount of the fluorescent compound in cytoplasm and nucleus. In Figure 3b small areas were marked to obtain an approximate value of local concentration.

3. LATERAL DIFFUSION

Principles employed in studies of molecular mobility are illustrated in Figure 4. In the displayed case a fluorescently labeled dextran of 150 kD molecular mass was dissolved in a viscous medium (30% wt/vol Ficoll 400, Pharmacia, Upsala, Sweden) and the solution was spread as an approximately 1 micrometer thick layer between a slide and a cover slip. Fluorescence scans were taken before and at various times after local photolysis by the krypton laser beam. We have yet to complete the algorithms by which scans can be evaluated in terms of the appropriate transport coefficients. However, as a first step software was written which generates intensity conture maps. A preliminary evaluation of such maps suggested that lateral diffusion coefficients can be correctly determined. The combination of laser scanning microscopy with fluorescence microphotolysis, here referred to as laser scanning micophotolysis, has far reaching perspectives, for instance for the study of anisotropic molecular motion in biological and artificial membranes. A pretentious goal is the 3-dimensional reconstruction of molecular transport in single cells at sub-micrometer resolution.

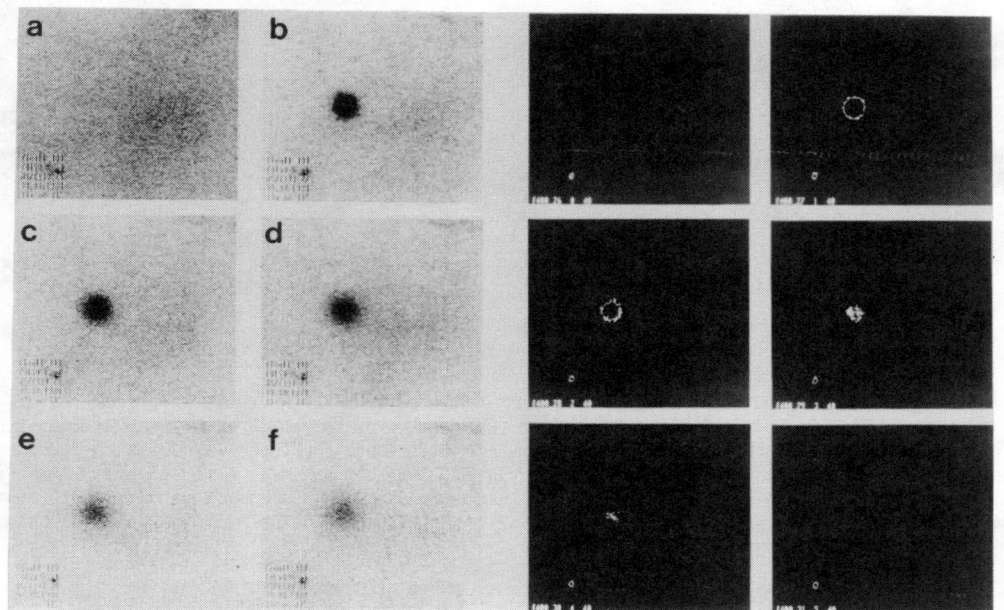

Figure 4. Lateral diffusion of a fluorescently labeled dextran in a viscous solution. Left hand side: Fluorescence x-y-scans of a thin fluid layer before (a) and 0.6 s (b), 3.2 s (c), 5.8 s (d), 8.4 s (e), 11.0 s (f) after photolysis, respectively. Right hand side: Intensity conture map of the fluorescence scans. Only one conture, corresponding 0.5 of the pre-photolysis intensity was drawn.

4. COMPLEMENT TITRATION OF ERYTHROCYTE GHOSTS

Proteins which insert into cell membranes to form transmembrane pores are widespread weapons in cellular warfare[4]. The classical example is the complement system. It involves about 20 different proteins. These interact in a complex reaction cascade which eventually leads to the formation of a large pore in the plasma membrane of the attacked cell and to cell death by osmotic lysis. In erythrocyte suspensions complement-induced lysis can be conveniently followed by the appearence of hemoglobin in the supernatant. For cells which do not contain large amounts of soluble chromoproteins no general method to follow complement lysis was available. We developed such a method which is based on the ability the LSM to generate thin optical sections.

As a first step the spatial resolution of optical sectioning by the LSM was tested. For this purpose sheep erythrocyte ghosts (i.e. isolated hemoglobin-depleted red blood cell membranes) were suspended in the solution of a fluorescently labeled 10-kD-dextran. A sample was prepared by spreading a small volume of the suspension between a glass slide and a cover slip. The specimen was inspected at high resolution (100-fold objective lens). Figure 5 displays a fluorescence x-z-scan of such a specimen. It may be recalled that an x-z-scan is a vertical optical section of the object. It is reconstructed from a series of line scans in the image (x-y-) plane during which the object is shifted along the optical (z-) axis. In the case of Figure 5 the z-increment was generated by a piezo element and amounted to 0.197 micrometer. In Figure 5 the glass-water interface and a ghost attached to it is seen. The ghost is impermeable for the fluorescent dextran. It can be recognized that the ghost lost its physiological biconcave shape and became a sphere. From the uncertainty in the position of the glass-water interface the resolution in z-direction can be estimated to be about 0.5 micrometer.

Figure 5. Vertical section (fluorescence x-z-scan) through an erythrocyte ghost suspended in a solution of a fluorescently labeled 10 kD dextran. The ghost is attached to the glass-water interface. Marker represent 10 micrometer.

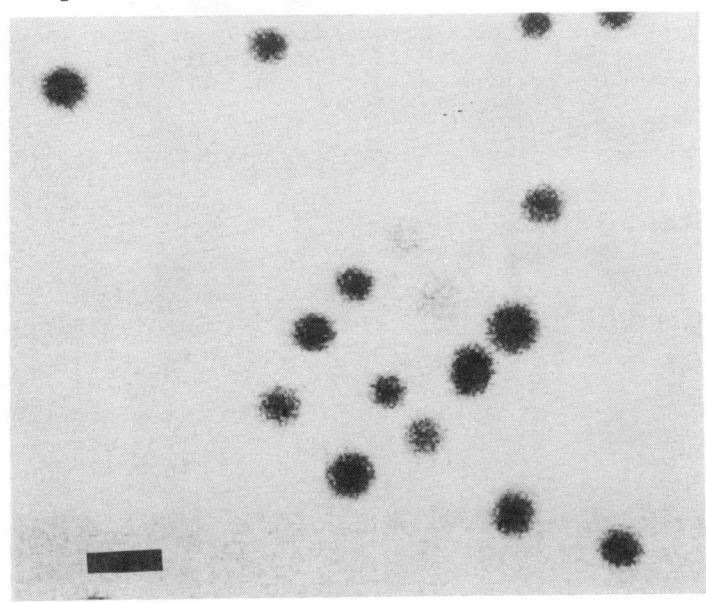

Figure 6. X-y-scan of a suspension of erythrocyte ghosts in a solution of a fluorescently labeled 10-kD-dextran. Marker represent 10 micrometer.

Our LSM-method for obtaining complement titration curves was worked out with erythrocyte ghosts but should be of general applicability. Resealed ghosts were prepared from sensitized sheep erythrocytes. Ghosts were preincubated with human serum depleted of complement component C8. Ghosts were titrated with pure C8 and mixed with solutions of fluorescein-labeled dextrans. Samples were inspected in the LSM and a pair of scans in the through-light and fluorescence mode, respectively, was taken. The through-light scan, especially after improvement by the DFC-method (see above and Figure 2), permitted to identify ghosts regardless of their permeability for the employed dextran. In the fluorescence x-y-scan (Figure 6) impermeable ghosts appeared as dark spots against a bright background whereas permeable ghosts had an intensity approaching that of the background. These relations were quantified by intensity measurements. Histograms such as those displayed in Figure 7 revealed that the intensity of impermeable ghosts was about 0.2-0.4 relative to background intensity. Ghosts which had been made permeable for the dextran by a high C8 concentration had a relative intensity of about 0.6-1.0. This difference was sufficient to discriminate between permeable and impermeable ghosts and to determine their fraction in a mixed population. Titration curves were constructed by determining the fractions of impermeable and permeable ghosts at different C8 concentrations.

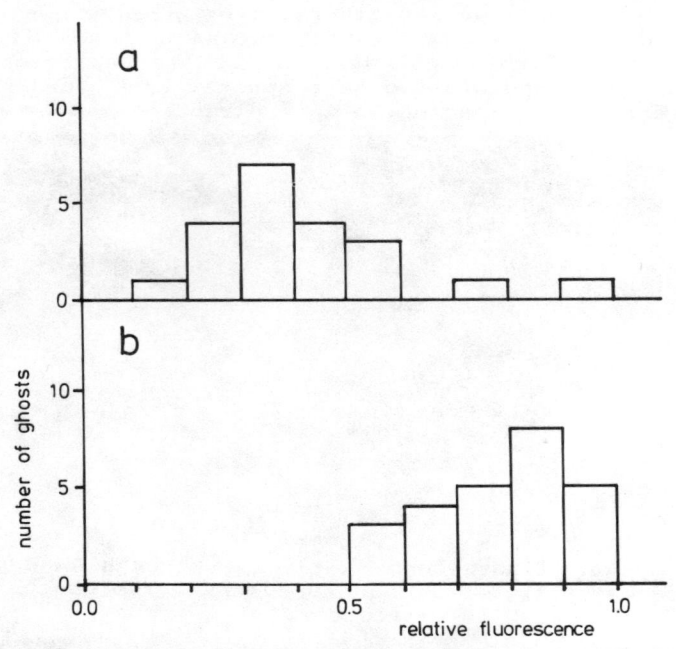

Figure 7. Intensity distribution in fluorescence x-y-scans as shown in Figure 6. Abszissa: Intensity of ghosts normalized to that of the solution. Ordinate: Number of ghosts. a) Distribution in absence of complement component C8. b) Distribution at a high C8 concentration.

Table 1. Kinetics of Nucleocytoplasmic Protein Transport as Derived from a Quantitation of Fluorescence Scans[a)]

Time [b)] (minutes)	Integrated Fluorescence (% of total cellular fluorescence) Cytoplasm	Nucleus	$F_{n/c}$ [c)]
I. Hybrid protein P4 containing wild-type nuclear localization sequence			
2.5	67	33	0.4
5.0	54	46	1.0
10.0	38	62	4.1
15.0	30	70	5.2
30.0	14	86	16.0
II. Hybrid protein P4 containing mutated nuclear localization sequence			
2.5	80	20	0.4
5.0	78	22	0.4
10.0	77	23	0.4
15.0	78	22	0.4
30.0	77	23	0.4

a) Values are mean of three measurements. Temperature was 20°C. b) Time after injection of the hybrid protein into the cytoplasm of HTC cells. c) $F_{n/c}$, nucleocytoplasmic fluorescence ratio derived from local intensities in the nucleus and a cytoplasmic region close to the nucleus.

5. NUCLEOCYTOPLASMIC TRANSPORT

In eukaryotic cells genetic information and protein biosynthesis are localized in different compartments, nucleus and cytoplasm. Therefore, an intensive and precisely scheduled exchange of matter between these compartments is required. The nuclear envelope has properties of a molecular sieve with a functional pore radius of about 50 Å. Thus, anorganic ions, metabolites and small proteins can pass between cytoplasm and nucleus by diffusion. However, the nucleocytoplasmic transport of macromolecules which exceed in size the functional pore radius requires specific transport mechanisms (for review, see Peters[3]).

Recently it has been discovered[5] that transport of large proteins from cytoplasm to nucleus is mediated by well-defined "signals" which consist of short streches of amino acid residues. Among such nuclear localization sequences (NLS) that of a viral protein, the T-antigen of the SV-40 virus, has been studied in greatest detail. The NLS of the T-antigen comprises residues 126-132 and consists of predominantly basic amino acids, namely Pro_Lys_Lys_Lys_Arg_Lys_Val. Residue 128 (lysine) plays a particulary important role because its exchange by a threonine or almost any other amino acid completely abolishes the targeting activity of the sequence. In order to further analyse functional aspects of the NLS we have generated a series of hybrid proteins which consists of T-antigen fragments and a large protein moiety, the E.coli β-galactosidase (Rihs and Peters, unpublished results). The proteins were constructed on the DNA level, expressed in E.coli, purified by affinity chromatography, labeled fluorescently, and injected into the cytoplasm of HTC polykaryons.

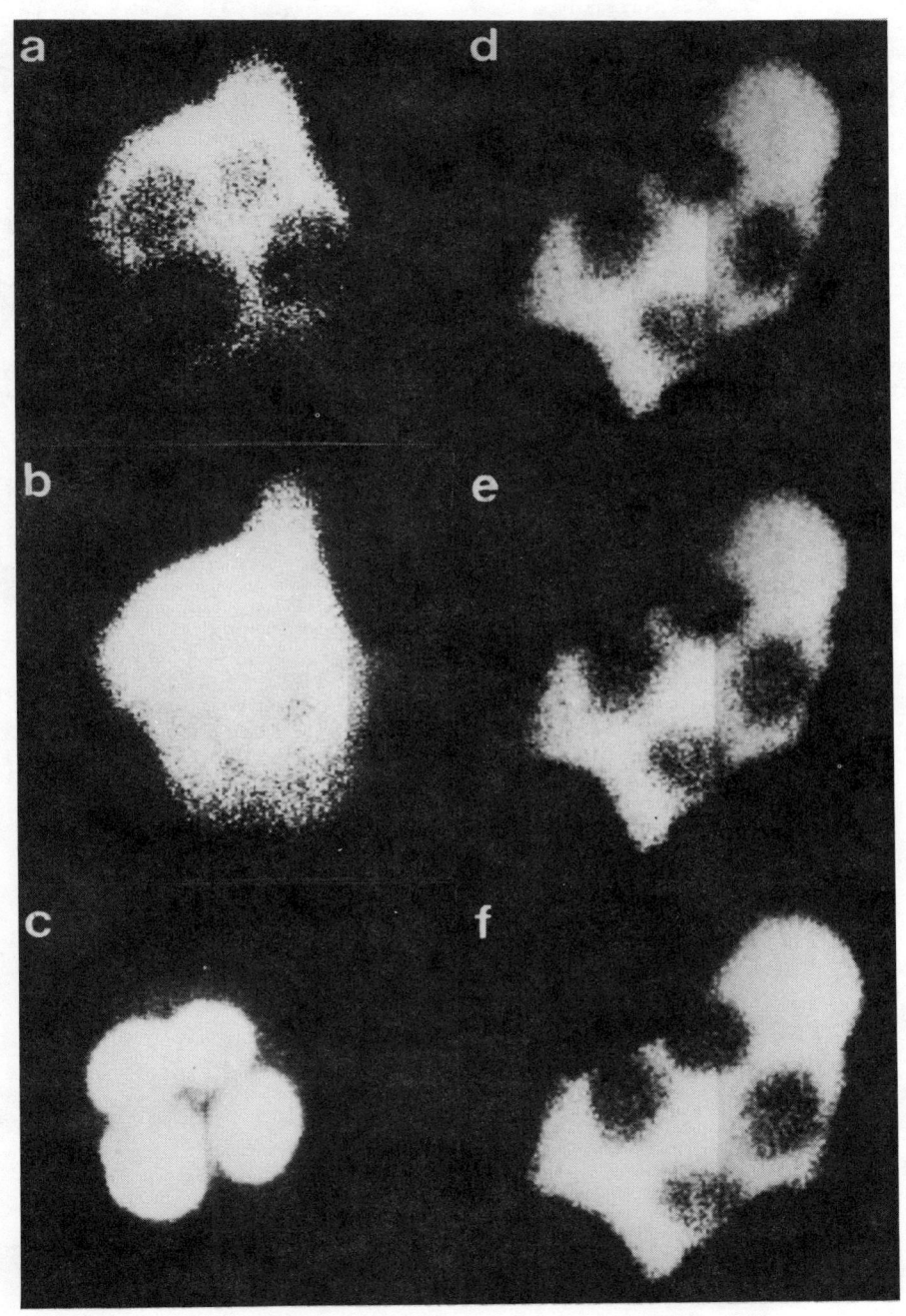

Figure 8. Nuclear targeting of a gentechnologically generated protein. Protein P4 containing the nuclear localization sequence of the SV-40 T-antigen was fluorescently labeled and injected into the cytoplasm of HTC cells. Fluorescence x-y-scans were taken 2.5 minutes (a,d), 5 minutes (b,e), and 30 minutes (c,f), respectively, after injection. a-c) Variant containing the wild-type nuclear localization sequence. d-f) Variant containing mutated sequence (threonine in position 128).

Figure 8 shows scans of HTC polykaryons injected with the hybrid protein P4. This protein contains T-antigen residues 113-135. If the NLS had its wild-type primary structure (i.e. a lysine in position 128) P4 was rapidly transported from cytoplasm to nuclei (Figure 8a-c). However, if the NLS was mutated to containing a threonine residue in position 128 the hybrid protein remained in the cytoplasm (Figure 8d-f).

The kinetics of nucleocytoplasmic transport were derived from sequences of fluorescence scans. Employing the method illustrated in Figure 3 both the integral fluorescence of cytoplasm and nuclei and the local fluorescence in small nuclear and cytoplasmic, perinuclear regions was determined. The results (Table 1) agree with the qualtitative impression given by Figure 8. Taking furthermore into account that the ratio of nuclear to cytoplasmic volume is 1/2 - 1/3 in HTC cells[6] the values of integral and local fluorescence are perfectly compatible with each other. Furthermore, kinetics of nucleocytoplasmic transport determined independently by laser microfluorimetry (Rihs and Peters, unpublished results) agreed with those obtained by laser scanning microscopy.

6. CONCLUSION AND PERSPECTIVES

Our results have shown that laser scanning microscopy can be used for quantitative purposes. It was therefore possible to merge the superior imaging qualities of the LSM with analytical methods such as fluorescence microphotolysis. We assume that combinations of this type will be of great value for cell biology and other disciplines.

7. ACKNOWLEDGEMENT

We should like to thank Mrs. Claudia Grosse-Johannböcke for excellent technical assistance. Support by the Deutsche Forschungsgemeinschaft is greatfully acknowledged.

8. REFERENCES

1. T. Wilson and C. Sheppard, Theory and Practice of Scanning Optical Microscopy, pp. 1-213, Academic Press, London (1984).
2. R. Peters, J. Peters, K.H. Tews, and W. Bähr, "A microfluorimetric study of translational diffusion in erythrocyte membranes," Biochim. Biophys. Acta 367, 282-294 (1974).
3. R. Peters, "Fluorescence microphotolysis to measure nucleocytoplasmic transport and intracellular mobility," Biochim. Biophys. Acta 864, 305-359 (1986).
4. S. Bhakdi and J. Tranum-Jensen, "Damage to mammalian cells by proteins that form transmembrane pores," Rev. Physiol. Biochem. Pharmacol., 107, 147-223 (1987).
5. D. Kalderon, W.D. Richardson, A.F. Markham, and A.E. Smith, "Sequence requirements for nuclear location of simian virus 40 large-T antigen," Nature (Lond.) 311, 33-38 (1984).
6. I. Lang, M. Scholz, and R. Peters, "Molecular mobility and nucleocytoplasmic flux in hepatoma cells," J. Cell Biol. 102, 1183-1193 (1986).

Confocal Fluorescence Microscopy of Epithelial Cells

Ernst H.K. Stelzer[*] and Robert Bacallao[+]

European Molecular Biology Laboratory (EMBL), [*]Physical Instr. Programme [+]Cell Biology Programme
Meyerhofstraße 1, Postfach 10.2209, D-6900 Heidelberg, West-Germany (FRG)

ABSTRACT

The investigation of epithelial cells with a confocal fluorescence microscope is only useful as long as the three dimensional structure of the sample is preserved. For many reasons it is not guaranteed that this goal can be achieved with fixed specimens. In order to perform relevant biological experiments one must find ways avoiding the "loss of information."

RESEARCH ON MDCK CELLS

Epithelial Madin-Darbey Canine Kidney (MDCK) cells have been the subject of intensive studies in the cell biology programme at the European Molecular Biology Laboratory (EMBL) for many years (Simons & Fuller, 1985). Grown on non-transparent filters these cells express their polarized nature strongly. Due to their thickness (15 - 18 µm) the out of focus contributions when observed in a conventional fluorescence microscope after labelling the sample with a fluorescent antibody are, in most cases, relatively strong and do not allow a detailed observation. The cells have, however, been observed by cell biologists in the electron microscope and have been the subject of experimental studies by biochemists (Hansson et al., 1986). During widespread investigations, in which several groups at the EMBL participated since 1984, the confocal fluorescence microscope has proven to be the ideal tool for investigating these cells (van Meer et al., 1987). The research interests at the EMBL include: distribution of filamentous actin, distribution of microtubules, transcytosis, distribution of glycoproteins and glycolipids and colocalization experiments.

SAMPLE PREPARATION

The steps through which one has to go during a sample preparation that can be used in a confocal fluorescence microscope may be sketched as follows: a) In the first step the scientist picks a preparation method from the literature that seems suitable for an experiment. This method is verified with a conventional fluorescence microscope, modified and verified again until the scientist has the image he is interested in. The quality criterium will therefore be morphological. b) The confocal fluorescence microscope comes along in the next step as a new method to check the structural quality of the preparation. This check will force another modification of the preparation method which again will be verified first in the conventional and then in the confocal fluorescence microscope. The scientist will have to go back as far as step one. The quality criterium may now be different, but it is still based on morphological data. c) This procedure will lead to a completely new preparation method which will be used to prepare the samples that are finally the subject of a quantitation experiment with a confocal fluorescence microscope. The quantitation experiment should be seen in contrast to the qualitative control with a confocal fluorescence microscope. While the latter case is supposed to return "good images" (from a biologists point of view) the first case is used for evaluation. d) The final step is the quantitation with the confocal fluorescence microscope which leads to a visualization of the three dimensional fluorescence intensity distribution (e.g. stereo pairs, x/z-images, extended views, titled views), to a three dimensional reconstruction (test against a three dimensional density distribution) or to a quantitative analysis as a function of the spatial (in three dimensions) distribution of the fluorescence intensity.

The new preparation procedure will also become a better starting point when a new question is raised. However, it will still be necessary to go through the cycle indicated above since one relevant biological paper should answer several questions it will become necessary to develop more than one procedure.

QUALITY CRITERIA

Most of the criteria that will evaluate the quality of a preparation procedure will be based on images recorded on a conventional fluorescence microscope. The idea here is of course to have a high contrast, i.e. ideally the structure of interest will be bright and the rest dark. These criteria will bias all labelling procedures towards samples that finally have a high contrast. While this goal may be achieved if the three dimensional structure is preserved it is very unlikely in epithelial cells. It may, however, be achieved under different circumstances: a) Whenever the cells shrink or collapse the difference in height decreases. The structures will spread and become thinner. Even a small depth of field could now visualize the whole sample. b) The target may also be present only in parts of the cell because it has been removed or because the dye labels the target in the cell unevenly due to its limited penetration capabilities. In these cases the source of the out of focus contributions has been removed.

An important problem with epithelial cells is that they are quite thick when they are grown on filters. When the structural information contained in densely packed areas that are present throughout the whole cell it can be impossible to resolve any details with a conventional fluorescence microscope. The filters not only reflect the excitation light they also tend to bind some of the antibodies. This is a problem when one tries to introduce the antibodies not only from the apical, but also from the basal side of the cell. On the other hand growing these cells on glass reduces their height and since the volume remains the same the structures of interest (e.g. microtubules) are spread laterally thus reducing the need for a small depth of field. These conditions are, however, unphysiological and should be avoided. Especially since the microtubules are organized differently in cells grown on glass (Bacallao unpublished results).

AVOIDING ARTEFACTS

Good preparation procedures return results that resemble the three dimensional in vivo structure of the cell. The simplest way to achieve this is to work with in vivo systems wherever that is possible. The other way is to find means of verifying that the structure has not changed. The confocal fluorescence microscope is one way to accomplish this. Another way is to measure the height of the living cell by focusing up and down in a conventional transmission microscope and to compare this data with that measured after the sample has gone through the preparation procedure. Ultrastructural investigations and cryo-sections may also become independent methods of verification. Standard embedding material such as Gelvatol and Moviol cause shrinkage of the cells (Bacallao et al., in prep.) while Glycerole with e.g. 50cell volume. Finally, the presure of the cover glass on the sample that may cause damage mechanically can be avoided by using spacers (e.g. drops of nail polish).

COLOCALIZATION EXPERIMENTS

Combinations of two (or more) labelling procedures are even more complicated since four factors influence the number of useful specimens: a) the quality of the first label, b) the quality of the second label, c) the preservation of the structure of the first target and d) preservation of the structure of the second target. One reason why the goal of accurate preservation is difficult to achieve is due to variations in the density distributions of the reagents used in the preparation of the sample. Another reason is that cells may grow differently according to where they have been seeded on the filter or on the cover slip. Verification has to be performed on two dyes and this can become a very time consuming task.

Another important problem is that different preparation procedures are used for each target. To label the membrane of the Golgi complex **and** the microtubules may be a conflicting goal if one would also like to preserve their location in space! This is understandable since many of the preparation procedures are quite drastic and remove e.g. almost all membranes in a cell so that an optimal access to microtubules is guaranteed.

CONCLUSIONS

Several conclusions can be drawn for the practical work with confocal fluorescence microscopes: 1) Have an excellent conventional fluorescence microscope available. Standard modern microscopes are excellent and reliable devices. Lenses with a high numerical aperture (more than 1.25) already provide a small depth of field which is usually sufficient for most purposes. Additionally, images recorded with this microscope provide a standard against which images derived from a confocal fluorescence microscope may be tested. 2) Use the confocal fluorescence microscope only when the third dimension is of any importance to the experiments. 3) Establish new quality criteria. These criteria may be based on different methods and should ask if the fixed sample resembles the structure of the cell as it was in vivo. The consequence may be that a lot of the established procedures turn out to be of no use for confocal fluorescence microscopy. 4) Develop preparation procedures that are designed to preserve the three dimensional structure of the cells. 5) Critically evaluate established preparation procedures as soon as colocalization experiments are to be performed.

The basic problem is that the criteria for structural quality will be based on morphologically derived data while the results that are derived from these experiments should lead to conclusions concerning a functionality.

It is also interesting to note that the confocal fluorescence microscope provides a path on which deliberate violations in terms of unphysiological conditions can be avoided.

In the long term, new experiments that investigate the three dimensional structures of cells must aim at using in vivo systems. Only these guarantee that the organelles, the cytoskeleton and the membranes are still in place. Such systems, however, require very fast confocal fluorescence microscopes, i.e. devices that work at a speed that is comparable to the speed with which one uses a conventional fluorescence microscope.

REFERENCES

1. R. Bacallao and E.H.K. Stelzer, "Preservation of biological specimens for observation in a confocal fluorescence microscope and operational principles of confocal fluorescence microscopy," in Methods in Cell Biology, A.M. Tartakoff, ed., (to appear) Academic Press, Orlando, Florida, USA (1989).
2. R. Bacallao, et al. (manuscript in preparation).
3. G.C. Hansson, K. Simons and G. van Meer, "Two strains of the Madin Darbey Canine Kidney (MDCK) cell line have distinct glycolipid compositions," EMBO J. 5, 483-489 (1986).
4. K. Simons and S. Fuller, "Cell surface polarity in epithelia," Annu. Rev. Cell Biol. 1, 243-288 (1985).
5. G. van Meer, E.H.K. Stelzer, R.W. Wijnaendts-van-Resandt and K. Simons, "Sorting of sphingolipids in epithelial (MDCK) cells," J. Cell Biol. 105, 1623-1635 (1987).

Inverted confocal microscopy for biological and material applications

*G.J. Brakenhoff, **R.W. Wijnaendts-van-Resandt, **J. Engelhardt, **W. Knebel and
and *H.T.M. van der Voort

*Department of Molecular Cell Biology, Section of Molecular Cytology,
University of Amsterdam, Pl. Muidergracht 14, 1018 TV Amsterdam, The Netherlands

**Heidelberg Instruments GMBH
Im Neuenheimer Feld 518, D-6900 Heidelberg, West Germany

ABSTRACT

Aspects of the inverted versus top view configuration for confocal microscopy are examined with the inverted-type clearly offering advantages for high-resolution imaging of certain types of biological specimen. The role and possibilities of the use of pinholes of variable size in confocal microscopy is discussed and data are presented for combinations of objectives and various sizes of pinholes.

1. INTRODUCTION

Confocal microscopy is becoming a well established technique for the investigation of the three-dimensional structure in biological and industrial materials. (Brakenhoff et al 1985, 1988, Carlsson, 1985, Wijnaendts van Resandt 1985, Steltzer, 1987). The basis for this success is the optical sectioning capability of this type of microscopy, which enables one to study the 3-D structure of intact specimens in their natural environment. The principle of confocal microscopy has been described before (Sheppard et al 1977, Wilson et al 1984). For first demonstrations of the improved imaging see Brakenhoff (1979) and optical sectioning Brakenhoff (1980) and Wijnaendts van Resandt (1985).

As in conventional microscopy one can realize confocal microscopy either in the normal, top-view configuration or in the inverted one. We examine some aspects of these approaches from the point of view of optical conditions for imaging and biological experimental possibilities. Essential parts of a confocal microscope are the illumination and detection pinholes. We show that if a microscope is equipped with a pinhole of variable diameter the sectioning power and the dimension of the optical sampling volume can be optimalized for specific requirements. Some data on the sectioning power are presented for various combinations of detector pinhole sizes and objectives.

2. INVERTED VERSUS TOP-VIEW APPROACH FOR CONFOCAL MICROSCOPY

Although in principle all the potentialities of confocal imaging can be realized in both modes, we think that for many specimens of biological and industrial origin, the inverted design offers a number of advantages. These stem from a combination of often occurring conditions in practical specimens and the optical demands for high resolution confocal microscopy. The latter requires that the optical path during image formation conforms as closely as possible to the design parameters of the used confocal objective lens(es). For confocal microscopy in biological specimens there are two basic factors which influence the 3-D depth imaging of the microscope. The first is the difference of the refractive index from the design value along the length of the optical path traversed in the specimen by the probing beam. The wave-front distortion caused by this effect is in the first order comparable to the one associated with spherical aberration (Brakenhoff, 1980). The second factor is that the confocal probe forming wave-fronts will be affected by scattering, absorbtion and refractive index irregularities in the specimen before reaching the probing depth. Both factors will affect the imaging to a greater degree as the depth of the specimen probed increases. It is therefore desirable to have the areas of greatest interest in the specimen near the glass interface on which the specimen is supported.

A typical high resolution oil immersion objective is designed for operation with the light path fully in a medium with a refractive index of 1.515. Part of the distance between objective an object is traversed in the cover glass (thickness 170 µm) and part in the immersion oil present in the working distance (typically 50-90 µm) between the objective and cover glass. For confocal microscopy optimal conditions are realized if such a thin 170 µm glass is used as support glass for the object, and the object is examined in the inverted mode. Then the first 10 to 20 µm above the support glass provide near ideal confocal imaging conditions with increasing degradation, depending on object, at larger penetration distances. This confocal depth range is usually sufficient for satisfactory examination of many biological objects like cells, cell cultures, tissues etc. This should be compared with the situation in the top-view method. There the specimen is lying on a 1 mm thick support glass and is then covered by the 170 µm thin glass through wich the object is viewed. The distance of this glass from the support glass will depend on the surface topology of the object and may be appreciable. Then areas of interest in the specimen are imaged through an unknown, and often rather long, length of optical path through a medium of differing refractive index.

A very important aspect of the inverted approach is accessibility of the specimen, supported on its 170 µm glass, from above. Basically the specimen does not have to be covered from above while being imaged from below. This means that during the confocal imaging stains can be added, chemical circumstances changed etc. Also in adapted culture chambers with thin glass bottoms the time development of live specimens can be followed with minimal interference of the confocal imaging process. This configuration can also be very well combined with micromanipulation techniques.

3. THE USE OF VARIABLE ILLUMINATION AND DETECTION PINHOLES

Confocal imaging can be looked upon as the interaction of the object with the product of the 3-D illumination an detection sensitivity distribution functions coinciding in the specimen. The shape of these distributions is determined by the optics used and the size of the used pinholes. The influence of the detection pinhole diameter on imaging has been investigated by Carlini et al (1987) and van der Voort et al (1988). The general trend is that with increasing size of the detection pinhole, first the lateral resolution falls off before the axial one. At such large pinhole sizes that the lateral resolution has become comparable to the one in conventional microscopy, still an appreciable degree of optical sectioning is present, which is then governed by geometrical arguments. Below data are presented on the sectioning power of some objectives. If a laser light source is available operating 100% in the TEM 00 mode than, strictly speaking, no illumination pinhole needs to be used. With good beam expansion optics a very satisfactory illumination wave front can be obtained and excellent use can be made of the available laser power.

Specific possibilities arise for confocal imaging if pinholes are used in the illumination and detection paths of which the size can be varied under control of the instrument. We specifically think of pinholes of which the diameter can be varied continuously from 5 µm up to 500 µm (Brakenhoff, to be published). These have the size ranges needed for placement at the image plane of objectives of 160 mm working distance. Such an optical set-up - or infinity corrected objectives with field lenses of limited long focal length - leads to a compact and stable confocal arrangement.

At the illumination side such a variable objective positioned in the illuminating laser beam can be used for controlling through diffraction, the divergence angle of the laser beam passing the pinhole. If we start with a low divergence laser beam, illuminating just a part of the entrance pupil of the confocal lens, we can by reducing the size of the illumination pinhole increase the filling degree of the confocal objective lens pupil from this starting point up to completely filled. This is equivalent to being able to change the Numerical Aperture (N.A.) of the illumination under instrument control. The N.A. of the illumination determines the lateral and axial FWHM's of the illumination distribution in the object through the relations, $FWHM_{lat} = 0.61 \lambda /N.A.$ and $FWHM_{axial} = 2\lambda /(N.A.)^2$ respectively (as adapted from (Born and Wolf). This optical arrangement for the illumination has the interesting characteristic that by varying the size of the pinhole, the dimensions of the illumination distribution can be changed but the illumination intensity in the optical probe stays in the first approximation constant. This is because the reduction of power passing the illumination pinhole (proportional to the b_i^2 with b_i the illumination pinhole diameter) is compensated by the concentration of this radiation into a specimen area smaller in proportion to the used NA squared. The capability of changing the optical probe dimensions can be employed in confocal microscopy for adapting the optical confocal sample size to the sampling density of the data collecting system, i.e. at low magnifications large size samples may be used etc. Beyond the point where the confocal objective is filled, a further reduction in the pinhole size will result in a fall-off of the illumination intensity at the specimen proportional to $(1/d_i)^4$, because both the power of the beam passing the pinhole and the fraction intercepted by the pupil are proportional to $(1/b_i)^2$. This provides a very effective control over the confocal illumination intensity by way of the pinhole diameter. Other methods of accomplishing this, like gray filters or polarizers are optically less attractive. Controlling the laser power directly often results in a noisy beam at low powers, as in many laser systems the beam power is only stabilized to an absolute and not a relative fluctuation level.

The detection pinhole can be chosen so that specific detection conditions are realized. Two ranges can be distinguished. In the first range where the filling factor of the pupil controls the illumination spot size in the object, we can match for optimal collection efficiency the width of the detection distribution to the one of the illumination. It is important to note first that then for detection the full aperture of the lens is used, thus assuring optimal signal collection efficiency, and second, that the widening in the lateral detection sensitivity distribution is for _geometrical_ reasons.

It can be shown that optimal lateral match of the pupil-controlled illumination distribution and the geometrically-controlled detection distribution occur when $d'_i = d'_d$ with d_d the detection pinhole diameter and the accent indicating the value of the diameters projected in the confocal spot in the specimen. For optimal axial match it can be shown that approximately the following relation has to be satisfied: $d'_d = 4\ 2\ tg\ (d'_i)^2/\lambda$ Operational conditions like the balance sought between lateral and axial resolution and the structure in the specimen investigated will determine the optimal setting under practical conditions.

In the second range, where the illumination pupil is filled, varying the detection pinhole can be used for balancing confocal resolution and sectioning power against signal collection efficiency as indicated by Carlini et al (1987) and van der Voort (1988).

4. PRACTICAL CONFOCAL AXIAL RESPONSES

We measured the axial response in fluorescence confocal microscopy for a number of different combinations of pinholes and objectives. The method is basically the same as used by Wijnaendts-van-Resandt et al (1985), where the optical sectioning power was determined as the FWHM of the differentiated confocal step-response to a fluorescent FITC solution in ethanol, situated above a glass interface. The possible influence of photo-bleaching could be eliminated by the evaluation of repeated measurements at various power levels. While the basic shape of the responses is in accordance with the Born and Wolff theory as used before (Wijnaendts-van-Resandt, 1985), we find some unexpected differences in the actual response width at small pinhole sizes. Table I shows the results of our experiments. For each objective used, the confocal setting of the pinholes was checked and if necessary optimalised. The measurements were done in a beam-scanning confocal microscope (off-axis) at relatively small scan amplitudes, i.e. using the center 10% of the field-of-view of the used lenses. Notable in the results is that there is clearly an optimal size (indicated with a *) beyond which a further increase of the pinhole diameter dramatically reduces the optical sectioning capability. This result is in accordance with the findings of Carlini et al (1987), van der Voort (1988). It is not clear why there is the difference in sectioning power at small apertures for both the 1.4 objectives and the relatively bad value for the 40x ,1.3 objective. At these small pinholes the sectioning power is diffraction limited and should be essentially the same for all three high power objectives. We have the impression that the objectives may be corrected to varying degrees of perfection and that this is the cause of the differences.

pinhole size (μm) (in image plane) at 160 mm	Objective			
	100x 1.4	63x 1.4	40x 1.3	25x 0.75
15	0.60	0.82	1.10	2.2*
35	0.62	0.87	1.37*	3.2
50	0.63	0.98*	1.75	
70	0.67*	1.18	2.52	5.3
200	1.65	2.24	3.95	11.8

Table I. The confocal optical sectioning power in μm of various objective/pinhole combinations.

5. CONCLUSIONS AND COMMENTS

The above considerations concerning the application of the inverted configuration to confocal microscopes and the use of the variable size pinholes is equally true for the on-axis case with object scanning (where beam is stationary on-axis) as well as for the off-axis one (where the beam is scanned). The variable size pinholes, discussed above, can be, in principle, directly controlled by the computer system operating the microscope. Then a system can be envisaged where, depending on scan amplitude and sampling frequency, the optimal optical sample size is automatically chosen. An added benefit is then that, as shown above, the beam intensity at various optical sample sizes is constant, which may be useful if specimen bleaching in fluorescence plays a role. Also other schemes can be contemplated for the use of the variable pinholes for automatic control of the detection signal and the signal-to-noise ratio in various circumstances. Above we have contributed some ideas how variable pinholes may be used in conjunction with each other to obtain optimal imaging in confocal microscopy under various circumstances. However as the measured data on real-life objective lenses indicate, the actual imaging of specific objectives has to be taken into account before these ideas are put into practice.

6. REFERENCES

1. G.J. Brakenhoff, P. Blom and P. Barends, J. of Micros. $\underline{117}$, 219-232 (1979).
2. G.J. Brakenhoff, J.S. Binnerts and C.L. Woldringh, In Scanned Image Microscopy ed. E.A. Ash. Academic Press, London (1980).
3. G.J. Brakenhoff, H.T.M. van der Voort, E.A. van Spronsen, W.A.M. Linnemans and N. Nanninga, Nature $\underline{317}$, 748-749 (1985).
4. G.J. Brakenhoff, H.T.M. van der Voort, E.A. van Spronsen and N. Nanninga, Scanning Microscopy, $\underline{2}$, 33-40 (1988).
5. A. Carlini and T. Wilson, Proc. SPIE 809, 97-100 (1987).
6. K. Carlsson, P.E. Danielson, R. Lenz, A. Liljeborg, L. Maylöf and N. Ashlund, Opt. Lett. $\underline{10}$, 53-55 (1985).
7. C.J.R. Sheppard and A. Choudbury, Optica $\underline{24}$, 1051 (1977).
8. E.A.K. Steltzer and R.W. Wijnaendts van Resandt, Proc. SPIE 809, 130-137 (1987).
9. H.T.M. van der Voort and G.J. Brakenhoff, Proc. SPIE, this volume (1988).
10. R.W. Wijnaendts van Resandt, J. Micros. $\underline{138}$, 29-34 (1985).
11. T. Wilson and C.J.R. Sheppard, Theory and Practice of Scanning Optical Microscopy, Academic Press, London (1984).

Applications of the microscope system LSM

Hans-Georg Kapitza, Volker Wilke

Carl Zeiss, D-7082 Oberkochen, West Germany

ABSTRACT

The new universal confocal LSM is a second-generation laser scanning microscope. This means, that laser scanning microscopy now made the transition from experimental set-up lab types to integrated workstations, where the manual handling of mechanical and optical components is left to the computer. The built-in microcomputer - now not only drives scanners and transforms signals into images but also controls directly the microscope functions. It turned out that this is a crucial step for making the LSM an universal instrument for widespread use in research and development. The switching from conventional microscopy to laser scanning modes and vice versa is performed by simply pressing keys. Not only images can be stored on the built-in hard disk but at the same time automatically the corresponding set of parameters: Even weeks or months after creating an image the settings of the instrument belonging to this image can be called from the operators panel by loading a parameter file which defines the laser line used and its intensity setting, nosepiece position, zoom factor, averaging conditions, microscopy mode (transmitted, reflected or fluorescence) and parameters for signal conditioning. Since the microscope stand is motorized at a high degree, the computer recreates automatically the exact conditions desired after dialing the number of the parameter file. In this way working with the LSM becomes not only reproducible, but also the user is freed from the handling of mechanical parts and typing commands on a keyboard. Finally the automatized LSM allows true remote control by a host computer necessary for the most demanding 3D-reconstruction. The characteristics pointed out so far are prerequisites for the daily use by microscopists in life science, semiconductor research, development and testing and materials research.[1-3]

1. DESCRIPTION OF THE MICROSCOPE SYSTEM LSM

The newly designed large research microscope stand of the LSM is prepared to apply any light microscopy technique at the same high level of performance. The optics are of the new ICS- (Infinity Color-corrected System) type. Scan optics are designed especially for this infinity type objective lenses. In this way without realignment one can use lenses ranging from 1.25 x up to 100 x on the motorized nosepiece. The optics in the detection paths are designed to give optimal throughput. Especially in fluorescence applications this is very impor-

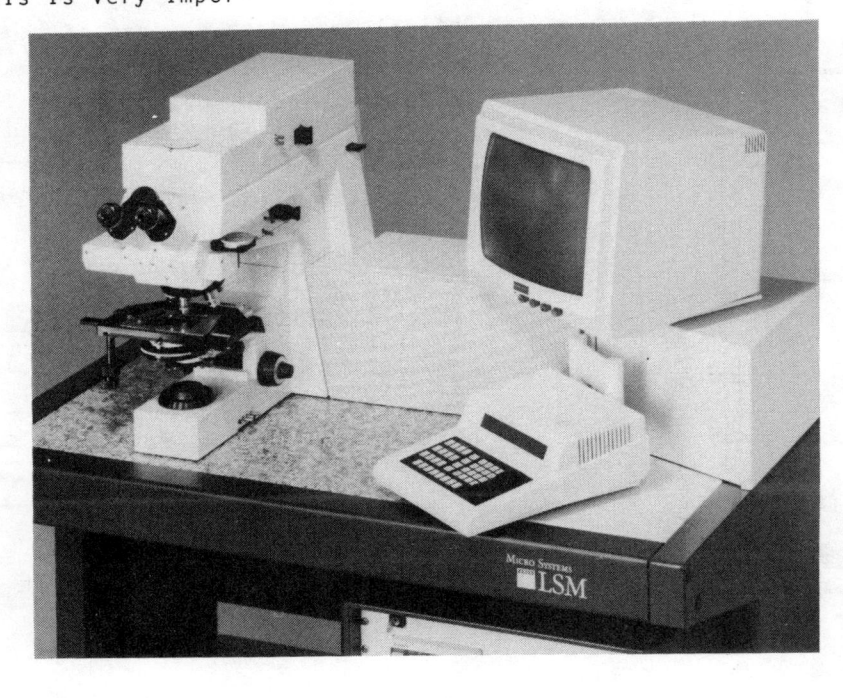

Figure 1.

The new microscope system LSM.

tant because the detection limit is set by the design of the detection channel rather than by laser power because of the bleaching of fluorophores. The stand furthermore is universal because external sensors (e. g. PMT), TV-cameras or external lasers can easily be adapted. A Helium-Neon-Laser (633 nm) is always built into the stand and provides a reference beam for external lasers. The external laser might be an Argon-Laser (488 nm / 514 nm), an Infrared Helium-Neon-Laser (1152 nm), or a laser provided by the user. For the 633 nm laser line an acousto-optical modulator can be integrated. All lasers used are aligned to the reference Helium-Neon-Laser and since the scan optics is chromatically corrected it delivers scan images of perfect raster coincidence at different laser lines. The scanning is performed by servo-controlled galvanometer scanners providing a scan-zoom factor of 1:10 without changing lenses. The whole optical system including the external lasers mentioned above is mounted to a common granite base plate giving protection against vibrations and providing long-term stability. The operation panel displays a permanent read-out of parameters and the images generated are observed using a RGB-monitor. The computer is controlled by the 80286 CPU and contains 20 Mbyte on hard-disk plus 720 KB on floppy-disk. Two serial interfaces RS 232 C and one parallel IEEE are standard. For motorized z-axis control (as an option) multiples of 0.05 µm can be selected for each step and series of images can be taken automatically. Also for motorized x- and y-control a scan stage can be added providing steps down to 0.25 µm. Positions in all directions are read automatically by the computer due to optical encoders used for each axis.

A videoprinter for quick documentation and and a 35 mm photo-camera taking high-quality images from a flat-screen, high resolution monitor are included. It should be mentioned that the LSM can be operated as a fully equipped research microscope in the "conventional" mode. This is a very helpful feature for the user, because one can connect the various light microscopy techniques to the results obtained in the laser scan mode.

2. APPLICATIONS

Laser scanning microscopy applications have been established mainly in three different fields:

A. Life sciences

B. Semiconductor Research, -Development, Testing

C. Materials research

Naturally the number of applications is permanently increasing because of the rising awareness of researchers about the possibilities of laser scanning microscopy.

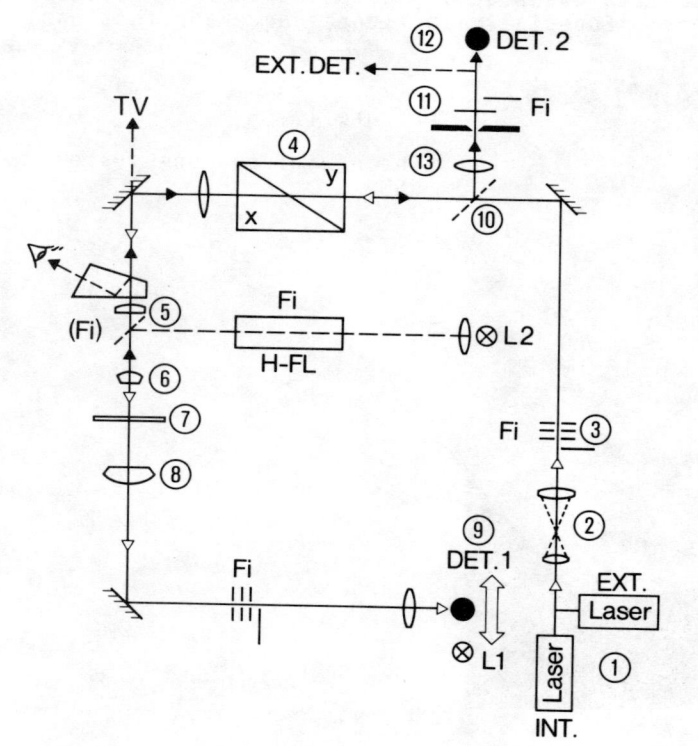

Figure 2.

Optical beam path diagram of the LSM:
1. Lasers
2. Beam expander optics
3. Motorized laser light attennuator
4. X-Y-Scanner unit
5. Tube lens
6. Objective lens
7. Specimen/stage
8. Substage condensor
9. Transmitted light detector
10. Beam splitter
11. Barrier filters
12. Reflected light/epifluorescence detector
13. Confocal spatial filter

L1 Transmitted light source,
L2 Incident light source for conventional microscopy.
Fi Filters
H-Fl Incident light illuminator

LIFE SCIENCE

In the life sciences, namely Cell Biology, Cytochemistry, Neurology, Genetics and Biophysics, at present fluorescence applications prevail. The quantitative fluorescence imaging benefits from the intrinsic high resolution seen even at low label concentration. The reason is, that in laser scanning microscopy the laser spot size always defines resolution. Therefore there is no degradation of images as observed usually with TV intensifier cameras. Also intensity shading and center-to-corner loss of resolution is unknown with the LSM. A distinct advantage in fluorescence images at very low optical magnification arises, because a clear nonshaded fluorescence image up to one centimeter in square can be created with one scan. With this approach for example large sections can be examined within seconds.

Multiple fluorescence imaging is easily performed with the LSM using 488 nm, 514 nm and 633 nm excitation and observing the corresponding emissions. Common fluorophores like Fluorescein, Rhodamine, Texas Red, Propidium Iodide, Acridine Orange, Phycoerythrin and others can be used. Even Feulgen stain or natural autofluorescence is offen suitable and shows outstanding stability. Most of the time the fluorophores used are linked to antibodies for exact targeting in biological samples.

All these fluorescence properties hold also for confocal fluorescence observations which allow to examine the three-dimensional structure of cells, eggs, embryos or similar living objects. In this particular application the most spectacular results have been achieved. The single confocal image is purified from the out-of-focus fluorescence by the confocal spatial filter in front of the detector. Since the LSM is designed for the use with ICS optics including a tube lens this spatial filter doesn't require adjustments for different laser lines or objective lenses. In this way even multiple confocal fluorescence images are easily and quickly obtained. Recent examples for the use of this technique are protein-transport studies in single cells and the observation of ongoing differentiation in embryos which is important in Developmental Biology experiments.

Figure 3. Confocal (upper) and non-confocal (lower) fluorescence images of cells.

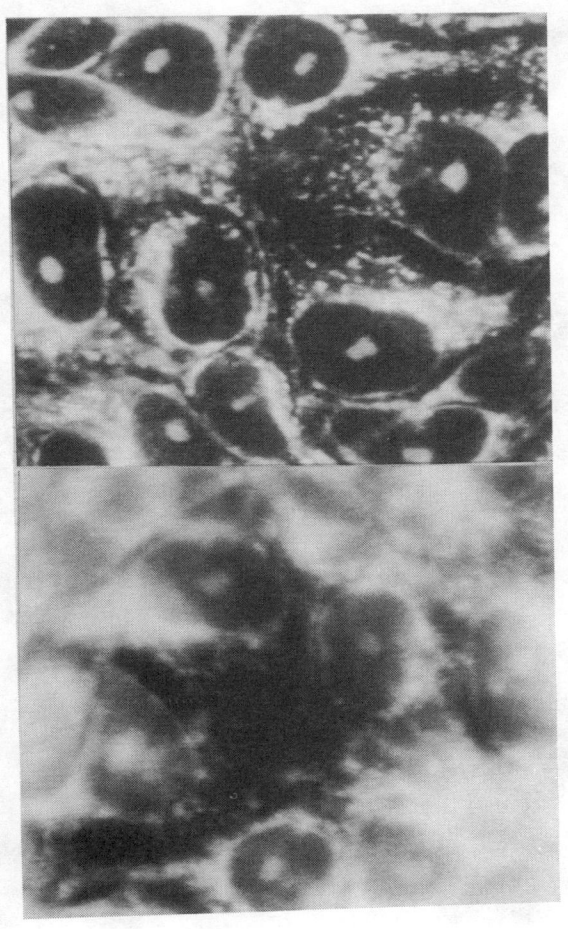

Figure 4. How the confocal spatial filter (7) works: Only light emanating from the focus plane (4, arrow) of the objective lens is allowed to pass the pinhole (7) without loss of intensity. In this way the detector (6) "sees" predominantly this one plane of the object.

TL: tube lens; Fi: filters.

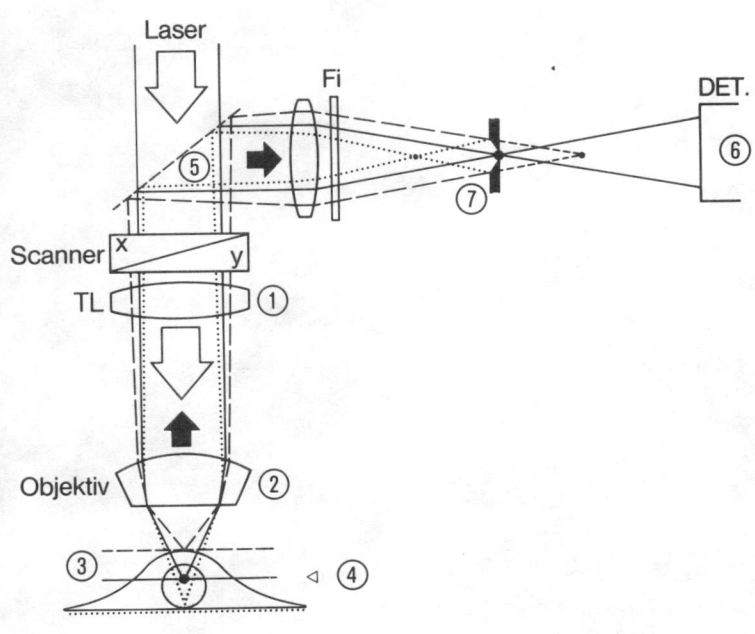

The confocal mode is not only restricted to fluorescence but generally available in all incident light techniques. This opens the way to observe reflecting particles of different kinds like immunogold (down to approach 5 nm), silver grains in autoradiography, Golgi stain for nervous tissue or even macromolecules able to scatter light much more than the surrounding medium. Good results have been achieved by using crossed polarizers in front of the detector and best results are given by the use of the Antiflex lens Plan-Neofluar 63x/1.25 which includes a quarter-wave plate for removing unwanted reflexes. Especially with immunogold staining in the confocal mode localization of filaments or viruses in cells can be performed.

In biological applications the transmitted light techniques are important because they are widely used to characterize or check the specimen. Also highest resolution studies are performed with transmitted light methods especially using the differential interference contrast (DIC) together with electronic contrast enhancement. Here the zooming capabilities of the LSM are helpful in the study of fine details near the resolution limit. And in contrast to optical zoom-systems there is no contrast change at various zoom-values. This is due to the fact that in scanning systems the scan angle is varied for different zoom values - there is no change in the optical system. Lateral resolution in transmitted light mode has been shown to be between 0.15 und 0.2 µm at good contrast. This marks the limit achievable with blue laser light (488 nm). Of course laser scanning itself doesn't increase resolution, but the electronic contrast enhancement shows details which are otherwise below the detection threshold.

Figure 5: Laser scanning fluorescence images of cells grown on glass surfaces. Left side: Overview; right side: Detail (x 5). The scan zoom permits wide variation of magnification (1:8 min.). Fluorescence label is fluorescein-conjugated antibody.

Objective lens: Plan Apochromat 63x/1.40 Oil

Excitation wavelength: 488 nm   Emission wavelengths: ≥ 515 nm

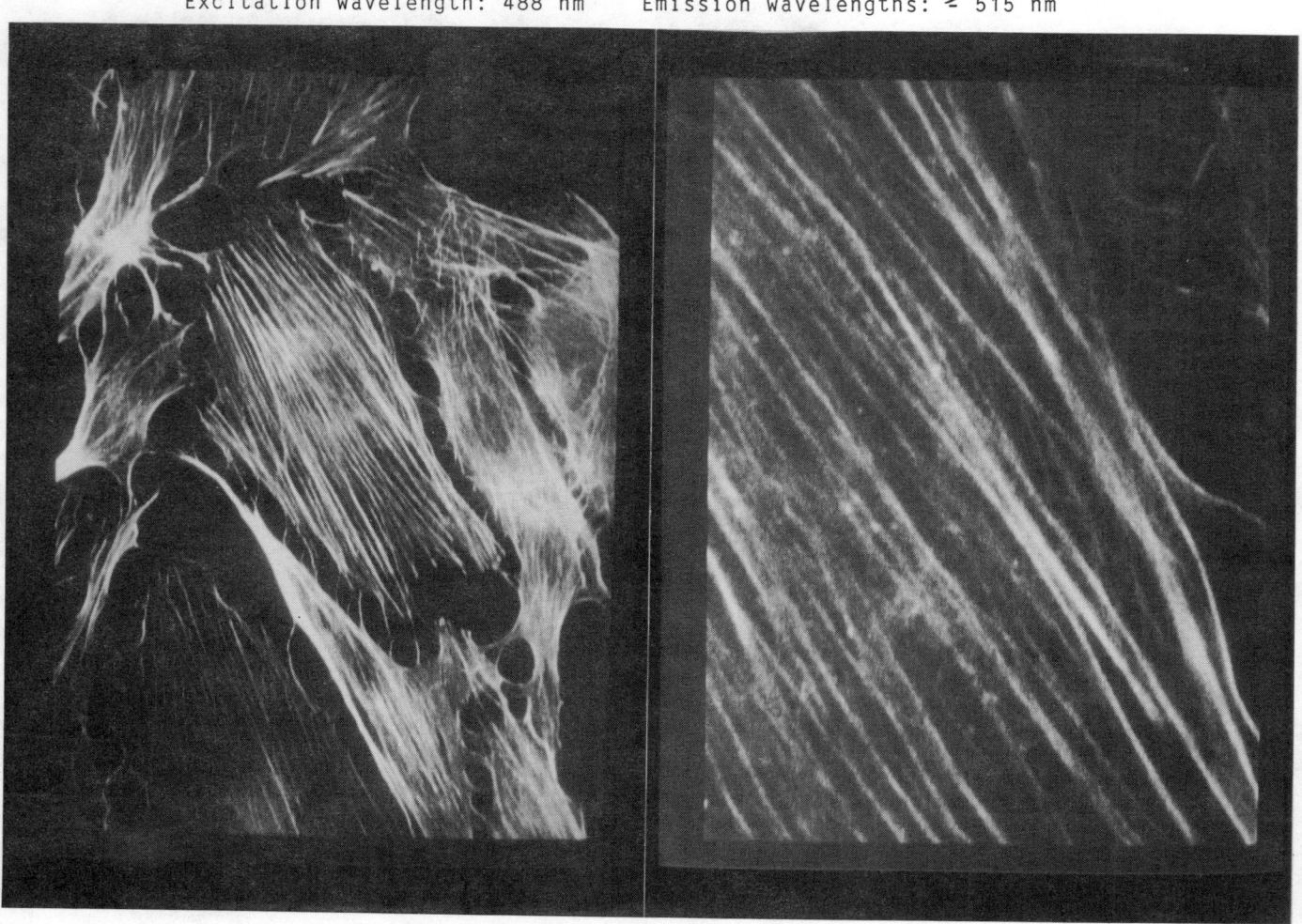

SEMICONDUCTOR RESEARCH, DEVELOPMENT AND TESTING

The most often used technique in this field is the OBIC method (Optical Beam Induced Current)[4]. The laser not only serves to image the structure of a device but also to stimulate photon-induced currents inside the sample under observation. The LSM 20/21 IR is equipped with a special specimen holder for clamping active devices to the stage. The pin selection for readout is done by software using multiplexers. Signal conditioning is easily perfected from the operator's panel in a similar fashion as the contrast/brightness-adjustments in microscopy mode. The resulting current pattern is displayed on the RGB monitor together with the reflected light image using pseudocolor coding. For quick evaluation an external computer may drive the selection of pins and resulting patterns can be downloaded to the host via parallel interface. In this way logical testing can be performed under remote control.

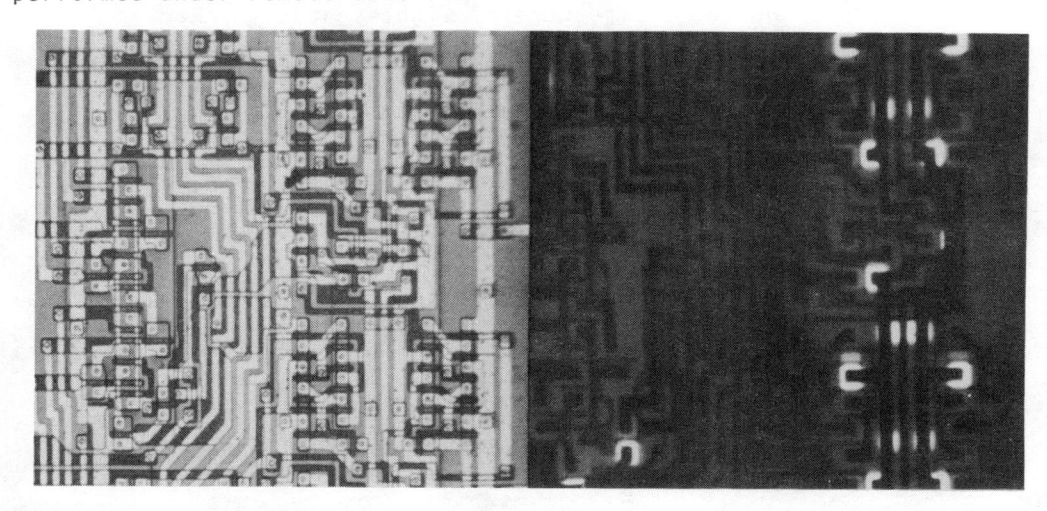

Figure 6:
OBIC image of an IC (right side) and the corresponding reflected light image (left side).

Both images are created simultaneously during scanning the IC and shown as a pseudocolor overlay on the RGB monitor.

The OBIC-applications so far included the detection of leakage currents, the latch-up analysis in CMOS devices and localization of electro-static damage (ESD). Not only integrated circuits but also laser diodes, photo diodes and solar cells can be examined. Since the stand provides ample space around the device to be tested, complex experimental setups can be realized e. g. including a prober station. The main advantages of OBIC as compared to EBIC (E for electron) applied in the elctron microscope are:

- OBIC is non-destructive
- OBIC doesn't require vacuum
- OBIC bears no need for removing insulating (passivation) layers from the device as necessary in the EM.

The IR-laser scanning can be done also in the confocal mode, which allows optical sectioning deep inside semiconductor material.
More recently the LSM has been used to carry out photoluminescence studies: In this application semiconductor materials can be imaged by detecting local variations of photoluminescence which depend on the local concentration of certain dopants[5]. The concentration map is changed by the diffusional built-up of Cotrell clouds of the dopant atoms trapped in defect sites. Photoluminescence excitation is done in the red (633 nm typically) and the detector can receive emitted photons up to approximately 850 nm.

Different methods can be used to detect "hot spots" in an active device by laser scanning. One is to use the polarized laser beam to observe a thin layer of liquid crystal pasted to the device surface: Upon heating in reflected light mode the change of optical properties within this layer is observed and "hot spots" are visualized as dark areas. The other method uses fluorescence in a similar way. Confocal scanning improves the results in either method.

MATERIAL SCIENCES

For material sciences the combination of the infinity corrected ICS-optics and the large electronic contrast enhancement factor has been further extended by confocal laser scanning in the reflected light mode. Besides, polarized light methods have been employed to visualize directly not only domains and Bloch walls but also Bloch lines. In this case a special one-sided darkfield mode has been used in transmitted light laser scanning[6]. More generally laser scanning microscopy is a wellsuited method to investigate the magnetic structures of memories, also those working on the base of magneto-optic effects. In more "classical" materials like metal alloys, ceramics and polymers it is the confocal reflected light mode, which leads to new results. There are three main areas where new results have been achieved:

- Investigations of the surface and the inner structure of light-scattering media

- studies on highly reflecting surfaces where normally only poor contrast is observed

- sub-surface analysis of visually opaque appearing materials.

Main interest in these methods comes from researchers working on polymers or emulsions (stray light problems) and from those in corrosion research (surfaces). Of course also compound materials like fiberglass-rein-forced polymers or ceramic-polymer hybrides are reveiling new details in confocal reflected light or fluorescence microscopy.

Figure 7.

Fiberglass-reinforced polymer.
Surface image shows structured matrix in confocal reflected light.

Objective lens: Epiplan-Neofluar
               50 x/0.85 Pol

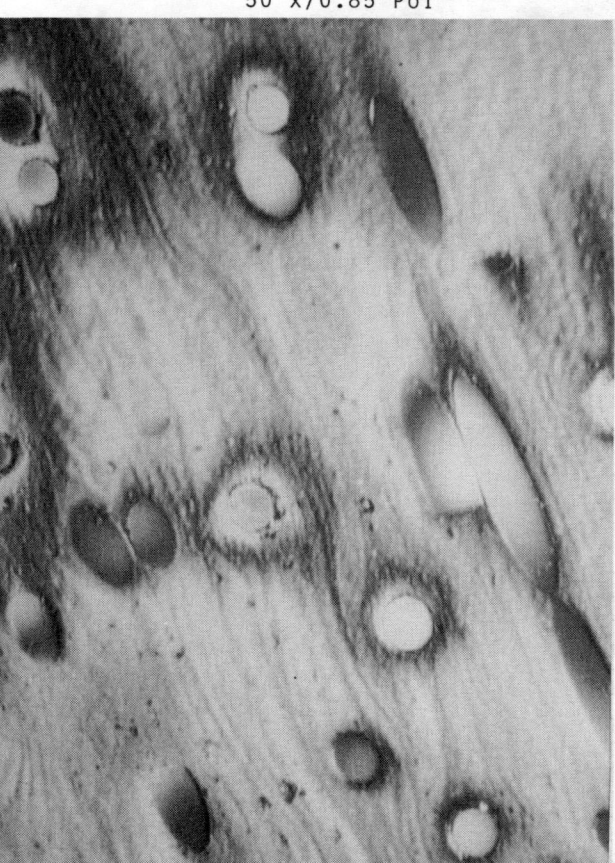

Figure 8.

Fiberglass-reinforced polymer subsurface confocal image taken 10 µm below the surface reveals fibers and matrix.

Objective lens: Epiplan-Neofluar
               50 x/0.85 Pol

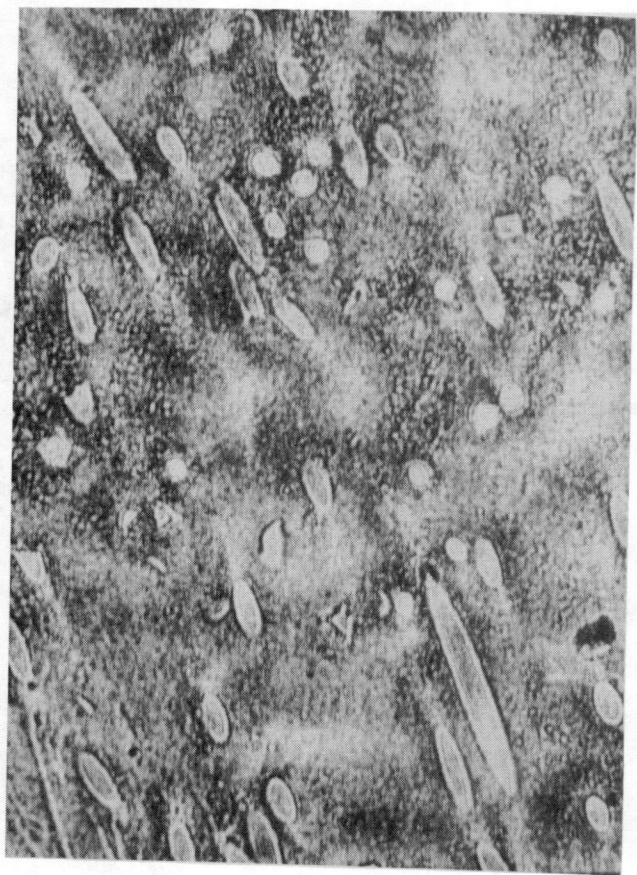

3. CONCLUSION

The new universal confocal LSM marks the beginning of a qualitative different epoch in laser scanning microscopy: Although the principle has been around for a couple of years and has demonstrated its power, the use of instruments has been restricted mostly to trained specialists. Now the second generation computerized LSM gives any microscopist the possiblitiy to use laser scanning routinely. Full computer control together with motors and servo elements make the LSM so easy to operate as a photomicroscope. In addition ICS optics and integrated optical design have been demonstrated to give best results in terms of sensitivity, resolving power and contrast enhancement. This type of instrument turns out to become the bridge between classical light microscopy and electron microscopy.

4. REFERENCES

1. T. Wilson and C. Sheppard, "Theorie and Practice of Scanning Optical Microscopy". Academic Press, New York (1984).

2. V. Wilke, "Optical scanning Microscopy - The Laser Scan Microscope", Scanning 7, 88-96 (1985).

3. A. Boyde, "Confocal optical microscopy", Microscopy and Analyses 1, 7 - 13 (1988).

4. J. Otto, E. Plies and J. Quincke, "Schaltungsanalyse in IC's mit Rasterlasermikroskop", VDI Berichte 659, 381-394 (1987).

5. J. Marek, A. G. Elliot, V. Wilke and R. Geiss, "High resolution scanning photoluminescence characterization of semi-insulating GaAs using a laser scanning microscope", Appl. Phys. Lett. 49, 1732-1734 (1986).

6. A. Thiaville, L. Arnoud, F. Boileau, G. Sauron and J. Miltat, "First direct optical observation of Bloch Lines in bubble garnets", IEE Trans. Mag. (1988).

ECO1
SCANNING IMAGING

Volume 1028

SESSION 5

Semiconductor Microscopy

Chair
Tony Wilson
Oxford University (UK)

Photoluminescence and Optical Beam Induced Current Imaging of Defects

P. D. Pester and T. Wilson

Oxford University, Department of Engineering Science,
Parks Road, Oxford OX1 3PJ, England.

ABSTRACT

The minority carrier distribution in a semi-infinite semiconductor due to light focussed onto the surface via a high numerical aperture lens is given. The contrast obtained in both Photoluminescence and Optical Beam Induced Current defect imaging is derived as a function of the illuminating lens numerical aperture. It is shown that the maximum resolution attainable in both techniques is comparable.

I. INTRODUCTION

In order to extract meaningful results from Photoluminescence (PL) or Optical Beam Induced Current (OBIC) investigations of semiconductors it is clearly necessary to have a good physical model for the generation of the excess carriers by the exciting beam.

Most early analyses [1,2] assumed that the generation of the excess carriers took place at one point either on, or just below, the surface of the semiconductor. To try and model the absorption of the incident beam by the semiconductor more closely some authors [3,4,5] have introduced an array of point sources decaying exponentially into the semiconductor, and others [6] have taken account of the beam width by introducing a Gaussian lateral profile. However, both of these cases are only approximations to the true light intensity [7] within the semiconductor due to light focussed onto the surface via a high numerical aperture lens, as is the case in the Scanning Optical Microscope (SOM).

Therefore, in the following, we calculate the minority carrier distribution injected into a semi-infinite semiconductor, characterized by both surface and bulk recombination rates, due to a highly convergent light beam focussed through a high numerical aperture lens. The resulting minority carrier distribution is then used to predict the contrast obtained, as a function of numerical aperture, due to a point defect in the semi-infinite sample in both the PL and OBIC examination techniques.

II. THE INJECTED MINORITY CARRIER DISTRIBUTION

The steady state distribution of holes, $\Delta p(r,z)$, induced in an n-type sample in the low injection limit can be found [8] by solution of the continuity equation [9],

$$\nabla^2 \Delta p(r,z) - \frac{\Delta p(r,z)}{L^2} + \frac{g(r,z)}{D} = 0, \qquad (1)$$

where D and L are the minority carrier diffusion coefficient and diffusion length respectively and $g(r,z)$ is the generation term which is proportional to the light intensity within the semiconductor. For a highly convergent light beam focussed onto the semiconductor surface $g(r,z)$ is given as [7],

$$g(r,z) = g_0 A \exp(-\alpha z) \left| \int_0^1 J_0(v\rho) \exp\left(-\frac{iu\rho^2}{2}\right) \exp\left(-\frac{\beta u \rho^2}{2}\right) \rho d\rho \right|^2 \qquad (2)$$

where v and u are normalised optical coordinates [10] given by,

$$v = \left(\frac{2\pi}{\lambda}\right) r \sin\theta \qquad u = \left(\frac{2\pi}{\lambda}\right) z \sin^2\theta. \qquad (3)$$

In Equations (2) and (3) λ is the wavelength of the incident light, $\sin\theta$ the numerical aperture of the objective lens, β is related to α, the intensity absorption coefficient via $2k\beta = \alpha$ with k being the wavenumber, ρ is a radial variable in the objective pupil plane, g_0 is a constant taking into account surface reflectivity and quantum efficiency and the premultiplying factor A is a normalisation constant given by,

$$A = \frac{\alpha k^2 \sin^4\theta}{2\pi \ln\left[1 + \frac{\sin^2\theta}{2}\right]}, \qquad (4)$$

chosen such that,

$$2\pi \int_0^\infty \int_0^\infty g(r,z)\, r dr dz = g_0. \qquad (5)$$

If we choose to consider a semi-infinite semiconductor bounded by a surface, in the $z = 0$ plane, which is characterized by a surface recombination velocity, v_s, the boundary conditions on the solution of equation (1) can be written,

$$\left.\frac{\partial}{\partial z}\Delta p(r,z)\right|_{z=0} = s\,\Delta p(r,z), \qquad (6a)$$

and,

$$\left.\Delta p(r,z)\right|_{z\to\infty} \to 0, \qquad (6b)$$

where $s = v_s/D$ is a normalised surface recombination velocity. Figure 1 shows isometric surface plots of the distribution, $\Delta p(r,z)$, found from equations (1), (2) and (6), as a function of position in the semiconductor for the case of red light ($\lambda = 0.632\mu m$) focussed onto silicon with a surface recombination velocity of $v_s/D = 1\mu m^{-1}$ and a bulk diffusion length of $L = 5\mu m$. The intensity absorption coefficient is taken as [11] $\alpha = 0.4\mu m^{-1}$ and the objective lens numerical aperture is (a) $\sin\theta = 0.1$, (b) $\sin\theta = 0.5$ and (c) $\sin\theta = 0.9$.

As would be expected the peak in the distribution is less well confined radially at low numerical apertures, however, it is also clear that the distribution penetrates further into the semiconductor implying that it is also less well confined axially.

The effect of the surface recombination velocity can be seen in figure 2 where the minority carrier distribution is plotted for the case when $\alpha = 0.4\mu m^{-1}$, $L = 5\mu m$, $\sin\theta = 0.6$ and $v_s/D \to 0$, $v_s/D = 1\mu m^{-1}$ and $v_s/D \to \infty$. It is clear that for any non zero value of v_s the distribution always peaks some distance below the semiconductor surface [6], this is a consequence of the increased recombination rate on the surface when $v_s \neq 0$.

(a)   (b)   (c)

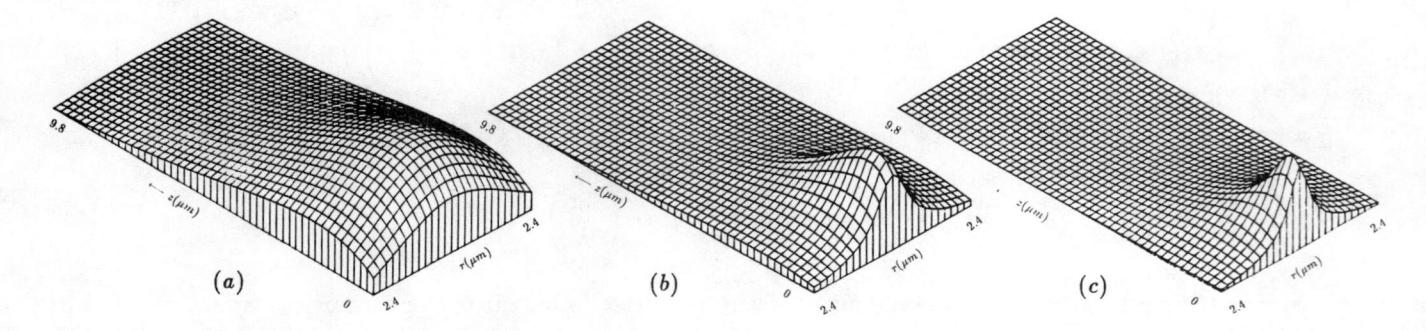

Figure 1. Isometric surface plots of the minority carrier distribution, $\Delta p(r, z)$ injected by red light focussed onto the surface of silicon having a bulk diffusion length of $L = 5\mu m$ and a surface recombination velocity of $v_s/D = 1.0\mu m^{-1}$, for the objective lens numerical aperture, (a) $\sin\theta = 0.1$, (b) $\sin\theta = 0.5$ and (c) $\sin\theta = 0.9$.

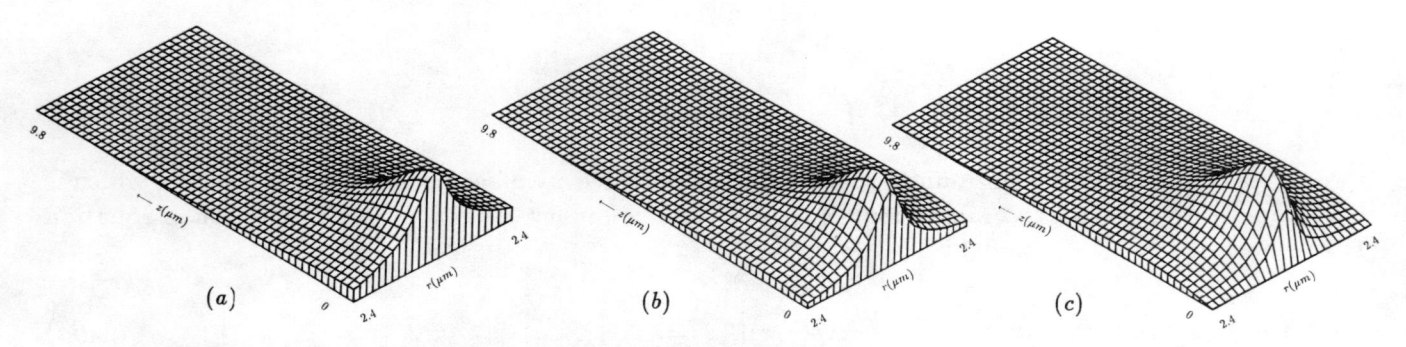

(a)   (b)   (c)

Figure 2. The minority carrier distribution for red light in silicon when the objective lens numerical aperture $\sin\theta = 0.6$, the diffusion length $L = 5\mu m$ and (a) zero surface recombination velocity, (b) a surface recombination velocity of $v_s/D = 1\mu m^{-1}$, and (c) infinite surface recombination velocity.

III. IMAGING OF ELECTRICALLY ACTIVE DEFECTS

One of the main attractions of the scanning PL and OBIC techniques is the ability to image electrically active defects close to the semiconductor surface. The contrast, $C_{\text{pl}}(\xi, a)$ as a function of beam position, ξ, and defect depth, a, obtained as a point defect in a semi-infinite semiconductor is image using the scanning PL technique has been calculated to be [8],

$$C_{\text{pl}}(\xi, a) = \Theta(\alpha, L, v_s, \sin\theta, a)\, \Delta p(\xi, a), \qquad (7a)$$

where,

$$\Theta(\alpha, L, v_s, \sin\theta, a) = \frac{\gamma_0}{g_0} \frac{\left\{1 - \left(\frac{sL}{(1+sL)}\right)\exp(-a/L)\right\}}{\left\{1 - \left(\frac{sL}{(1+sL)}\right)\frac{\ln\left[1 + \frac{\alpha L \sin^2\theta}{2(\alpha L + 1)}\right]}{\ln\left[1 + \frac{\sin^2\theta}{2}\right]}\right\}}. \qquad (7b)$$

Here γ_0 is the *strength* of the defect and g_0 is the *strength* of the exciting beam. Similarly it is possible to calculate the contrast, $C_{\text{obic}}(\xi, a)$, as a point defect in a semi-infinite Schottky barrier diode is imaged in the OBIC technique as,

$$C_{\text{obic}}(\xi, a) = \Phi(\alpha, L, \sin\theta, a)\, \Delta p(\xi, a), \qquad (8a)$$

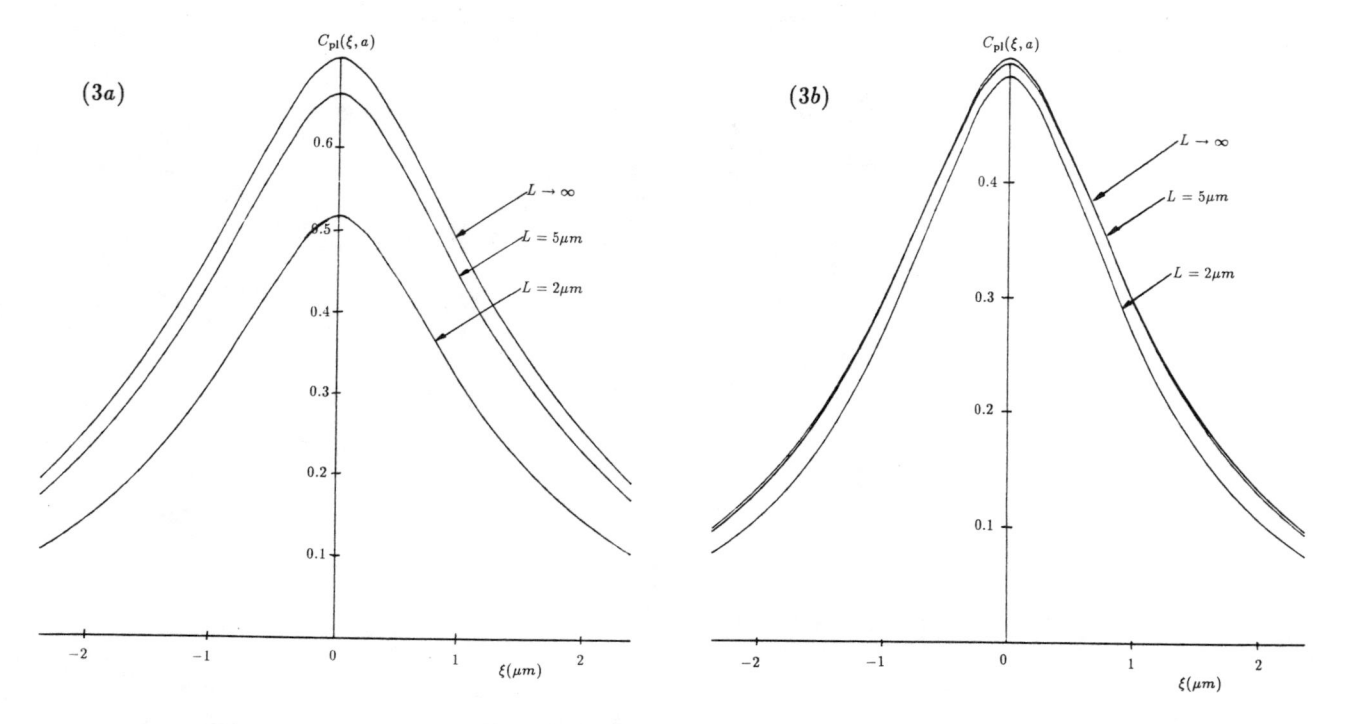

Figure 3. The contrast obtained in the PL technique when a defect at a depth of $a = 2\mu m$ in a semi-infinite semiconductor is imaged using red light focussed through a lens with a numerical aperture of $\sin\theta = 0.6$. The surface recombination velocity is taken as (a) $v_s/D \to 0$ and (b) $v_s/D \to \infty$.

where the function Φ is given as,

$$\Phi(\alpha, L, \sin\theta, a) = \frac{\gamma_0}{g_0} \left\{ \frac{\ln\left[1 + \frac{\sin^2\theta}{2}\right]}{\ln\left[1 + \frac{\alpha L \sin^2\theta}{2(\alpha L + 1)}\right]} \exp\left(-\frac{a}{L}\right) \right\}. \qquad (8b)$$

It is clear that, knowing the minority carrier distribution $\Delta p(r, z)$, it is straightforward to find both contrast functions.

Specalizing to the PL case we show, in figure 3, the PL contrast for a defect at a depth of $a = 2\mu m$ when illuminated with red light through a lens with a numerical aperture of $\sin\theta = 0.6$. It is interesting to note that the $L \to \infty$ approximation is more applicable at lower values of diffusion length in the $v_s/D \to \infty$ limit than in the $v_s/D \to 0$ limit. From the form of the expressions given in equations (7) and (8), where the contrast functions differ only in premultiplying factors, it is clear that the resolutions predicted for the PL, considering the infinite surface recombination velocity limit owing to the presence of the surface junction, and OBIC techniques are similar.

The effect of the illuminating lens numerical aperture on the resolution of the PL technique can be investigated by plotting the half-width of the contrast curves as a function of numerical aperture. Figure 4 shows such a plot for the case when $\alpha = 0.4\mu m^{-1}$, a diffusion length of $L = 5\mu m$ and a surface recombination velocity, $v_s/D = 1.0\mu m^{-1}$. From figure 4 we can see that, for a shallow defect where $a = 0.5\mu m$ the curves are as one might expect, that is the resolution increases with increasing numerical aperture until a limit is reached at around $1.1\mu m$, although for deeper defects the curves are quite different. It is clear that the resolution goes through a maximum, that is the half-width goes through a minimum, at an intermediate value of numerical aperture. This is an important point to note since,

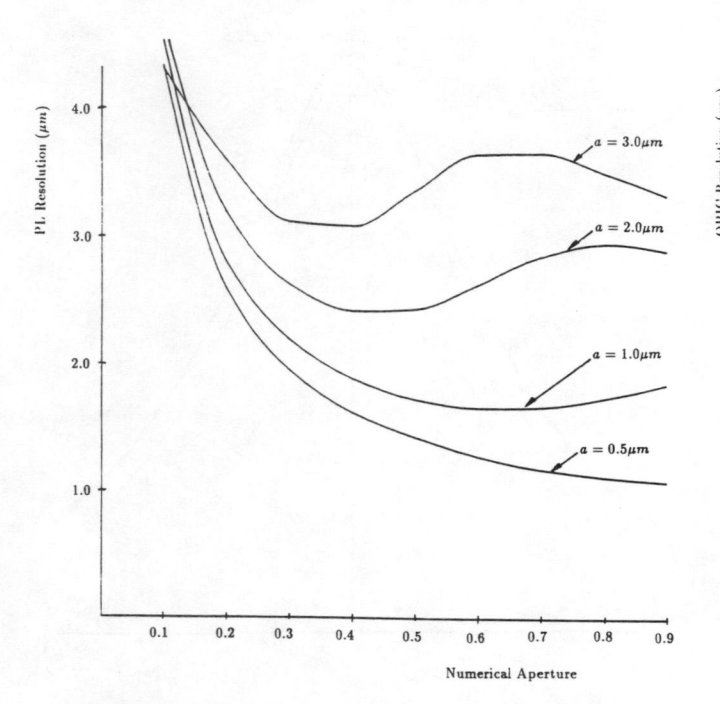

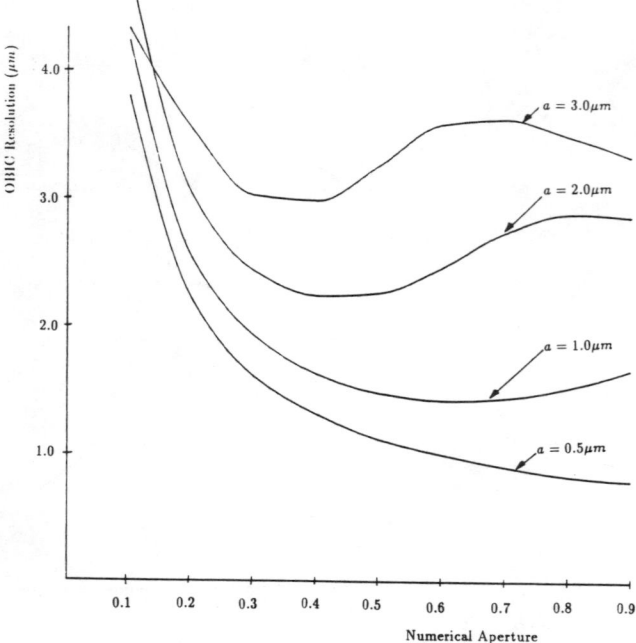

Figure 4. The resolution of the photoluminescence technique for the case when $\alpha = 0.4\mu m^{-1}$, a diffusion length of $L = 5.0\mu m$ a surface recombination velocity of $v_s/D = 1.0\mu m^{-1}$ and various defect depths.

Figure 5. The resolution of the OBIC technique for point defects at various depths in a silicon Schottky barrier diode. The case shown is for when $\alpha = 0.4\mu m^{-1}$ and the diffusion length is $L = 5.0\mu m$.

when imaging defects which are not close to the surface, it is not always wise to use a high numerical aperture lens to obtain optimum resolution.

Following the same procedure to investigate the resolution of the OBIC technique we obtain, for a defect in a semi-infinite Schottky barrier, resolution curves as shown in figure 5. The resolution of the OBIC technique clearly follows the same trends as in the PL case, although the maximum contrast here is seen have increased slightly to around $0.8\mu m$.

III. SCANNING PHOTOLUMINESCENCE IMAGES

Figure 6 shows two scanning photoluminescence images of InP obtained by using a red, $\lambda = 0.632\mu m$, HeNe laser focussed through a objective lens with a numerical aperture of 0.6, in a scanning optical microscope, to excite a photoluminescence signal of wavelength $\lambda = 0.920\mu m$. Reflected light images of the specimen showed no surface details at either low or high magnification. The photoluminescence images were obtained in transmission, with suitable filters being used to block the exciting wavelength, at room temperature using a silicon dectector. Figure 6a is a low magnification image showing a network of defects in the InP sample, in particular a pair of line defects crossing at an angle of about 50 degrees to each other can be clearly seen in the centre of the image. In figure 6b an increased magnification photoluminescence image is shown of the cross-over point of the two main features, from which it is clear that what appears to be a single line defect in figure 6a is actually a pair of parallel line defects and that, as suggested in figure 4, the resolution of the PL technique can be as good as $1\mu m$.

(6a)   (6b)

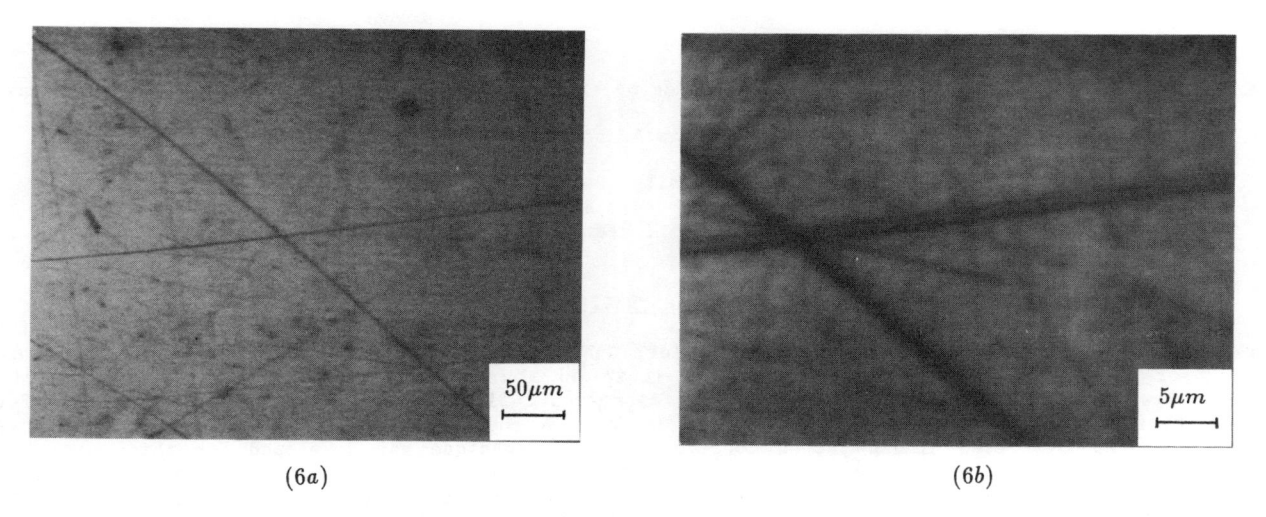

Figure 6. Scanning photoluminescence images of InP showing defect networks, (a) is a low magnification image and (b) is a high magnification image.

IV. CONCLUSIONS

We have shown the form of the minority carrier distribution injected into a semi-infinite semiconductor due to light focussed onto the surface through a lens with finite numerical aperture. We have found that the distribution is less well confined both radially and axially at low numerical apertures.

The resolution obtained when imaging a point defect in both the scanning PL and OBIC examination techniques was shown as a function of illuminating lens numerical aperture. Limits on the resolution of the two techniques in the cases considered were predicted to be around $1.1 \mu m$ for scanning PL and $0.8 \mu m$ for OBIC. Experimental scanning photoluminescence images were presented where it was shown that, in good agreement with the theory, a resolution of about one micron is attainable.

V. REFERENCES

(1) F. S. Goucher, *Phys. Rev.*, **81**, 475, (1951)
(2) F. S. Goucher, G. L. Pearson, Sparks, Teal and W. Schockley, *Phys. Rev.*, **81**, 637, (1951)
(3) C. Hu, C. Drowley, *Sol. St. Elect.*, **21**, 965, (1978)
(4) T. Wilson, P. D. Pester, *J. Appl. Phys.*, **61**, 2307, (1986)
(5) A. Sinha, S. K. Chattopadhyaya, *Sol. St. Elect.*, **19**, 3345. (1976)
(6) T. Wilson, P. D. Pester, *Phys. Stat. Sol.(a)*, **97**, 323, (1986)
(7) T. Wilson, E. M. McCabe, *J. Appl. Phys.*, **58**, 2638, (1986)
(8) P. D. Pester, D.Phil. Thesis, University of Oxford, (1988)
(9) W. Van Roosbroeck, *Phys. Rev.*, **91**, 282, (1953)
(10) M. Born and E. Wolf, *Principles of Optics*, (Pergamon, Oxford, 1975)
(11) W. C. Dash, R. Newman, *Phys. Rev.*, **99**, 1151, (1953)

Minority carrier lifetime mapping in gallium arsenide by time-resolved photoluminescence scanning microscopy

Thomas A. Louis

Heriot-Watt University, Physics Department,
Riccarton, Edinburgh EH14 4AS, United Kingdom
Tel. (031) 449 5111, Fax (031) 451 3088

ABSTRACT

The minority carrier lifetime in gallium arsenide (GaAs) can be determined from time-resolved photoluminescence near the optical band edge, around 872 nm (1.42 eV) at room temperature. Spatial resolution of the order of a few micron is required for investigating microscopic fluctuations of he minority carrier lifetime in GaAs materials and in small, highly integrated devices. A photoluminescence lifetime spectrometer (PLS) based on the time-correlated single photon counting (TCSPC) technique was developed for this purpose. The instrument uses only solid state components, i.e. pulsed diode laser excitation and single photon avalanche diode (SPAD) detection, and is capable of measuring photoluminescence decay time constants of the order of 10 ps with 3 μm spatial resolution. Operation at a repetition rate above 50 MHz reduces the data collection time for a complete decay curve to a few seconds and thus permits one and two-dimensional scans to be done. The design of the PLS is discussed and first results from LEC grown GaAs substrates are presented.

1. INTRODUCTION

The minority carrier lifetime τ and the related minority carrier diffusion length L are important parameters for characterizing the quality of GaAs material and for modeling the performance of GaAs devices. τ and L are related through the minority carrier diffusion constant D ($= L^2/\tau$).

In the simplest case of a surface free sample under low excitation conditions, where the excess minority carrier density is small compared with the majority carrier density, linear non-radiative recombination is the dominating recombination mechanism and the externally observable photoluminescence decay time τ_{PL} is equal to the minority carrier lifetime τ, which is a bulk parameter. Measurement of τ_{PL} then directly yields τ. More generally, the problem of determining the bulk parameter τ is equivalent to the problem of measuring the observable quantity τ_{PL} and interpreting it in terms of the bulk parameters τ, D and the surface/interface recombination velocity S. The presence of surfaces/interfaces will always tend to lower the effective minority carrier lifetime τ_{eff} below the bulk value τ. Strictly speaking, τ_{PL} and τ_{eff} are meaningful only when the PL decay is a simple exponential. As this is rarely true, we shall avoid the use of τ_{PL} and τ_{eff}. Instead we shall refer to time-resolved photoluminescence (TRPL) throughout, implying that τ (and maybe other material parameters) can be derived from the measured TRPL data.

In the technologically most important GaAs substrates, LEC grown undoped or In-doped SI-GaAs substrates, τ and L are of the order of several 100 ps[1] and below 4 μm respectively. This value of τ is far below the value of the bulk radiative lifetime, τ_r, and is dominated by the non-radiative lifetime for recombination at deep levels, τ_{nr}, plus localized recombination at dislocations and the surface. Commercially available SI-GaAs substrate materials show strong inhomogeneities in the distribution of defects across the wafer due to difficulties in controlling the growth process. These substrate inhomogeneities usually affect epitaxially grown layers and are known to influence the electrical parameters of simple GaAs devices, such as field effect transistors (FETs), when directly implemented on the substrates.

Topographical methods, displaying information graphically as one-dimensional scans or two-dimensional maps, have revealed spatial variations of many material properties across SI-GaAs wafers both on a macroscopic (W-shape) and a microscopic scale (cellular structures, microscopic lines). A number of different topographical methods, such as based on CW photoluminescence (CWPL) and near infrared absorption (NIR)[2-4], CW and time-resolved cathodoluminescence (CWCL, TRCL)[5], etch pit density (EPD), X-ray and electrical measurements such as electron/optical beam induced current (EBIC/OBIC), have been reported in the literature. Correlation of topographical images obtained with various methods has been used to discriminate between bulk and surface effects and identify certain types of defects. However, interpretation of the observed quantities in terms of material parameters is often ambiguous. More detailed information can be expected from mapping the photoluminescence lifetime with high spatial resolution, but this has so far been impossible due to the lack of sufficiently sensitive instrumentation.

Experimental methods used for TRPL can be distinguished by their means of exciting excess carriers, i.e. photoexcitation (laser beam) or impact excitation (electron beam), and by the observable quantity used for measuring the excess carrier recombination rate, i.e. luminescence (TRPL, TRCL), microwave absorption or an induced electrical signal (OBIC, EBIC). They are further classified as being either time domain or frequency domain methods. Time domain methods use pulsed excitation and measure the time decay of the signal, frequency domain methods use an intensity modulated CW excitation source and determine the phase shift in the modulated signal.

TRPL techniques have become very popular, mainly because picosecond laser pulses are now readily available. The analytical interpretation of TRPL data is also simpler than in TRCL, because in TRPL the excitation is well-defined, hence there are less transitions involved when carriers return from the excited state(s) to the ground state. TRPL has been investigated with a variety of experimental techniques, using streak camera systems[1], non-linearity of the PL intensity[6], sum frequency generation[7] and time-correlated single photon counting (TCSPC)[8-10]. The characteristics of the detection system determine time-resolution, sensitivity, wavelength range of operation and simplicity of use.

TCSPC is known as by far the most sensitive technique for measuring fast fluorescence decays in the UV-VIS region, but it has so far not been extensively used for measuring TRPL in semiconductors due to the lack of adequately fast single photon detectors with high sensitivity in the near infrared. Recent improvements in multialkali photocathodes have produced photomultiplier tubes (PMTs) for use in TCSPC up to about 930 nm[11]. The timing properties of micro-channelplate photomultipliers (MCP-PMTs) have also continuously improved, fastest devices reaching response widths of around 25 ps (FWHM). A novel type of detector, the single photon avalanche diode (SPAD) in connection with an active quenching circuit[12-14], has a response width below 45 ps[15] and good sensitivity in the NIR up to about 950 nm. Performance comparison of a MCP-PMT with a SPAD detector has recently demonstrated that a TCSPC setup with an overall instrumental response width of 50 ps (FWHM) with a MCP-PMT or 70 ps (FWHM) with a SPAD detector can resolve fluorescence decay times of the order of 10 ps with ± 2 ps accuracy[16].

In this paper we report, for the first time, TRPL measurements on a TCSPC system using only solid state components for excitation and detection, namely a pulsed diode laser source and a SPAD detector. The paper is structured as follows : Section 2 defines the general performance requirements for TRPL measurements in GaAs, section 3 describes the new photoluminescence lifetime spectrometer (PLS), section 4 shows first results from measurements on a GaAs substrate and section 5, conclusions, discusses future trends and applications for the PLS system. Frequently used abbreviations and symbols are listed in section 6.

2. MINIMUM PERFORMANCE REQUIREMENTS

It is generally not possible to analyze the PL signal with very high time resolution, spatial and spectral resolution simultaneously, not even by signal averaging over a large number of events. The characteristics of the experimental technique limit the sensitivity that can be obtained with a practical instrument. All methods for measuring TRPL allow trade off between temporal, spatial and spectral resolution, at least to a certain extent. Hence, for a given temporal, spatial and spectral resolution, the sensitivity of the technique defines the lower end of the excitation density range in which TRPL can be investigated. It is clearly desireable to be able to measure TRPL under low density excitation, where material parameters such as τ and L are constants, independent of the excess carrier density.

We shall briefly describe the minimum performance requirements for doing spatially resolved TRPL measurements in GaAs before deciding on the method of choice.

2.1. Time resolution

The shortest minority carrier lifetimes are measured in Cr-compensated GaAs substrates, less than 50 ps. The technologically more important LEC grown undoped SI-GaAs substrates have values of τ of the order of several 100 ps and in best epitaxial material τ may reach many µs. TRPL measured with high spatial resolution in the immediate vicinity of localized strong recombination centers is also expected to show a very rapid decay, resulting in decay time constants much shorter than the presently measured spatial averages obtained with low spatial resolution techniques. A TRPL technique suitable for high resolution spatial mapping should therefore have a time resolution of at least 10 ps and a large dynamic range.

2.2. Spatial resolution

The smallest scale on which fluctuations of the PL decay time are expected is given by the carrier diffusion length, less than 5 µm in SI-GaAs. Therefore the ultra high spatial resolution provided by a scanning electron microscope (SEM), which, at first sight seems an inherent advantage of SEM based techniques, such as intensity modulated EBIC or TRCL, is not actually required. Carriers excited by a finely focussed electron pulse immediately diffuse and "smear out" the TRCL signal.

In all TRPL techniques, spatial resolution is limited by diffraction to about 1 µm. Even here, depending on the actual diffusion length, carriers excited by a highly focussed laser beam will diffuse laterally and into the bulk. This results in a decay of the excitation density in the center of the excited volume. Even though this may not change the total number of excited carriers within the - usually much larger - field of view of the detector, at higher excitation densities, in the non-linear regime dominated by bimolecular recombination, this results in an undesirable additional "decay" component in the TRPL data. In order to avoid lateral diffusion related effects, it is therefore preferable to obtain spatial resolution by limiting the field of view of the detector whilst homogeneously exciting a sufficiently large sample area. In a similar fashion, the effect of bulk diffusion and surface recombination can be reduced by choosing an excitation wavelength near the band edge thereby increasing the penetration depth.

2.3. Spectral resolution

High spectral resolution is not essential for determining the near band edge photoluminescence decay time. If the sensitivity of the experimental technique allows to measure the sample even at room temperature, excitonic transitions and the band-acceptor transition in p-doped material are strongly broadened and cannot be resolved from the band-band transition in bulk GaAs anyway. Normally, an interference filter with 10 nm bandwidth centered at the photoluminescence peak plus a suitable long pass filter would be sufficient to suppress scattered excitation light, provided that polarization and spatial discrimination is used as well (see section 3.1.). Only when the excitation wavelength is very close to the band edge and therefore close to the photoluminescence signal does spectral resolution become important. It may then be necessary to use a monochromator instead of filters.

2.4. Method of choice

Sum frequency mixing techniques provide very high time resolution but involve a low-efficiency conversion process, are therefore not very sensitive and require averaging over finite sample volumes in order to obtain a large enough PL signal at a reasonable excitation density.

The non-linear pump-probe technique is similar to the above in as far as it can provide very high time resolution when using a non-linear optical gate for detection. In order to obtain a good enough sensitivity for doing highly spatially resolved TRPL measurements in GaAs, however, an avalanche photodiode, boxcar integrator and lock-in averaging was found necessary, which resulted in a reduced time resolution of approx. 500 ps.

Synchroscan streak cameras, although more sensitive than sum-frequency mixing, non-linear pump-probe techniques and single sweep streak cameras, still require high excitation densities and cooling of samples to produce sufficient photon flux. Streak cameras suffer from basically the same problems as do MCP-PMTs, namely the rapid drop in spectral sensitivity of multialkali photocathodes in the near infrared above 800 nm. Hammamatsu synchroscan streak cameras, e.g., also suffer from saturation effects, visible as a broad shoulder on the rising edge of a pulse, which limit the useful dynamic range to typically less than 50:1.

We thus conclude that the above mentioned methods are unsuitable for high resolution spatial mapping TRPL measurements in GaAs. TCSPC is, at present, the only method with good enough sensitivity to be used in such an application. Provided a good kinetic model is available for fitting TRPL data by reconvolution analysis, TCSPC provides a time resolution comparable to synchroscan streak camera systems, which is sufficient to investigate quasi-equilibrium carrier dynamics in GaAs on a picosecond time scale.

3. DESIGN OF THE PHOTOLUMINESCENCE LIFETIME SPECTROMETER (PLS)

The main components in a TCSPC setup are the pulsed light source, the optical spectrometer, the single photon detector, the timing electronics, the computer and the software for reconvolution analysis. Standard fast TCSPC setups used in photochemistry and photobiology are usually based on a synchronously pumped and cavity dumped picosecond dye laser system, a spectrometer in conventional L-geometry with a grating monochromator in the detection channel and a fast MCP-PMT detector. The PLS, specifically designed for high spatial resolution applications in TRPL of semiconductors, differs mainly in the design of the spectrometer and, as result of the different nature of the samples, its ability to take full advantage of the new SPAD detectors.

3.1 Spectrometer

The spectrometer is based on a commercial light microscope with optics corrected for infinite tube length. This allows to insert the optical routing module (ORM) shown in fig.1 into the microscope column without affecting the microscope image quality. The ORM contains two crossed polarizing beamsplitter cubes (PBS1,PBS2) to couple excitation light into and luminescence photons out of the microscope beam path. (Note that fig.1 shows the two perpendicular axes of source beam, Y, and luminescence beam, X, in the plane both in the same plane for simplicity).

The excitation light is collimated to form a parallel beam of 0.9 mm diameter by lens L1 and positioned such that it passes through the unobstructed area of the reflecting objectives small convex mirror. With an objective of large numerical aperture (N.A.≈0.95) the beam is aligned to be incident upon the sample close to Brewster angle in order to reduce specular reflection off the surface. The excitation spot size is diffraction limited to about 4.5 µm in the plane of best focus (fig.2, focal plane 1). Axial adjustment of L1 is used to shift the plane of best focus off the sample surface and expand the excitation area to about 6 µm diameter in the focal plane of the detector (fig.2, focal plane 2). The magnification given by the focal length ratio of L2 and the reflecting objective and the diameter of the stop aperture determine the detector's field of view. These values can easily be varied according to the experimental requirements. In the experiments reported here, the magnification was 1.8, the stop aperture 5 µm diameter (= active area of smallest SPAD), hence the detector field of view was about 3 µm diameter. Due to the large refractive index of GaAs, n=3.7 at 872 nm, the PL wavelength, only the small fraction of the isotropically emitted PL photons within a narrow solid angle of 17° can escape from the bulk. Reflection at the surface, the microscope objectives limited solid angle of acceptance the detector's field of view further limit the photon luminescent flux selected for detection.

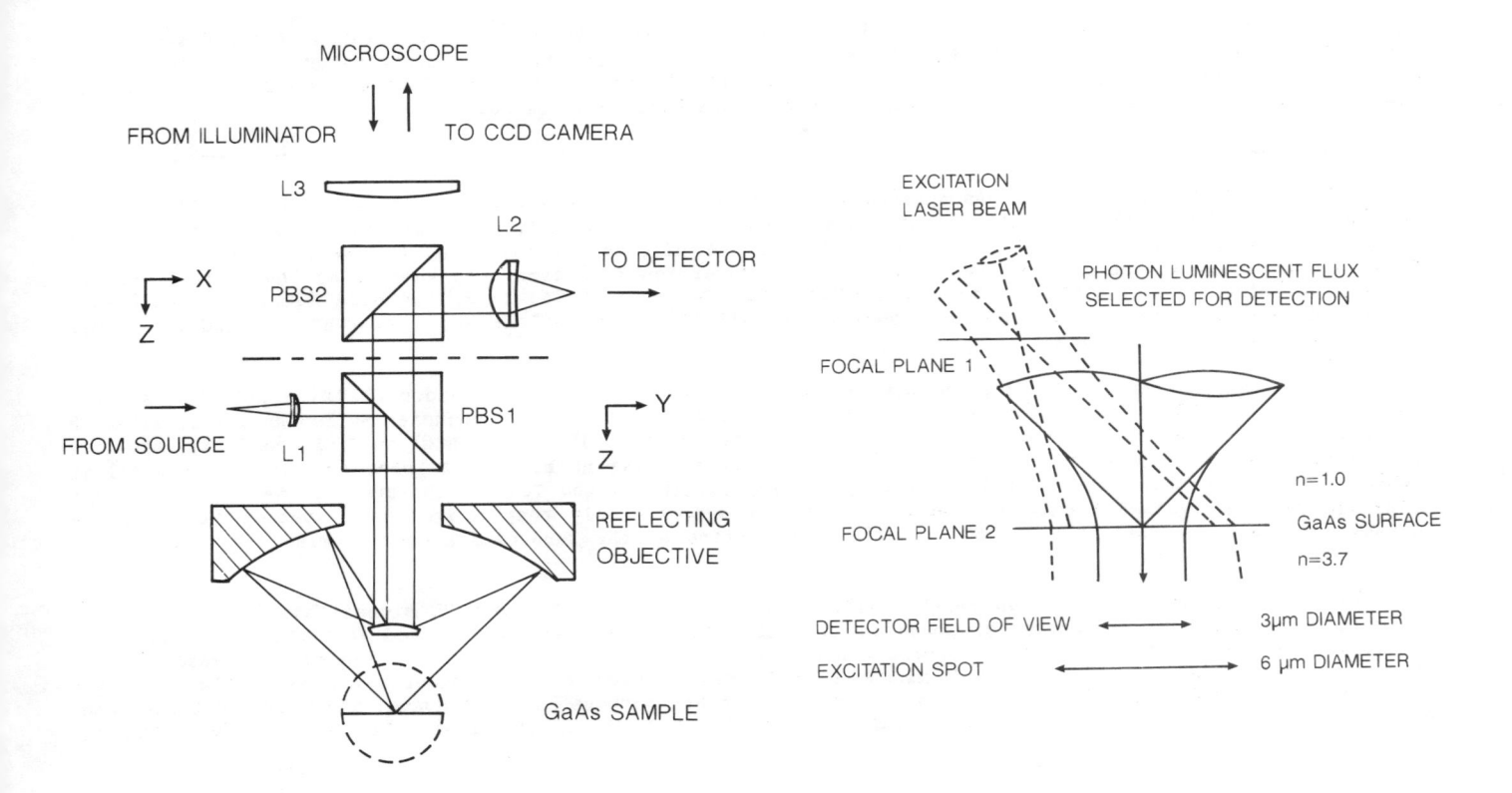

Fig.1 : Optical routing module of the
Photoluminescence Lifetime Spectrometer

Fig.2 : Beam profile near the GaAs sample surface
(enlarged view of encircled section in fig.1)

The ORM makes use of the fact that PBS1 and PBS2 separate the two polarization components (transmission of light in the p-plane and 90° reflection in the s-plane) only within a narrow range around the design wavelength λ_0, typically 0.9 to 1.1 times λ_0. As the the PBSs are centered at the excitation wavelength and the peak of the PL spectrum, 785 nm and 870 nm respectively, visible light below 700 nm passes both beamsplitters unhindered. Both PBSs are slightly tilted or bevelled (not shown in fig.1) in order to direct the inevitable strong reflections from the microscope's vertical illuminator at the PBS's narrowband AR-coated surfaces out of the CCD camera's field of view. For the purpose of sample inspection, focussing and positioning the laser probe, the reduced field of view thus obtained is perfectly acceptable.

Spatial resolution in the object plane is defined by the stop aperture in the detection channel. The stop aperture is formed either by the active area of the detector itself, as is the case e.g. when a small SPAD detector is directly attached to the spectrometer, or by the core diameter of an optical fiber. This is the case for a pigtailed SPAD detector, which can then be conveniently detached from the spectrometer and placed close to the active quenching circuit (AQC) in the instrumentation rack.

Suppression of scattered excitation light posed a serious design problem, as the axes along which excitation light, photoluminescence signal, microscope illumination and reflection from the sample travel coincide. Reflection off the sample surface was reduced by choosing the polarization and the angle of incidence of the excitation light such that it was close to the Brewster angle. Within the ORM, all optical surfaces along the excitation beam path were AR coated at the excitation wavelength and the components either tilted or the surfaces bevelled. As a result, residual reflections from surfaces in the parallel beam region of the ORM form spots, "ghost images" of the excitation source, which are well separated from the image of the luminescent sample and discriminated by spatial filtering at the stop aperture. Suppression of diffuse scattered light is achieved by using interference filters and long bandpass filters.

Tilting/bevelling of components was found to be crucial in discriminating against multiply reflected luminescence photons, which results in replicas of the decay superimposed onto the main decay, with peak amplitudes typically of the order of 1% the main peak height and shifted in time (see fig.4). This is unacceptable, because the S/N ratio in the peak can be as high as 10^3-10^4 and these features are clearly resolved from the noise.

An enormous practical advantage results directly from the high spatial resolution requirement and the high sensitivity of the TCSPC technique : The optical power requirements of the system are so low, that a pulsed laser diode can be used for excitation instead of a large picosecond laser system. The PLS system can take full advantage of this by handling all optical signals to and from the spectrometer via optical fiber. All the well known problems associated with running a large laser system, installation and alignment, start-up time, shutdown for maintenance, high running costs etc. are hence eliminated.

3.2. Timing electronics

The short PL decay times in GaAs substrates allow to take full advantage of the high repetition rate (up to 100 MHz) of commercially available pulsed laser sources. Theoretically, this permits data collection rates of over 200 kcps (0.2% of 100 MHz) without risking pulse pile-up. In practice, the maximum data collection rate is limited to around 100 kHz due to the finite conversion dead time even when using the fastest commercially available time-to-amplitude converter (TAC), the Tennelec TC 563 (min. 2µs dead time) and analog-to-digital converter/multichannel analyzer combination (ADC/MCA), the Silena Mod. 7423/UHS ADC and Mod. 7328 MCA/buffer (3µs fixed dead time).

However, strong non-linearities in the response of commercial TACs to random START signals are already observed when the TAC's STOP input is triggered with periodic signals at frequencies as low as 1 MHz, although these devices are designed to operate with random signals up to $3 \cdot 10^7$ cps (30 MHz) on STOP and START (specification for EG&G Ortec Mod. 567 TAC). For this reason, cavity dumping has so far been considered essential in TCSPC. This has certainly not stimulated widespread application of the TCSPC technique, as the need for cavity dumping, usually available only in connection with a synchronously pumped dye laser system, is clearly a disadvantage w.r.t. other TRPL methods capable of operating at the fundamental mode-locking frequencies of a simple Nd:YAG or ion laser.

So far, reduction of the high repetition rate signal to the much lower PL signal rate by some form of coincidence gating, has, although proposed in the literature, not been widely used in TCSPC experiments. We propose here the simple two stage anti-coincidence gating circuit shown in fig.3. The circuit makes use of four cascaded constant fraction discriminators (CFD1-CFD4), such as the commercially available Phillipps Scientific Mod. 714 QUAD CFD, and eliminates the need for any purpose built hardware. The only additional components required are three variable coaxial delay lines (DEL2-DEL3) and one fixed delay coaxial cable (DEL1).

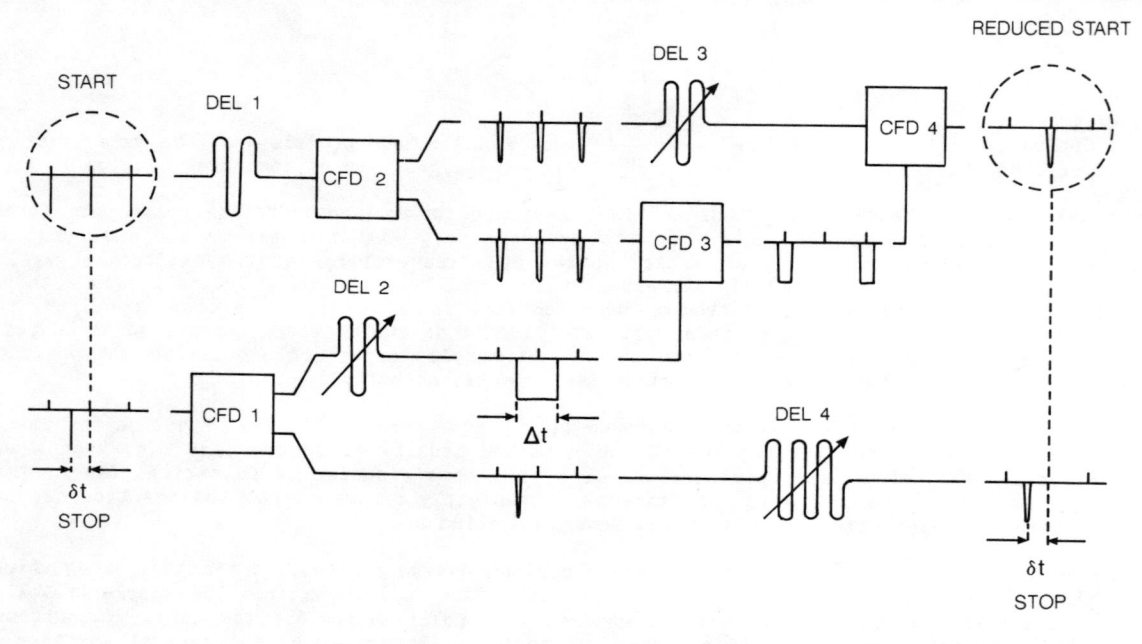

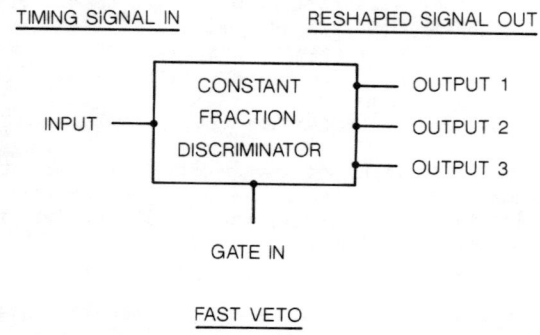

Fig.3 : Two stage anti-coincidence gating circuit based on a QUAD CFD with fast VETO (inset on the right) for reducing the high repetition rate TAC START signal to the low TAC STOP signal rate.

Three successive START pulses out of a continuous train are shown in fig.3. In this particular example, the 2nd START pulse corresponds to the single STOP pulse, which is delayed w.r.t. START by δt. (The statistical distribution of δt over a large number of events and discretization in the TAC/ADC/MCA chain produces a histogram equivalent to the PL decay.) The START pulse is deliberately delayed w.r.t. STOP by DEL1. This allows the GATE pulse triggered by the STOP pulse in CFD1 to overlap the leading edge of the shaped output from CFD2 present at the input of CFD3. The width of the GATE pulse is precisely adjusted to be Δt, the START pulse spacing. Synchronization of the reshaped START output from CFD2 and the GATE pulse from CFD1 is achieved by varying the delay DEL2. The START pulses corresponding to STOP pulses are eliminated in the output from the anti-coincidence gated CFD3. The CFD3 output is used to anti-coincidence gate CFD4 which then produces the desired REDUCED START pulse train from the shaped, delayed and synchronized second output of CFD1.

Any TCSPC system operating at high repetition rates (> 1 MHz) will benefit from the described circuit in three ways :

1. The integral non-linearity of the TAC, due to the high repetition rate periodic input, disappear.
2. The TAC can be operated in normal, i.e. non-inverted mode.
3. The useable time window Δt can be placed anywhere within the full TAC range.

As a result, the integral non-linearities of typically ±20% at repetition rates (STOP) above 10 MHz are reduced to the values specified for operation up to 30 MHz with random input (START and STOP), typically less than 0.2%. In addition, with most MCAs using Wilkinson type ADCs for converting the TAC output, the overall conversion rate is strongly improved when the decay peak is placed into a low channel. At very high repetition rates, up to 100 MHz, the width of the useable time window is Δt is increased significantly, up to 50%, by shifting the window away from the beginning of the TAC range, which is affected by the fixed dead time.

3.3. Data analysis

The goodness of fit obtained to the experimental decay data by reconvolution analysis determines the improvement of time resolution over the instrumental response width of the TCSPC setup. A perfect fit, however, can only be obtained with a correct kinetic model, taking into account material parameters, measurement conditions and sample geometry. In other words, attempts to fit experimental data by "brute force", e.g. by simply fitting a number of exponential functions, will not normally result in good fits as assessed by the criteria value of chi-square and distribution of residuals.

As an example we shall look at the correct analytical solution to the one-dimensional transient diffusion equation for a GaAs substrate, assuming very high surface recombination velocity (S > 10^6 cm/s)[17] :

$$I_{PL}(t) = A_0 + A_1 \cdot \exp(t/\tau) \left\{ A_2 \cdot W(t/\tau_2) + A_3 \cdot W(t/\tau_3) \right\} \tag{1}$$

where the function W is defined as

$$W(t/\tau_i) \equiv \exp(t/\tau_i) \cdot \mathrm{erfc}(\sqrt{t/\tau_i}) \quad (i=2,3) \tag{2}$$

erfc being the complimentary error function. τ is the bulk minority carrier lifetime, defined as

$$\tau = (1/\tau_r + 1/\tau_{nr})^{-1} \tag{3a}$$

with τ_r the radiative, τ_{nr} the non-radiative lifetime.

$$A_1 = N_0/\tau_r \tag{4a}$$

where N_0 contains known parameters, such as the number of photons in the excitation pulse, i.e. the excitation density. $I_{PL}(t)$ in eq.(1) varies linearly with N_0 and A_1, τ are independent of time t at low excitation density, which results in dominantly non-radiative recombination. τ_2, τ_3 depend on the minority carrier diffusion constant D, the absorption coefficient at the excitation wavelength, α, and the absorption coefficient for recombination radiation, β, at the PL detection wavelength[18]:

$$\tau_2 = (\alpha^2 \cdot D)^{-1}$$
$$\tau_3 = (\beta^2 \cdot D)^{-1} \tag{3b}$$

The corresponding amplitudes are given by

$$A_2 = \alpha \cdot (\alpha^2 - \beta^2)^{-1}$$
$$A_3 = -\beta \cdot (\alpha^2 - \beta^2)^{-1} \tag{4b}$$

and A_0 is the usual noise background. If we assume that N_0, α and β, hence A_2 and A_3, are known, the remaining terms in eq. (1), A_1, τ, τ_2, τ_3 contain the radiative lifetime τ_r, the non-radiative lifetime τ_{nr} and the minority carrier diffusion coefficient D as the only free parameters. These can then be simultaneously fitted by reconvolution analysis.

The analysis of large amounts of decay data produced in scanning applications has to be automatically carried out by a "master program" requiring little or no user interaction. This is realistic, given that data from neighbouring gridpoints are very similar and therefore good starting values are available for every new cycle of the reconvolution program. The fact that well-defined statistical criteria are available for automatic assessment of the quality of the reconvolution fit may be considered an inherent advantage of TCSPC w.r.t. other techniques, where even fitting of a simple exponential is normally done graphically and always requires strong user interaction.

4. EXPERIMENTAL RESULTS

Figure 4 shows a typical instrumental response curve for the PLS as initially obtained, before careful optimization.

The satellite peaks before and after the main peak are due to multiple reflections of light scattered in the optical routing module. Spatial filtering by the small detector active area, 5 µm diameter, eliminated these peaks after all optical components had been slightly tilted w.r.t. the optical beam bath.

The noise background is not flat, as would be expected, because of the strong integral non-linearity of the TAC produced by the 50 MHz repetition rate at the STOP input. The TAC was operated in inverted mode, therefore data shown in fig.4 is time reversed. This non-linearity disappears after reducing the high repetition rate to the signal rate with the help of a coincidence gating circuit similar to the one shown in fig.3.

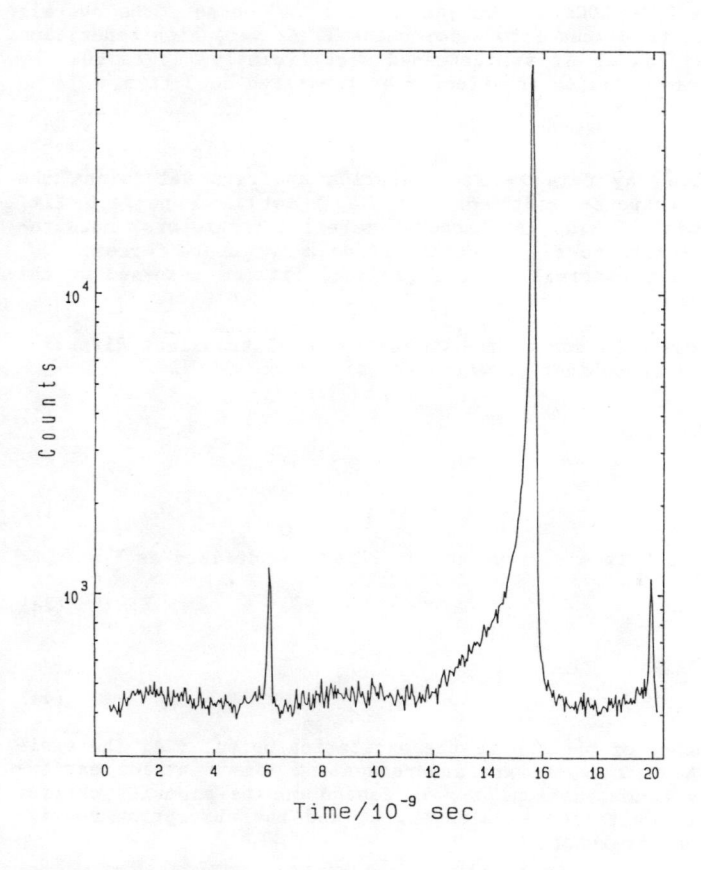

Fig.4 : PLS instrumental response before optimum alignment of stop aperture, TAC operated in inverted mode at 50 MHz STOP rate.

Fig.5 : PLS instrumental response (+), TRPL data from LEC grown n-doped GaAs:Si substrate (x) three-exponential fit (-) and residuals (below).

Figure 5 shows a typical TRPL data from LEC grown n-doped GaAs:Si substrate after annealing. Excitation wavelength was 785 nm, detection wavelength 830 nm with a filter of 10 nm bandwidth. The 5 µm diameter SPAD detector was operated uncooled, at room temperature. The instrumental response width was 68 ps (FWHM) with the SPAD detector operated at a bias voltage about 0.5 V above breakdown.

A three-exponential decay model

$$I_{PL}(t) = A_0 + A_1 \cdot \exp(-t/\tau_1) + A_2 \cdot \exp(-t/\tau_2) + A_3 \cdot \exp(-t/\tau_3) \tag{5}$$

was fitted to the experimental data in order to give a rough phenomenological description of the decay. The fitted time constants τ_i, relative amplitudes A_i and quantum efficiencies η_i are

$$\tau_1 = 49 \text{ ps} \quad A_1 = .14 \quad \eta_1 = 16\ \%$$
$$\tau_2 = 405 \text{ ps} \quad A_2 = .055 \quad \eta_2 = 52\ \% \quad (6)$$
$$\tau_3 = 1630 \text{ ps} \quad A_3 = .0086 \quad \eta_3 = 32\ \%$$

This three-exponential model is of course not a proper physical description of the photoluminescence decay but was chosen simply for convenience. The correct analytical solution given in section 3.3 had not been implemented in the reconvolution analysis yet at the time of this publication. Yet, the dominant second decay component with a time constant of 405 ps agrees well with the results obtained in streak camera measurements of the same sample. The latter were carried out at lower temperature and higher excitation density and are therefore not directly comparable. The non-exponential character of the decay, with a fast initial decay component and an additional slow decay component in the tail, also agrees with the prediction of the analytical model.

The value of chi-square, XSQ=1.70, and the residual distribution, however, clearly indicate a non-perfect fit. This can be seen as a demonstration of the sensitivity of the TCSPC method and reconvolution analysis, because, by merely looking at the fit to the data in fig.5, the three-exponential model would be considered as a rather good description of the decay !

The noise background was rather high, $A_0 = 242$, and the signal-to-noise ratio low, S/N = 40, because the 830 nm filter was not matched to the peak of the photoluminescence at 872 nm and therefore the overall signal rate was low, only 10,000 cps.

The total photon luminescent flux, I_{PL}, can be estimated by multiplying the internal quantum efficiency of the sample, η_{QE}, the collection efficiency, η_{COL}, which takes into account the geometric loss factors due to limited detector field of view and limited solid angle of the collection optics, the spectrometer efficiency, η_{SPEC}, including the losses due to polarizational and spectral discrimination, the external quantum efficiency of the detector, η_{COL}, the number of photoexcited carriers generated per excitation pulse, N_{EXC}, and the repetition rate of the source, f_{REP} :

$$I_{PL} = \eta_{QE} \cdot \eta_{COL} \cdot \eta_{SPEC} \cdot \eta_{DET} \cdot N_{EXC} \cdot f_{REP} \quad (7)$$

Typically, with $\eta_{QE} \approx 0.4\%$ (SI-GaAs substrate at R.T.), $\eta_{COL} \approx 0.05\%$ (as in fig.2), $\eta_{SPEC} \approx 10\%$ (10nm bandwidth filter), $\eta_{DET} \approx 10\%$ (1st generation epitaxial SPAD, non-etched), $N_{EXC} \approx 5*10^5$ (5mW$_{peak}$, 30ps (FWHM) at 785nm) and f_{REP}=50MHz for a diode laser, a photon luminescent flux of approx. $5*10^4$ sec^{-1}, or, 10^{-3} photons/pulse is obtained. For a detector field of view of 3 µm and an excitation depth of 1 µm, i.e. an observed sample volume of approx. 7 µm^3, the corresponding initial excess carrier density is of the order of 10^{15}-10^{16} cm^{-3}. Under these conditions, recombination is dominated by linear non-radiative recombination even in low-doped materials and all material parameters, τ, L, D, can be taken as constant, independent of excitation density.

5. CONCLUSIONS

We have shown that it is possible to measure picosecond TRPL in GaAs with 3 µm spatial resolution by using the TCSPC technique. The PLS instrument based on this technique is sensitive enough to allow the use of a pulsed diode laser as the excitation source. It was demonstrated that, by use of a suitably modified commercial light microscope, the small active area of the SPAD detector is not a disadvantage w.r.t. large area photocathode fast single photon detectors such as MCP-PMTs. The low power handling requirements of the PLS system allow extensive use of optical fibers for interconnecting source, detector and sample microscope, thereby completely eliminating alignment problems.

As a result, and in contrast to any other experimental setup reported for TRPL measurements in semiconductors so far, the TCSPC based PLS system reported here is suitable for routine operation in an industrial environment. Applications currently envisaged are on-line quality control of laser diodes, photodiodes and optoelectronic integrated devices.

Further work will be aimed at setting up a software library providing kinetic models for fitting to TRPL decay curves for various types of samples. A commercial PLS system is being developed in collaboration with Edinburgh Instruments Ltd., Edinburgh, UK.

6. ACKNOWLEDGEMENTS

The author wishes to thank A. Goetzberger, Fraunhofer Institut für Solare Energiesysteme, Freiburg, F.R.G., for accepting this work as part of a PhD thesis. D. Smith and A. Walker, Heriot-Watt University, physics department, Edinburgh, U.K., have encouraged further development and funded the first PLS prototype through a SERC grant. Technical and logistic support from Edinburgh Instruments Ltd. in setting up the PLS prototype is gratefully acknowledged. Sergio Cova, Istituto di Fisica del Politecnico, Milan, Italy, generously provided the SPAD chips and the active quenching circuit. Without his helpful collaboration, encouraging comments and continuous interest in the PLS development from the very beginning, the project would not have been so successful.

7. ABBREVIATIONS AND SYMBOLS

TCSPC :   time-correlated single photon counting
PLS :   photoluminescence lifetime spectrometer
ORM :   optical routing module
SPAD :   single photon avalanche diode
MCP-PMT : micro-channelplate photomultiplier
TRPL :   time-resolved photoluminescence
TAC :   time-to-amplitude converter
CFD :   constant fraction discriminator
MCA :   multichannel analyzer
ADC :   analog-to-digital converter
DEL :   coaxial delay line
τ :   (minority) carrier lifetime
L :   (minority) carrier diffusion length
D :   (minority) carrier diffusion coefficient

8. REFERENCES

1. K. Leo and W.W. Rühle, "Influence of annealing on free carrier lifetime and deep level luminescence in semi-insulating GaAs", Proceedings MRS meeting, Strasbourg (1987)
2. J. Windscheif and W. Wettling, "GaAs wafer investigation by near infrared transmission and photoluminescence topography techniques", Inst.Phys.Conf.Ser.No.8: Chapter 4, presented at the Int.Symp. GaAs and Related Compounds, Las Vegas (1986)
3. J. Windscheif and W. Wettling, "Combined photoluminescence and near-infrared transmission imaging for GaAs wafer inspection and process control", in Defect Recognition and Image Processing in III-V compounds II, E.R. Weber, ed., Elsevier Science Publishers, Amsterdam (1987)
4. W. Wettling and J. Windscheif, "Direct and fast comparison of near-infrared absorption and photoluminescence topography of semiinsulating GaAs wafers", Appl.Phys. A 40, 191-195 (1986)
5. A. Steckenborn, H. Münzel and D. Bimber, "Cathodoluminescence lifetime pattern of semiconductor surfaces and structures", Inst.Phys.Conf.Ser.No.40: Section 4, presented at Microsc.Semicond.Mater.Conf., Oxford (1981)
6. A. Von Lehmen and J.M. Ballantyne, "Investigation of the nonlinearity in the luminescence of GaAs under high-density picosecond photoexcitation", J.Appl.Phys. 58 (2), (1985)
7. Jagdeep Shah, "Ultrafast luminescence spectroscopy using sum frequency generation", IEEE J.Quantum Electron., Vol.24, No.2 (1988)
8. J.S. Weiner and P.Y. Yu, "Free carrier lifetime in semi-insulating GaAs from time-resolved band-to-band photoluminescence", J.Appl.Phys. 55 (10), (1984)
9. J.E. Fouquet and A.E. Siegman, "Room-temperature photoluminescence times in a $GaAs/Al_xGa_{1-x}As$ molecular beam epitaxy multiple quantum well structure", Appl.Phys.Lett. 46 (3), (1985)
10. J.A. Kash, J.H. Collet, D.J. Wolford and J. Thompson,   Phys.Rev.B 27, 2294 (1983)
11. D.J.S. Birch, G. Hungerford, B. Nadolski and R.E. Imhof, "Infrared time-correlated single photon counting", Edinburgh Instruments technical bulletin, Fluorescene No.6 (1987)
12. A. Andreoni, S. Cova, R. Cubeddu and A. Longoni, "Solid-state detector for single-photon measurements of fluorescence decays with 100 picosecond FWHM resolution", in Picosecond Phenomena III, Springer (1982)
13. S.Cova, A. Longoni, G. Ripamonti, "Active quenching and gating circuits for single photon avalanche diodes (SPADs)", IEEE Trans. Nucl. Sci., NS-29, 599-601 (1982)
14. S. Cova, G. Ripamonti, A. Lacaita, "Avalanche semiconductor detector for single optical photons with a time resoltuion of 60 ps", Nucl. Instrum. Methods, A253, 482-487 (1987)
15. S. Cova, private communication (1988)
16. T.A. Louis, G.H. Schatz, P. Klein-Bölting, A.R. Holzwarth, G. Ripamonti and S. Cova, "Performance comparison of a single photon avalanche diode with a micro-channelplate photomultiplier in time-correlated single photon counting", Rev.Sci.Instrum., 59 (7), (1988)
17. R.K. Ahrenkiel, D.J. Dunlavy and T. Hanak, "Photoluminescence lifetime in heterojunctions", Solar Cells, 24, 339-352 (1988)
18. for a detailed discussion of the effects of reabsorbed recombination radiation (RRR) see e.g. O.v. Roos, "Influence of radiative recombination on the minority carrier transport in direct band-gap semiconductors", J.Appl.Phys., 54 (3), 1390-1398 (1983)

Topography of GaAs/AlGaAs heterostructures using
the lateral photo effect

P.F. Fontein, P. Hendriks, J. Wolter

Faculty of Physics, Eindhoven University of Technology,
5600 MB Eindhoven, the Netherlands.

A. Kucernak, R. Peat, D.E. Williams

Harwell Laboratory, U.K. Atomic Energy Authority,
Oxfordshire OX11 0RA, United Kingdom

ABSTRACT

We studied the lateral photo effect in GaAs/Al$_x$Ga$_{1-x}$As heterostructures both theoretically and experimentally. We observe a linear dependence of the photo voltage as a function of the position of the light spot. In our model this corresponds to a recombination length of the spatially separated electrons and holes longer than the length of the sample (1 mm). Deviations of this linear dependence are a direct indication of inhomogeneities in the conductive properties of the two-dimensional electron gas at the interface of the heterostructure. Results are shown in which long (1 mm) but very narrow cracks are seen.

1. INTRODUCTION

Recently GaAs/Al$_x$Ga$_{1-x}$As heterostructures have become of great interest because of their high electron mobility (used in e.g. the high electron mobility transistor, the HEMT). However, in spite of this high electron mobility these heterostructures are not necessarily homogeneous with respect to electron concentration and resistivity. Knowledge of the homogeneity and defects therefore is of great importance, on the one hand to interpret electrical transport experiments, on the other hand to improve the yield of device fabrication. In this paper we describe a method to investigate the homogeneity of GaAs/AlGaAs heterostructures using the lateral photo effect. It is a non-destructive technique with a spatial resolution of approximately 0.3 μm.

The lateral photo effect was first observed in p-n junctions[1]. In these junctions two contacts are made at one (p or n) side of the junction. When the junction is illuminated locally between the contacts, a perpendicular (normal) photo voltage occurs. Due to the local illumination this voltage established across the junction depends on position and has a maximum value at the illuminated spot. As a consequence a voltage drop in the lateral direction appears and this is called the lateral photo effect. The effect is not unique for a p-n junction but can also be observed on a heterostructure[2]. This is the topic of this paper.

In the following paragraphs we will subsequently describe the experimental set-up, a model of the lateral photo effect and some experimental results.

2. EXPERIMENTAL SET-UP

The experimental set-up consists of a commercially available scanning optical microscope (Biorad Lasersharp). We used the system in both the scanning optical microscope (SOM) mode and optical beam induced current (OBIC) mode. The OBIC mode was used to measure the lateral photo effect. The sample (see section 3) is moved under a fixed spot of light (λ=628 nm) on a scanning table. The voltage across the two contacts connected to the sample is measured and stored digitally. This voltage is also represented on a video screen as an intensity, so a map of the lateral photo effect is obtained directly. A light area on the screen corresponds to a positive, a dark area to a negative voltage across the contacts, zero is in between. The spot size is approximately 1 μm and one complete scan takes about 7 s.

3. DESCRIPTION OF THE LATERAL PHOTO EFFECT

A theoretical description of the lateral photo effect in p-n junctions has been given by Lucovsky[3]. Whereas Lucovsky describes the lateral photo effect in case one side (p or n) of the sample is an equipotential, we will drop this restriction. However, we will only study a rectangular shaped sample of width b that is illuminated across this width b with a line of light of width 2a. So one (lateral) dimension drops out of the problem.

The band structure of a GaAs/AlGaAs heterostructure is given in Fig. 1. The two-dimensional electron gas (2-DEG) is confined at the interface of the GaAs and the AlGaAs. Illumination takes place from the AlGaAs side of the sample. This thin layer (with energy gap larger than the energy of the incoming photons) is transparent to the incoming light. So the creation of an electron hole pair occurs only in the GaAs (note that the electron hole pair creation in the extremely thin GaAs cap layer is neglected). The internal electric field of the heterojunction separates the electrons and holes and a photo voltage V develops

Fig. 1

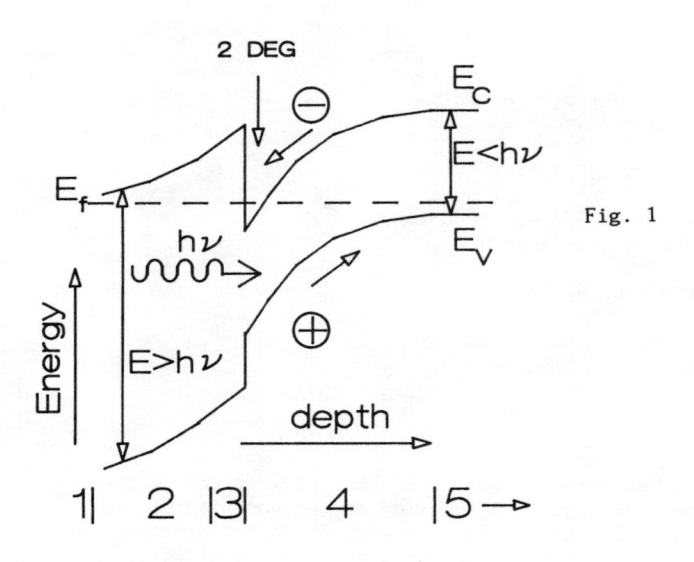

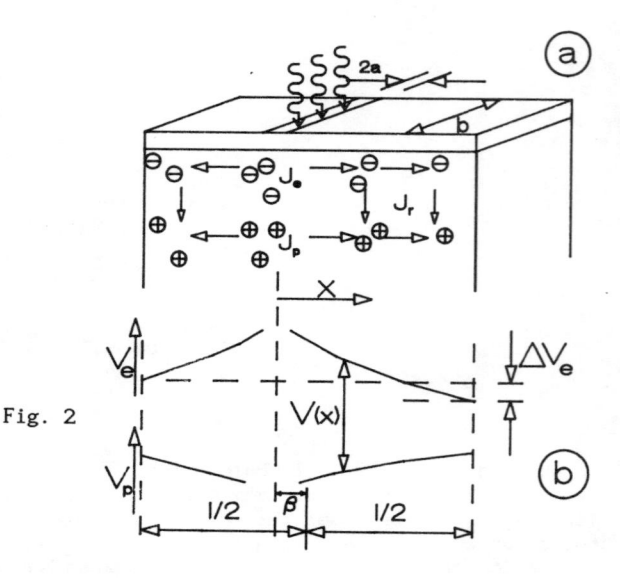

Fig. 2

Fig.1 The band structure of a GaAs/AlGaAs heterostructure. The two-dimensional electron gas is at the interface of the two materials. Indicated are the different layers of the structure. Visible are (1) a GaAs cap layer of 100 Å, which prevents the Al from oxidation, (2) a Si-doped AlGaAs layer of 500 Å from which the electrons in the two-dimensional electron gas originate, (3) an undoped AlGaAs spacer layer of 180 Å, which separates the electrons from their donors thus giving a high mobility and (4) the GaAs layer of 1 μm on (5) the semi-insulating GaAs substrate (slightly p-type). The electron concentration of the heterostructure under study here is $6*10^{15}$ m^{-2}, the mobility 0.6 m^{2}V^{-1}s^{-1}, both at 300 K.

Fig. 2 (a) A schematical picture of the lateral photo effect. Visible are the creation of electron hole pairs and the subsequent current flow. The sample width is b, the width of the line of light 2a. The currents J_e, J_p and J_r are indicated.
(b) The potentials V_e, V_p and the resulting potential difference ΔV_e are schematically indicated. The sample length is l, the distance to the centre of the sample β.

perpendicular to the junction. If V_e and V_p denote the voltages in the 2-DEG and in the p-side of the GaAs respectively we get $V(x)=V_e(x)-V_p(x)$, with x=0 at the centre of the line of light. Associated with V(x) is a recombination current $J_r(x)$ perpendicular to the interface. We use the linearized dependence

$$J_r(x) = J_0 q V(x)/kT \tag{1}$$

for this recombination current, with q the elementary charge, k the Boltzmann constant, T the temperature and J_0 a constant depending on the details of the heterostructure. Apart from this recombination current there are also currents J_l in the lateral direction because of the x-dependence of V_e and V_p, see Fig. 2. These currents obey the relation

$$dV_i/dx = -\rho_i J_{li}, \tag{2}$$

with index i=e for the electrons in the 2-DEG and i=p for the holes in the p-side of the GaAs, ρ is the resistivity. Because J_{le} has to be equal to J_{lp} in a stationary situation we may write $V = V_e(1+\rho_p/\rho_e)$. It is now possible to show that

$$dJ_{le}/dx = -J_r = -(J_0 q V_e/kT)(1+\rho_p/\rho_e). \tag{3}$$

Combination of Eq. (1) and Eq. (2) leads to the differential equation

$$d^2V_e/dx^2 - \alpha^2 V_e = 0, \tag{4}$$

with $\alpha^2 = (\rho_e+\rho_p)J_0 q/kT$. The solution in case of an infinitely long structure is simply

$$V_e = Ae^{-\alpha|x|}. \tag{5}$$

The constant A can be obtained directly from the equality of the generation and the recombination of the electrons and holes. Let g2ab be the total number of electron hole pairs generated per unit time. Then the combination of Eqs. (1) and (5) and the assumption that the recombination in the illuminated zone is small with respect to the total recombination leads to

$$(J_0q/kT)(1+\rho_p/\rho_e)\int_{-\infty}^{\infty} Ae^{-\alpha|x|}dx = (J_0q/kT)2A/\alpha = 2qga. \qquad (6)$$

So $A = (1+\rho_p/\rho_e)ga\alpha kT/J_0 = \rho_e gaq/\alpha$. The solution obtained for V_e is also the solution for a rectangular sample of finite length with total recombination in the end contacts. Using Eq. (5) we write for the voltage ΔV across such a sample with length l that is illuminated at a distance β from the centre of the sample, see also Fig. 2:

$$\Delta V = 2Ae^{-\alpha l/2}\sinh(\alpha\beta). \qquad (7)$$

In case no recombination occurs at the contacts a different solution is obtained. Consider again a sample of length l. The boundary conditions at $\beta=+l/2$ and $\beta=-l/2$ become $J_l=0$, so $dV_e/dx=0$ there. We can write for V_e that

$$V_e = A \sum_{i=-\infty}^{\infty} (e^{-\alpha|x-\beta+i2l|} + e^{-\alpha|x+\beta+l+i2l|}), \qquad (8)$$

which obeys the imposed boundary conditions, as can easily be verified. The constant A is obtained in a way similar to Eq. (6) and is again equal to $\rho_e qga/\alpha$. The calculation of the voltage ΔV across the sample results in

$$\Delta V = 2Ae^{-\alpha l/2}\sinh(\alpha\beta) \; 2(1+1/2 \; e^{-\alpha l/2}). \qquad (9)$$

Thus the β dependence is $\sinh(\alpha\beta)$ for both boundary conditions, see Eqs. (7) and (9). Only the magnitude of the voltage differs depending on $\alpha l/2$, varying between a factor of 2 to 3. The magnitude of the vertical (normal) photo voltage, however, depends strongly on the boundary conditions.

4. EXPERIMENTAL RESULTS AND DISCUSSION

A direct optical (SOM) image of the heterostructure used for the experiments is shown in Fig. 3 (a). Visible are the rectangular Hall bar (centre) of width 300 μm and length 2 mm. Connected to this Hall bar are six so called voltage probes or side arms with channels of about 20 μm wide. These are normally used to measure the voltages in a (Quantum) Hall experiment but are unconnected in our experiments. The two voltage contacts used to measure the lateral photo voltage are at the two ends of the Hall bar (not visible) and make contact to the 2-DEG. The material outside the Hall bar has been etched away down to just below the 2-DEG.

Fig. 3 (a)

Fig. 3 (b)

Fig. 3 (a) A direct picture of the Hall bar structure. The bar is situated vertically in the centre. The six side arms are also visible, three on each side. The two voltage contacts are not visible but just outside the picture (one above and one below).
(b) The lateral photo effect image of the structure, the orientation is the same as in Fig. 3 (a). Visible are the voltage gradients in the bar as well as in the side arms.

The result of a lateral photo effect experiment is shown in Fig. 3 (b). The orientation of the sample is identical to the orientation in Fig. 3 (a). Visible is a gradual change of the voltage from one end of the main channel to the other. This gradient has to obey the derived functional dependence of $\sinh(\alpha\beta)$ (Eqs. (7),(9)). All experiments we performed resulted in a linear dependence of the lateral photo voltage on spot position (neglecting inhomogeneities). So we have to conclude that $\alpha l \ll 1$, because in this case $\sinh(\alpha\beta)$ is approximately equal to $\alpha\beta$. That means that the recombination length of the spatially separated electrons and holes is much larger than the sample length. The resulting voltage has the order of magnitude of $2\rho_e gaq\beta$ which corresponds to the measured 10 mV if we take the following values: $2gab=10^{13}$ s^{-1} (a laser power of 0.01 mW reaching the sample and a conversion efficiency of 50 %), $\rho_e=1600$ Ω (the resistance of a square of 2-DEG), $\beta=10^{-3}$ m, $b=3*10^{-4}$ m. We neglected here of course the difference between illumination with a spot and with a line of light.

Also visible is a gradient of the voltage in the side arms. This gradient has a sign opposite to the one of the main channel. We can interpret this behaviour easily if we imagine a side arm disconnected from the main channel. If the (disonnected) side arm is illuminated there is no voltage difference between a contact at the main channel and the p-side of the GaAs directly underneath this contact, because there is neither a current flowing there nor a photo voltage built up at that position. Thus it is just as if the contacts of the main channel were placed at the p-side of the GaAs. The lateral photo voltage measured is thus the lateral photo voltage developed in the p-side of the GaAs underneath the side arm and thus of the mentioned opposite sign, see also Fig. 2. In reality, however, the side arm is not disconnected, so the importance of the effect depends on the resistance of the narrow channel from the side arm to the main channel.

Let us now look at some details. Structure is visible in the lower part of the main channel and the right upper side arm. The right upper side arm is shown again in Figs. 4 (a) and 4 (b). While nothing peculiar (except for some dust particles) can be observed in the SOM picture, we clearly see long 'scratch like' structures in the image of the lateral photo effect. These structures have also been observed by Hendriks et al.[4] with a liquid crystal technique. The origin of the cracks is not clear. While some of them seem to stop at the boundary of the 2-DEG others run straight through the substrate. This demonstrates that also the substrate is active with respect to the lateral photo effect. Probably the electric field piles up at the surface of the substrate due to surface states, making it photo active. Another possibility is that the cracks are photo active by themselves. The width of the cracks has been studied by a scanning electron microscope (SEM) technique too[4] and appears to be smaller than 0.1 μm. Since these cracks influence the electrical properties of the Hall bar under study it is of great importance to be aware of their presence and to know their origin in order to think of a way to avoid them. At the moment we can only speculate about their origin, but it seems to be clear that the cracks must be present in the substrate prior to the growth of the heterostructure.

In conclusion we state that the lateral photo effect has been proven to be a useful technique to study GaAs/AlGaAs heterostructures. Several interesting properties have been revealed, including the existence of long range cracks of unknown origin.

Fig. 4 (a) 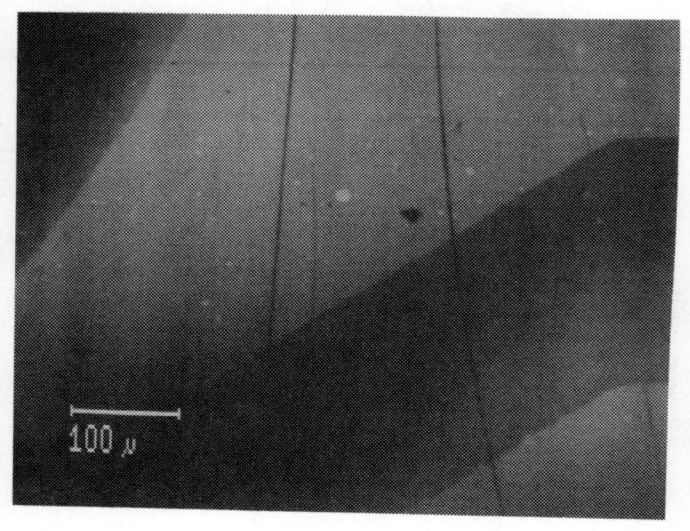   Fig. 4 (b)

Fig. 4 (a) A direct picture of the upper right side arm of Fig. 3 (a).
(b) The lateral photo effect image of the upper right arm of the structure. The orientation is the same as in Fig. 4 (a). Clearly visible are narrow cracks.

5. ACKNOWLEDGMENTS

The authors like to thank M. Hayles and M. Spijkers for the use of their SOM. Part of this work is supported by the Stichting voor Fundamenteel Onderzoek der Materie.

6. REFERENCES

[1] Wallmark T, Proc. I.R.E. 45, 474 (1957).
[2] Fontein P F, Hendriks P, Wolter J, Peat R, Williams D E, André J P, J.Appl.Phys., scheduled for september 1988.
[3] Lucovsky G, J.Appl.Phys. 31, 1088 (1960).
[4] Hendriks P, Kort K de, Horstman R E, André J P, Foxon C T, Wolter J, Sem.Sc.& Techn. 3, 521 (1988)

Scanning laser photocurrent spectroscopy of electrochemically grown bismuth sulphide films

Anthony R. Kucernak, Robert Peat, and David E. Williams

*UKAEA Harwell Laboratory, Material Developments Division,
Building 393, Didcot, Oxon OX11 0RA, UK*

ABSTRACT

The photoelectrochemical properties of semiconducting bismuth sulphide (Bi_2S_3) films (in contact with an aqueous electrolyte) grown on bismuth have been investigated. Using an Optically Beam Induced Contrast (OBIC) technique, it has been possible to image these films, and to identify local variations corresponding to grain boundaries and other recombination centres. It is seen that the form of these regions varies with the thickness of the film. Both object scanning and beam scanning have been investigated for generating images and these two methods are compared. Samples have also been investigated by Intensity Modulated Photocurrent Spectroscopy (IMPS). This method involves sinusoidal intensity modulation of the incident laser radiation at different frequencies, and the analysis of the resulting photocurrent response. Our extension of this method has been to map the IMPS response over the specimen surface in order to obtain specific information on local properties.

1 INTRODUCTION

Photocurrents generated by scanning a light spot across the surface of electrochemically grown films can be used to image various surface and near surface properties.

Figure 1 shows the semiconductor film interface. A contact is made to the semiconductor surface by using an aqueous electrolyte, and illumination is accomplished through this medium. Generation of charge carriers by photo excitation is followed by the separation of these charge carriers by the electric field. The magnitude of the photocurrent detected is dependent upon the generation and recombination rates, and the magnitude of the electric field.

The variation in energy across the semiconductor film interface is shown in the diagram for p-type materials. Holes produced by the incident illumination are separated by the field towards the metal contact apart from a small recombination current which will be discussed later. Electrons proceed to the surface where they can react with solution species ($O + e^- \rightarrow R$) or with the semiconductor (photoetching). Electrons may fill surface states present in the semiconductor film, and again react with solution species. Holes produced near the surface may transfer to these surface states and recombine with electrons that have filled these states. Recombination due to surface states decreases the photocurrent seen in an external circuit and also gives rise to a transient behaviour in the photocurrent. Thus the variation in concentration of surface states is one reason for localised changes in contrast.

Stimming[1], has reviewed the photoelectrochemistry of passive metal films. On a metal electrode the sign and magnitude of any photocurrent is dependent on a number of characteristics of the photoactive layer:

* Whether it is a p-type or n-type semiconductor, or insulator (films on valve metals).

* The film thickness.

* The concentration and nature of trap states and surface states.

* Whether the current is produced by photoexcitation in the film or photoemission from the underlying metal into the film.

Thickness is usually the predominant factor for very thin films as they usually do not absorb all of the incident light.

Whereas standard electrochemical techniques used for the study of electrochemically grown semiconductors usually sample the entire electrode, the techniques that we have developed and present here are spatially resolved, and produce information on the positional variation of the parameters we are studying.

Our studies have involved looking at films that are electrochemically grown on top of metals. These products and the processes that lead to them are of wider interest too, since they may be applicable to other systems, examples being:

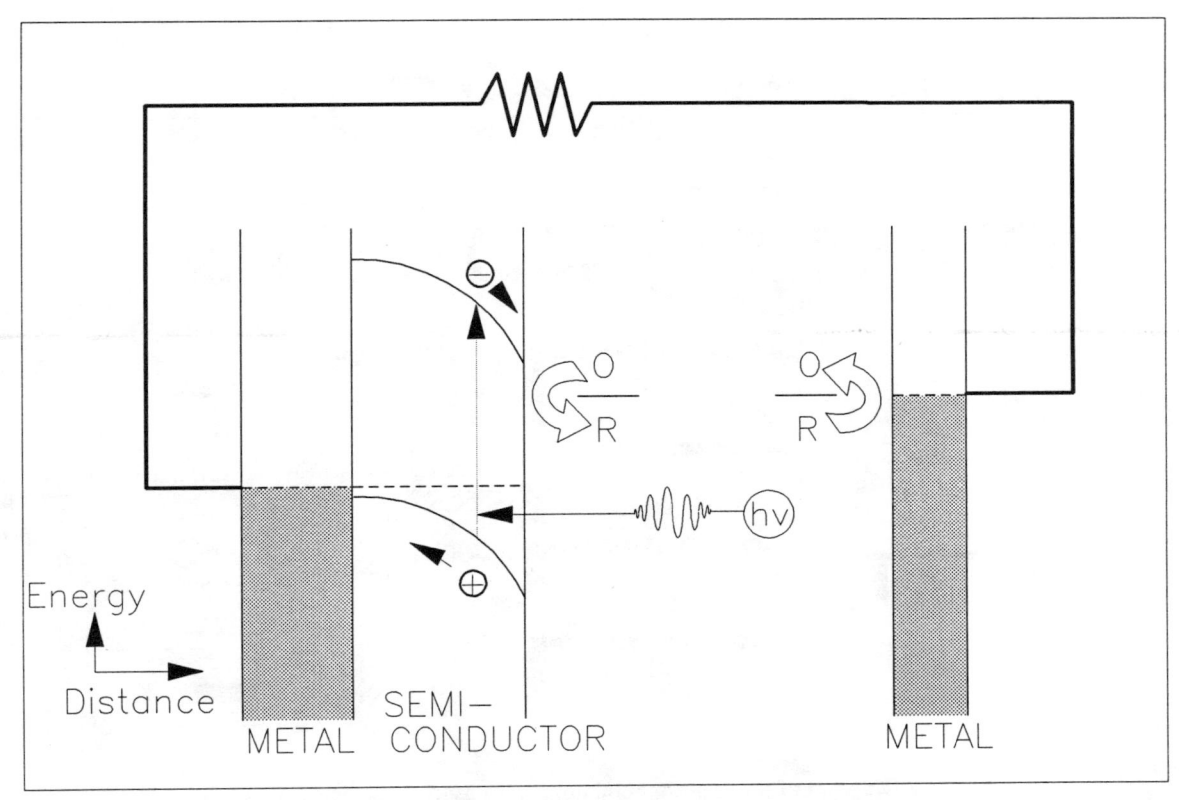

Figure 1: Diagram showing the aqueous/semiconductor/metal film interface

* <u>Corrosion:</u> During corrosion processes the metal surface invariably becomes covered with a film. Either native as for example with the passive oxide film on iron or deliberate by introduction of inhibitors to the electrolyte phase.

* <u>Semiconductor Manufacture:</u> Electrochemical processes are used in some sections of the semiconductor industry (e.g. in the production of liquid junction solar cells).

* <u>"Pure" Electrochemical Processes:</u> e.g. anodisation, electrodeposition etc.

The time dependence of the induced photocurrent may be determined by the technique of IMPS (Intensity Modulated Photocurrent Spectroscopy). The theory of IMPS for the case of a thin semiconductor film on a metal has been described previously[2] and compared with the experimentally derived data for the formation of the passive film on iron. In this paper it is shown that the IMPS technique can be applied using a focussed laser spot to obtain a spatially resolved profile of the transient photocurrent response. Under certain circumstances it is possible to decouple the values of these elements and thus obtain the local variation of the film resistivity.

2 EXPERIMENTAL

The arrangement of our system is such that we can choose either to raster the laser spot over the surface of our sample (keeping it still), or to scan the sample whilst keeping the laser beam steady.

For the systems we have looked at (comprising of samples which have been electrochemically grown in situ), we have usually employed a beam scanning technique (where the laser beam is deflected over the sample using a pair of mechanically controlled mirrors). The reasons for using this technique as opposed to an object scanning method are:

* The electrochemical cells in which we grow the samples in are relatively bulky and this limits the maximum rate which we can scan the sample due to the inertia of the combined system.

* The systems we look at are aqueous electrochemical ones, and we would wish to limit the amount of stirring and convection which goes on within the cell.

* The mounting and removal of an electrochemical cell would be more difficult, especially with regard to the possibility of spillages and other minor "accidents".

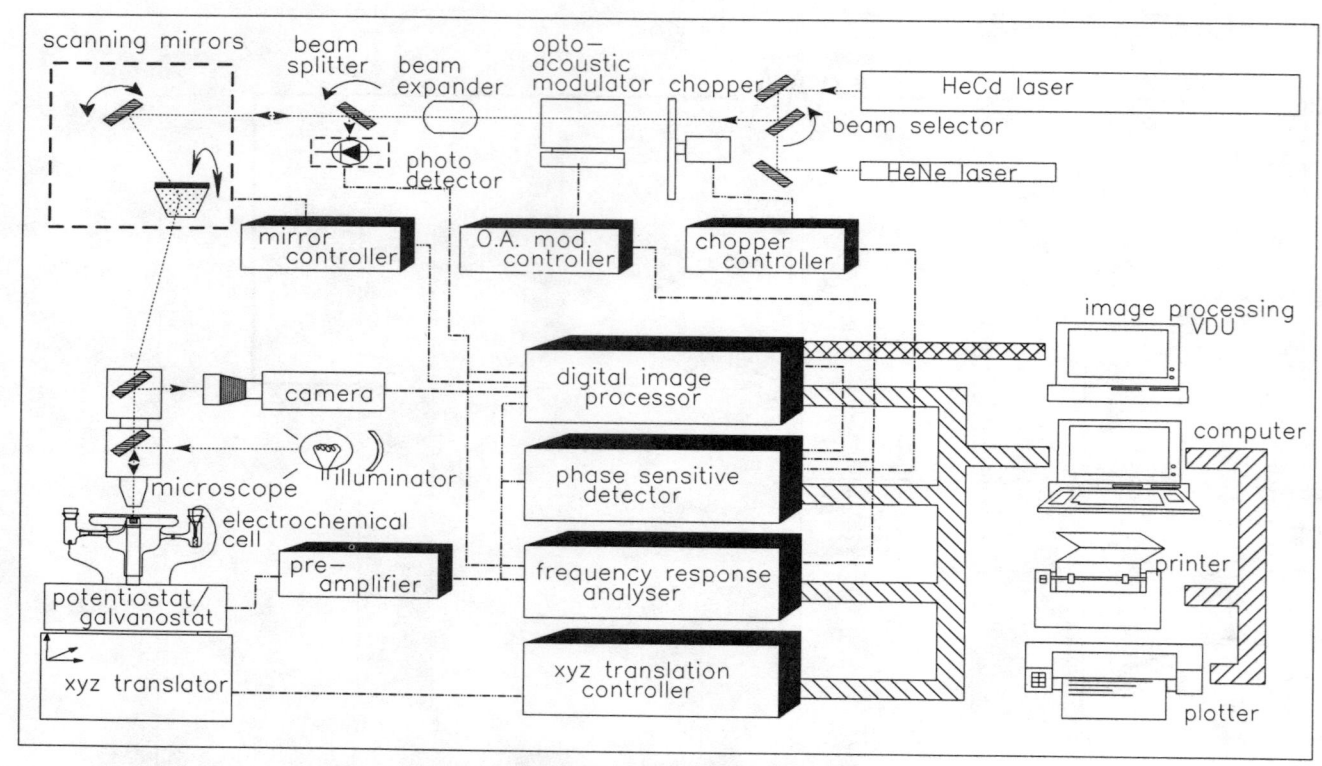

Figure 2 Diagram of the experimental apparatus used

Figure 2 schematically illustrates the experimental apparatus used in these experiments. Photocurrent is induced in samples using light from one of two laser sources: either a 3 mW Helium-Neon (λ = 632.8 nm), or a 10 mW Helium-Cadmium (λ = 464nm). The laser sources can be easily interchanged, and the beam intensity modulated by either a mechanical chopper (low frequency, square wave intensity profile), or by an acousto-optic modulator (high frequency, sine or square wave intensity profile), and scanned over the sample using a pair of orthogonally mounted mirror galvanometers. The intensity profile may be monitored via a beam splitter and a photodiode, and this signal provides a reference signal for the Phase Sensitive Detector (PSD) or Frequency Response Analyser (FRA). The controlling signals for the galvanometer mirrors are produced by the framestore used to acquire the image. The accessible scan speeds range from 3ms to 200s per line; allowing a minimum acquisition time of 3s per complete frame. The laser beam is focused through a microscope objective onto the sample, and provision is made to view the sample through the microscope via a video camera which is also connected to the framestore.

The framestore acquires images at a resolution of 512 x 512 pixels, each pixel being digitised to 8-bit accuracy. It also provides the ramp signals used to control the mirror galvanometers. Input for the framestore is either a video signal from the camera which displays a non laser scanned image of the sample as seen through the microscope, or a signal representing some parameter induced by the laser beam scanning the surface. This may be a photoacoustic emission (thermo acoustic imaging), a photon emission (photoluminescence), or reflected light.

The sample may be positioned under the microscopy by an xyz translator, providing a 10 μm step resolution in the xy plane perpendicular to the laser beam and 0.25 μm resolution in the z direction. Instead of using the mirror galvanometers to scan the laser beam, it is possible to scan the sample using this translator, although this is a much slower process, partly because of the need of a 1-2s dwell time at each point to allow vibrations to die down.

A Research Machines VX40/XL with an HP BASIC Language Processor (HP 82321A) were used to control the experiments.

Either a potentiostat or galvanostat was used to control the electrochemical growth of the semiconducting film by holding the potential or current (respectively) constant.

Both the potentiostat and galvanostat were purpose built, and optimised for high bandwidth because with the maximum scan speed, each point is illuminated for as little as 6 μs. Their output was passed through a preamplifier (PAR Model 113) before being passed to the framestore. The electrochemical cell was mounted directly on top of the potentiostat (which in turn was mounted on top of the xyz translator), this being done to reduce the length of leads. A Hi-Tek Instruments PPR1 wave generator was used to generate ramps and establish controlling potentials.

The electrochemical cell used is of the "piston" type, a diagram of which is given in figure 3. The sample was mounted on a threaded brass button before being inserted into a teflon sleeve, gaps between the sample and the sleeve being filled with epoxy

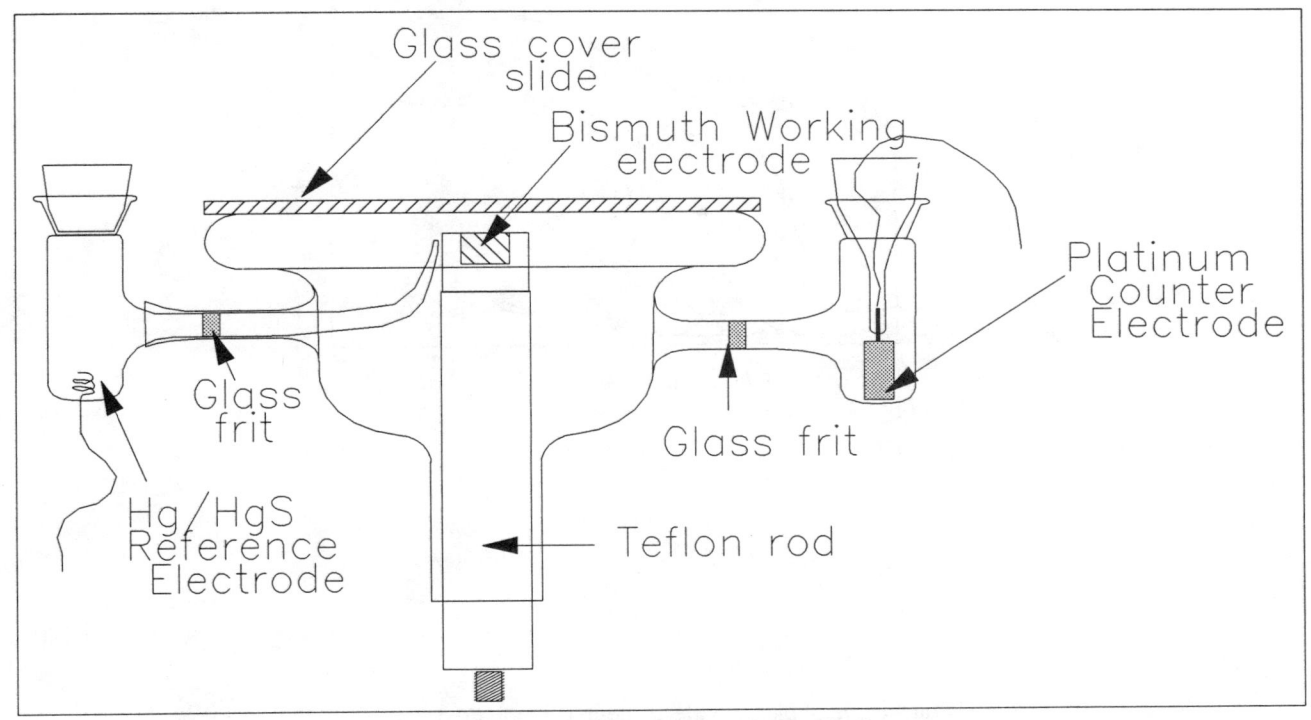

Figure 3: Diagram of the electrochemical cell used

resin. This was then screwed on to a rod surrounded by a teflon sleeve which was machined to closely fit the glass cell. The arrangement thus arrived at could be moved within the cell enabling the sample to be positioned either closer or further away from the window at the top of the cell. A Hg/HgS(red) ($E^{\circ} = -0.69V$) electrode was used to provide a reference potential, and all potentials are quoted referenced to it. A platinum gauze was used as the counter electrode. Both reference and counter electrodes were separated from the main solution by glass frits.

99.999% Bismuth shot was used to make the electrodes, and Analar grade chemicals were used throughout. Before each experiment, the electrodes were polished to a mirror finish with 0.05 μm Alumina, and etched by dipping in 0.5% nitric acid for 2 seconds before thoroughly washing. The bismuth sulphide film was either grown potentiostatically by holding the potential at 0.5 V for 10 minutes, or by the method of Peter[3].

3 RESULTS AND DISCUSSION

3.1 Electrochemistry of bismuth in a sodium sulphide solution

A standard technique for studying electrochemical processes is cyclic voltammetry. Figure 4(a) shows the current-voltage curve for a bismuth electrode in 0.5 mol.dm^{-3} sodium sulphide solution, in which the potential is ramped from -0.50V to 1.0V and back again at a sweep rate of 10 mV/s. The main features to note are:

* A small anodic peak present on the forward sweep just past 0.0V attributed to the formation of 6-8 monolayers of Bi_2S_3 by a mechanism of successive nucleation of two-dimensional layers.

* A much larger anodic peak at ca. 0.60V is again attributed to the formation of Bi_2S_3 attributed to the nucleation and growth of three dimensional centres on the metal electrode.

* A number of cathodic peaks seen on the reverse sweep at voltages less than 0.0V due to the reduction of the film to regenerate the native metal electrode.

It is possible to deduce information concerning the semiconductor properties of the electrochemically grown compounds. In order to do this, chopped light is used to illuminate the electrode while the voltage is ramped as in the above experiment. Detecting the component of the current that is modulated at the same frequency as the illuminating light using a PSD allows the Photocurrent Voltage curve to be obtained. This is shown in Figure 4(b). Laser light from a Helium Neon laser was used, and this light was chopped with a mechanical chopper at 420 Hz. Points to note are that the photocurrent does not start increasing till well after the film has started growing, and that it disappears before the film has started to be reduced off the electrode. Reasons for the first point noted above are that some reordering of the film is necessary before it becomes efficient at producing

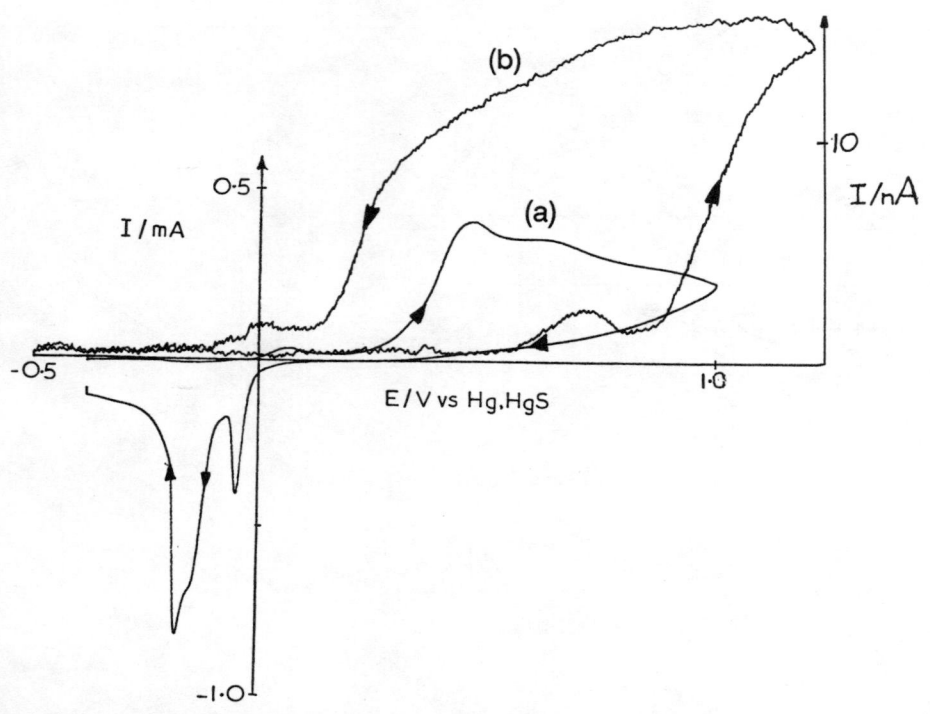

Figure 4: (a) Current Voltage curve of bismuth electrode in 0.5 mol.dm-3 sodium sulphide solution; (b) Photocurrent Voltage Curve

photocurrent and also at higher anodic potentials there is the possibility of the formation of some bismuth oxide (Bi_2O_3). Once the film is grown it behaves as a typical n-type semiconductor in which the photocurrent decreases rapidly as the junction is biased towards the flatband potential.

Both of the techniques presented above are bulk techniques: they sample the entire electrode surface and only present an average property of the system. Because the films we are studying are not single crystal of fixed orientation, and thus not spatially uniform, the OBIC technique can be used to to display their heterogeneous properties.

3.2 OBIC of electrochemically grown thin bismuth sulphide film

Figure 5(a) shows an optical micrograph of a bismuth electrode after preparation but before any bismuth sulphide film was grown. Notice the pronounced grain structure visible on this electrode which is 2mm in diameter. A 1 μm thick bismuth sulphide film was then grown on the surface by polarising the electrode to +0.40V for 10 minutes. Under visual examination the film appeared coloured due to the interference of light reflected off the front face of the film, and the metal/film interface. The film thus produced was then imaged using the OBIC mode.

Figure 5(b) shows the OBIC image for the electrode described above, illumination coming from a Helium-Neon laser, (λ = 632.8 nm), and the potential was controlled at 0.30V. The presence of the underlying metal's grain structure is clearly visible as a contrast difference. Note that there has been a slight rotation of the images with respect to one another.

The contrast in this image is due to local variations in photocurrent across the film. The lighter regions corresponding to higher amounts of photocurrent.

The thickness of the semiconducting film is different on different metal grains and is small compared to the absorption coefficient for bismuth sulphide. Hence the amount of light absorbed and thus the photocurrent produced profiles the local thickness variation. For thin films light which is not absorbed as it passes through the film, reflects off the underlying metal electrode, and then passes through the film again (where there is the possibility of further absorption). Films grow to different thicknesses due to variations in parameters that affect the rate of growth, such permeability of the film to ions and substrate orientation.

3.3 OBIC of electrochemically grown thick bismuth sulphide film

Figure 6(a) shows a micrograph of the bismuth metal electrode used in this experiment; again, before any film was grown. It was obtained by looking at light from the laser reflected off the metal electrode as the laser was scanned over the surface (type II, non confocal). Again a film was grown on the surface as described above, but this time the film produced was much thicker, appearing black under visible light illumination (i.e. absorbing all incident light).

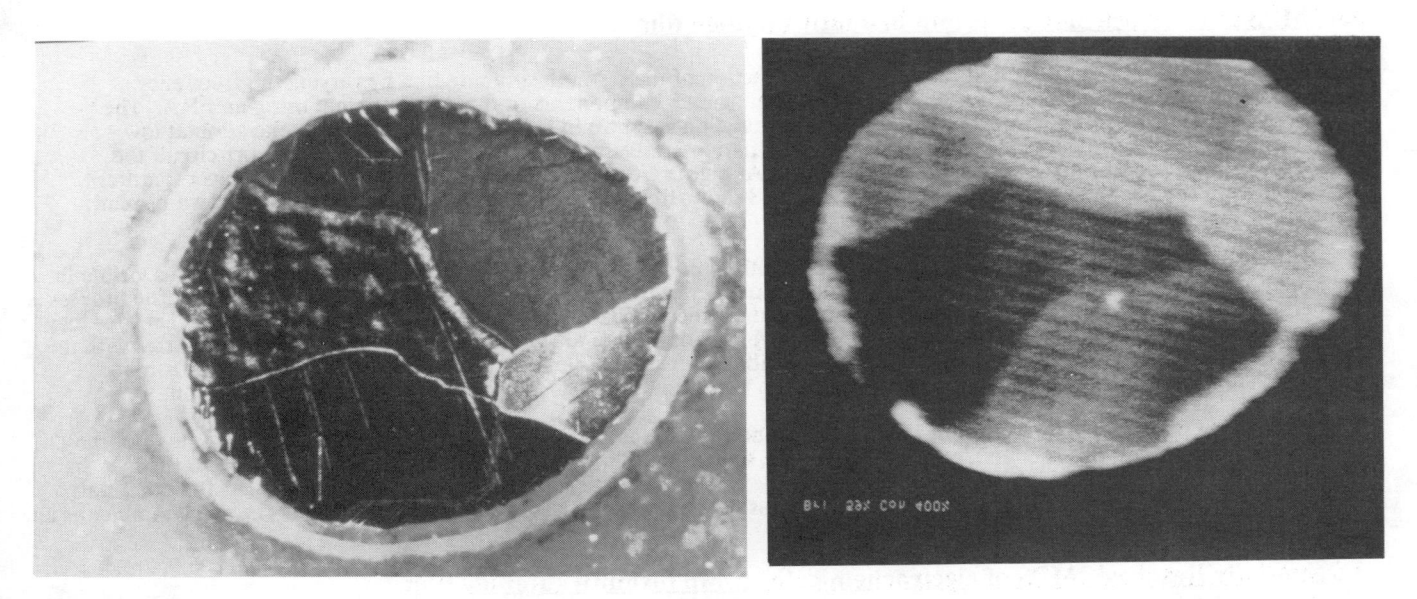

Figure 5: (a) Optical micrograph of bismuth electrode showing grain structure (2mm diameter); (b) OBIC image of a thin bismuth sulphide film grown on (a) showing the presence of the underlying metal's grain structure.

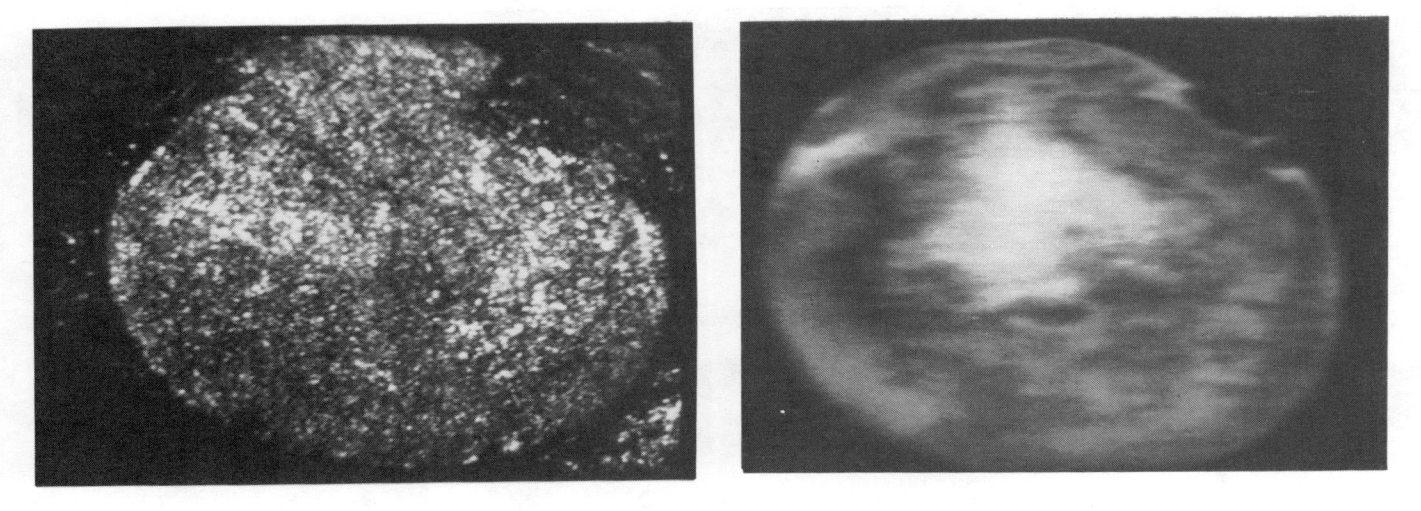

Figure 6: (a) Scanning laser reflected light micrograph of bismuth electrode showing rough surface (1.4mm diameter); (b) OBIC image of the thick bismuth sulphide film grown on (a).

The corresponding OBIC image is shown in figure 6(b). Again using a Helium Neon laser for illumination, and again the potential applied to the working electrode was controlled at 0.30V. Since the OBIC and reflected light images were both performed in situ on the same sample, it is possible to superimpose these two images in the framestore, and compare them. Little correlation is seen between the structure of the electrode and contrast seen in the OBIC image. It can thus be concluded that contrast is not due to the thickness of the film, but rather some other mechanism (e.g. concentration of recombination centres) must be the cause.

3.4 IMPS of electrochemically grown bismuth sulphide film

We have also looked at the photocurrent transfer function of the bismuth sulphide film by varying the frequency of illumination of incident light and recording the real and imaginary components of that photocurrent using an FRA. The theoretical prediction for the semiconductor/electrolyte junction is shown in figure 7. It consists of a relaxation at low frequencies in the positive quadrant which provides information on the rate constant for processes that short-circuit the photocurrent as is the case for a recombination process. At higher frequencies a semicircle appears in the lower quadrant associated with the RC time constant of the film. Figure 8 shows the IMPS spectrum of a galvanostatically grown bismuth sulphide film at different potentials.

The form of these curves should be compared to the theoretical plot[2] given in figure 7. A relaxation process is visible in the upper quadrant but the time constant given by the frequency at the maximum of this semi-circle indicates that the process is slow compared with the millisecond time regime observed for recombination processes in other materials. However, this may be distorted by the dominating effect occurring in the lower quadrant due to the high capacitance of the film in series with the resistance of the system. This resistance contains a contribution from the film and a relatively small component from the solution. The low frequency intercept on the real axis is a measure of the DC photocurrent obtained from the film.

Comparing the trend in the different IMPS shown, one can see that as the potential is decreased, the magnitude of the DC photocurrent decreases; this is equivalent to the response seen in the Photocurrent Voltage curves, and is due to the fact that the applied voltage approaches the flat band condition the effective charge separation process is reduced. At the potentials where these experiments are performed, the film undergos no electrochemical reactions and is acting in the same manner as an ordinary semiconductor Schottky barrier.

3.5 Spatially Resolved IMPS of electrochemically grown bismuth sulphide film.

Spatially resolved IMPS spectra were obtained by focussing the laser beam to a spot and rastering the sample underneath the beam.

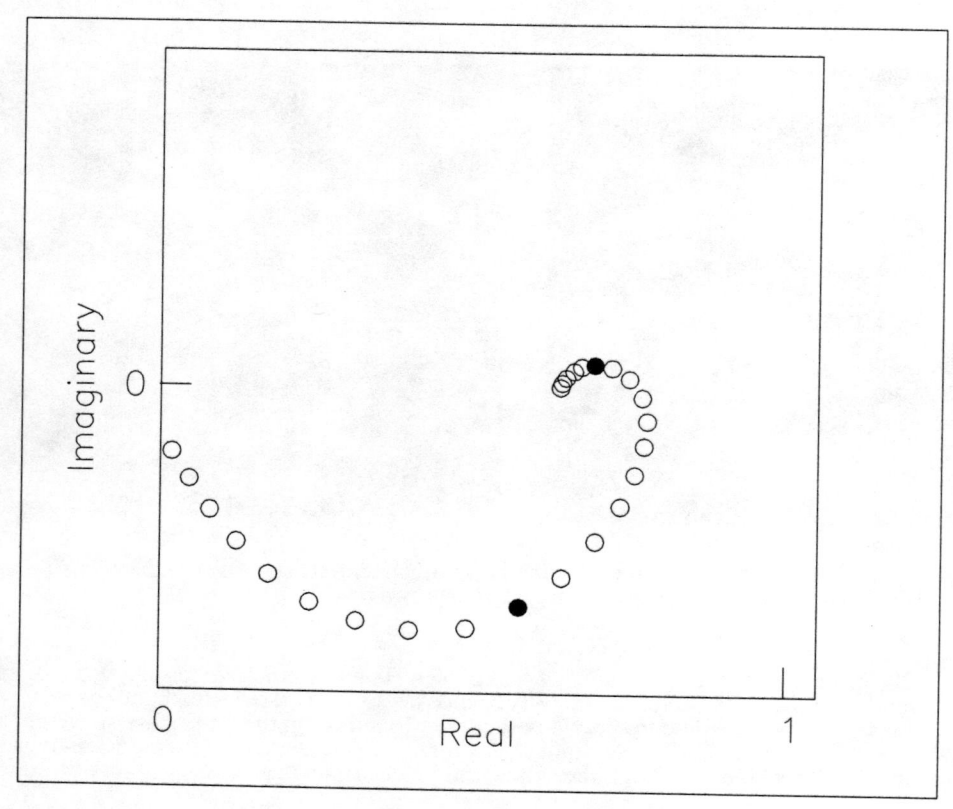

Figure 7 Theoretical photocurrent response resulting from the RC time constant of the cell.

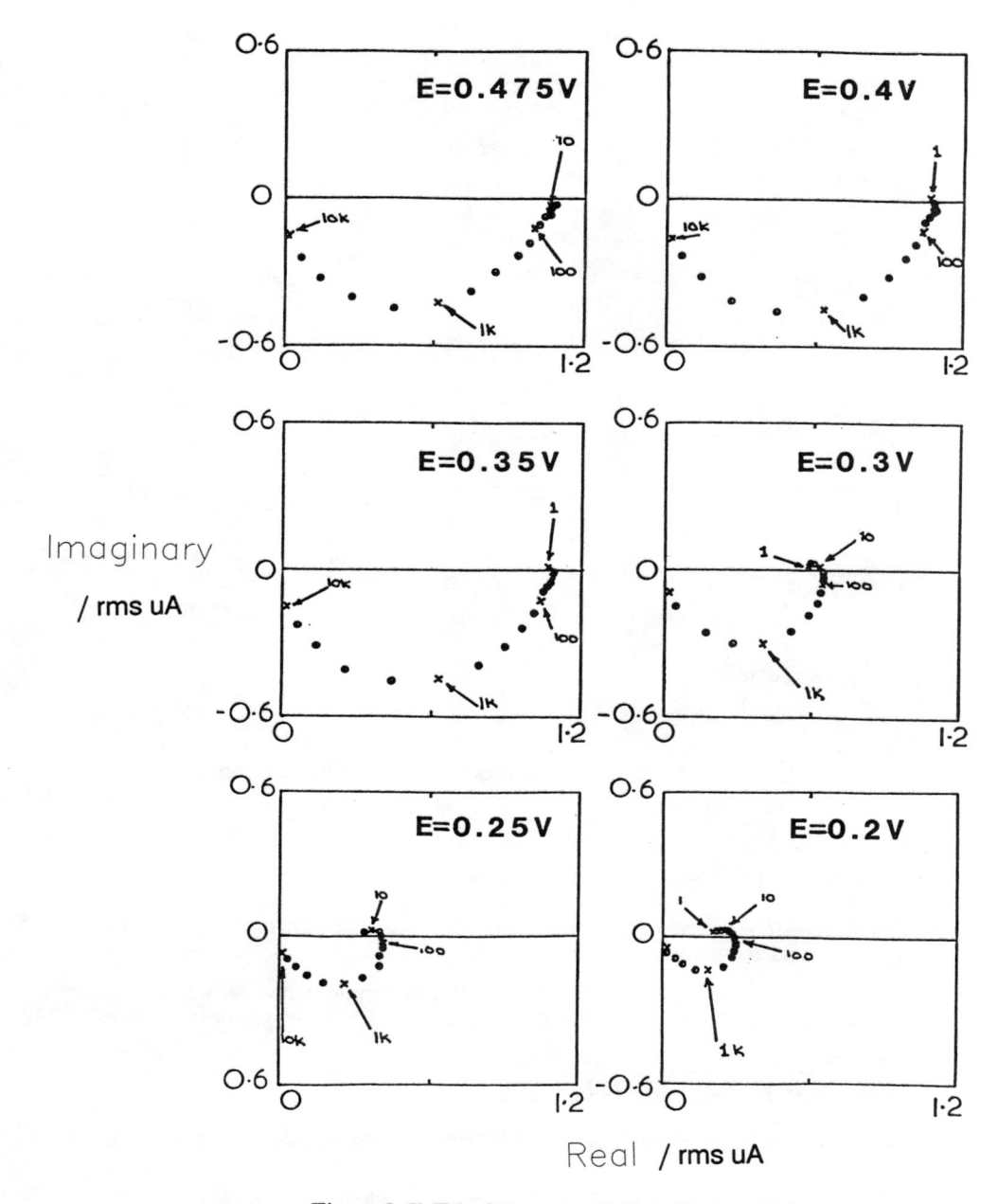

Figure 8: IMPS of bismuth sulphide film at different potentials.

The film was the same one as imaged in figure 6, the area scanned over was 1000 μm in the x direction (across the page), and 400 μm in the y direction (down the page) at 20 μm intervals i.e. 50(x) x 20(y) spectra were run. The equipment used for running the IMPS spectra were the same as used in the "normal" (non-spatially resolved) IMPS experiment listed above. A total of 10 frequencies were sampled from 1000 Hz to 1 Hz using a logarithmic scale with three frequencies per decade.

For normal IMPS, the illumination is defocussed across the entire electrode, and both the resistance and capacitance measured are of the entire interface. If we focus the beam down to a single spot and measure an IMPS spectrum, then the capacitance we measure is still that of the entire film ($R_{film} \gg R_{solution}$ i.e. a macro property), but the resistance we measure is the point resistance of the sample at the spot that we are illuminating.

Figure 9 shows these spatially resolved IMPS spectra, with the real and imaginary responses plotted at different frequencies. Note that Imaginary response has been multiplied by a factor of (-1). The maximum real photocurrent observed was 300 nA, and the maximum imaginary was 150 nA.

Because there is considerable variation in the response of the film at a spatially resolution higher than our sampling resolution, the plots are not predominantly smooth. This gives the appearance of noise in these figures. However, regions are present that are relatively smooth (e.g. the lines closest to the observer), and there is correlation between some features on successive lines.

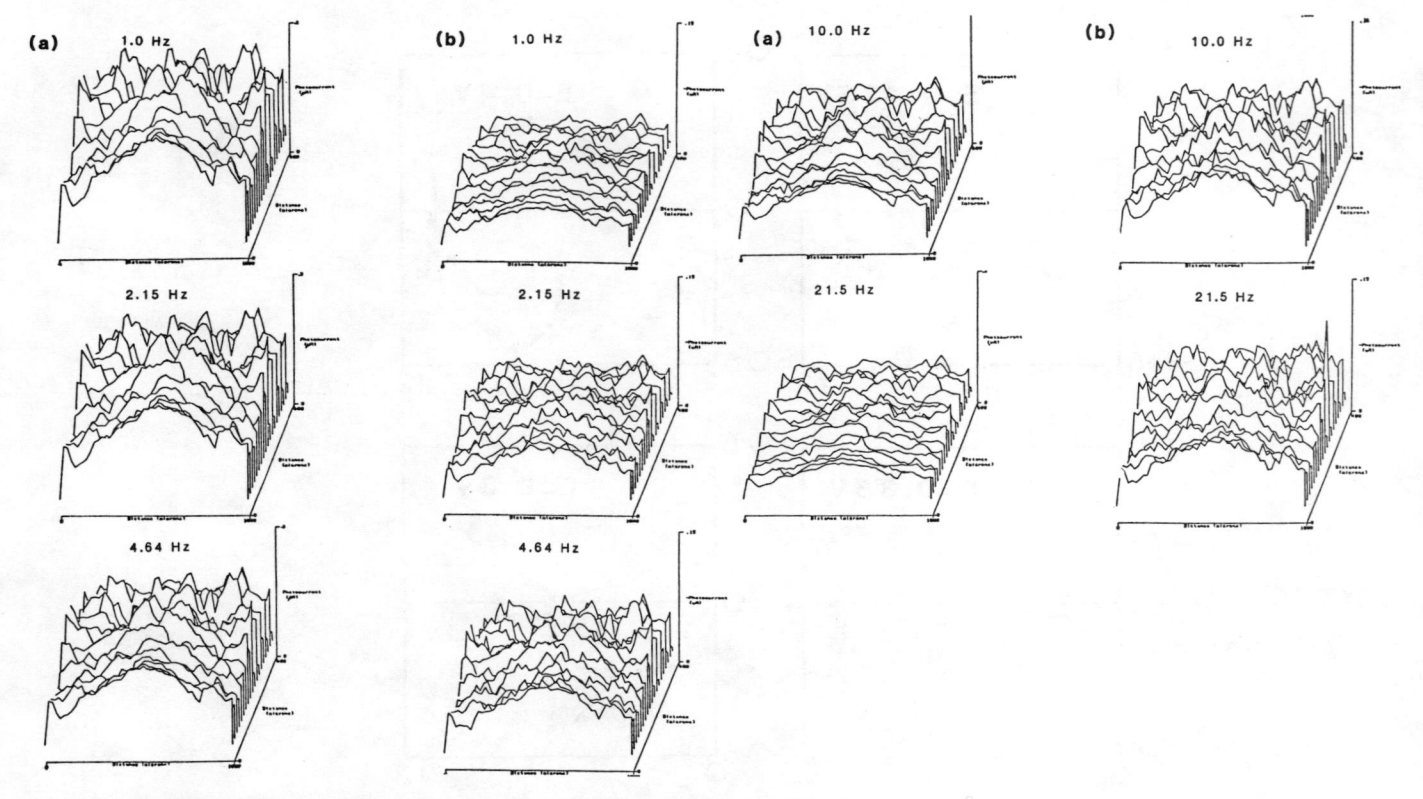

Figure 9: Spatially resolved IMPS spectra, (a) Real and (b) -Imaginary responses plotted at different frequencies

4 CONCLUSION

We have shown that it is possible to image electrochemically grown films using variations in photocurrent to provide contrast. Contrast mechanisms include film thickness and density of surface states.

A new technique has been developed by extending the standard technique of IMPS to include spatial variation. It allows the spatially resolved acquisition of the photocurrent transfer function. Using this technique, it should be possible to look at the spatially resolved film resistance.

5 ACKNOWLEDGEMENTS

This work has been funded under the Underlying Research Budget of the United Kingdom Atomic Energy Authority.

REFERENCES

1. U. Stimming, "Photoelectrochemical studies of passive films", Electrochim. Acta, Vol. 31, pp. 415-429. 1986.
2. R. Peat, and L.M. Peter, "A study of the passive film on iron by intensity modulated photocurrent spectroscopy", J. Electroanal. Chem. 228 (1987) pp 351-364.
3. L. Peter, "The photoelectrochemical properties of anodic Bi_2S_3 films" J. Electroanal. Chem. 98, 49-58 (1979)

Circuit analysis in ICs using the scanning laser microscope

J. Quincke, E. Plies, J. Otto

Research Laboratories of Siemens AG,
Otto-Hahn-Ring 6, D-8000 München 83, FRG

ABSTRACT

The laser probe of the scanning laser microscope generates electron-hole pairs in the IC semiconductor material which are separated at inversely-biased pn-junctions and can then be detected as photocurrent at external terminals. This OBIC (Optical Beam Induced Current) effect provides the physical basis for the localization and sensitivity measurements of the undesired latch-up in CMOS devices. The laser probe can also be used to measure logic-level time diagrams at the internal nodes of an IC. In addition, the scanning laser probe can be used to stimulate and localize malfunctions when the IC is operated at its marginal conditions.

1. INTRODUCTION

Chip verification and failure analysis of integrated circuits (ICs) often necessitate measurements in the interior of these ICs. This is done using fine mechanical measuring probes as well as the electron and laser probe /1/, the latter being the most recent of these tools. Used as the scanning probe of a scanning laser microscope (SLM), it is particularly suitable for practical circuit analysis.

2. SCANNING LASER MICROSCOPE

Fig. 1 shows a block diagram of an SLM by the Zeiss company /2,3/. The beam of a red He-Ne laser (632.8 nm, 5 mW) is initially expanded, then traverses firstly an acousto-optical modulator which is active only at specific operating modes and subsequently a half-silvered mirror. After the ensuing reflections at the two galvanometer mirrors for the x and y scans, the laser beam is focussed as a fine probe from the objective onto the object.

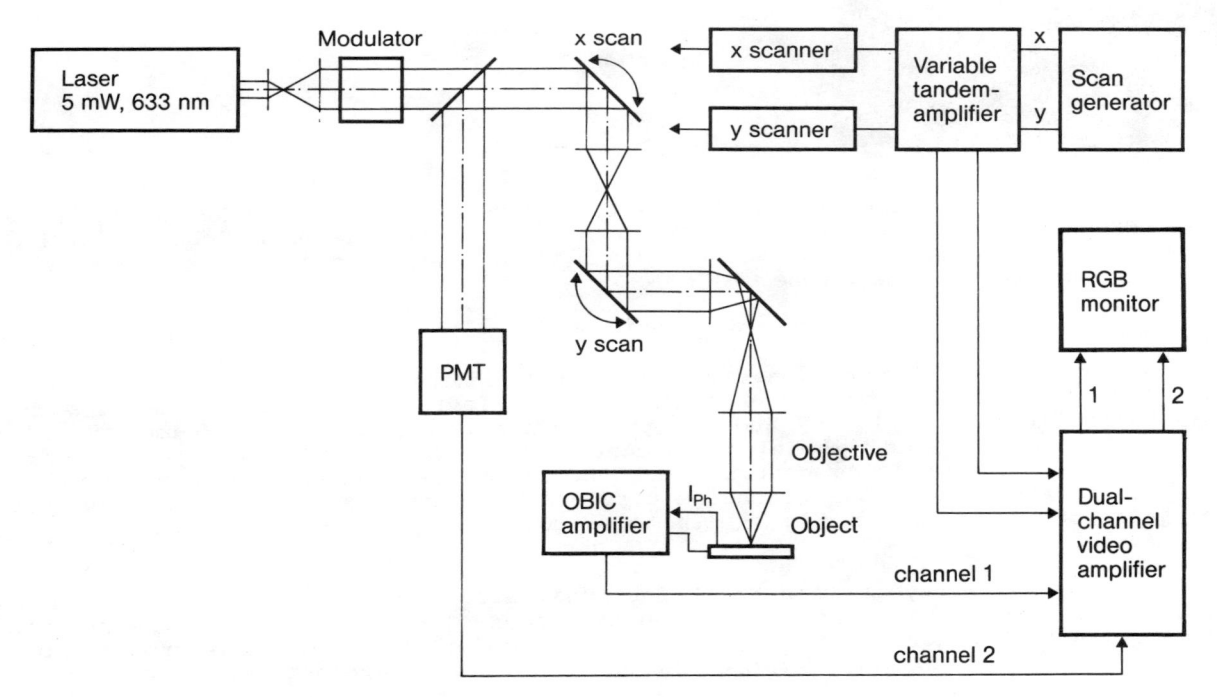

Figure 1. Function schema of scanning laser microscope LSM 42 by Zeiss. The second laser (1150 nm, 2 mW), which is coupled in after the modulator, is not included. PMT = photo-multiplier, OBIC = Optical Beam Induced Current, RGB = Red/Green/Blue.

In this process, probe diameters down to about a single wavelength can be attained with an objective of high numerical aperture. The position of the scanning laser probe is determined by setting the electrically-driven galvanometer mirrors, and the optical power at the object has a maximum value of 0.2 mW. The laser light reflected from the object is detected in the reverse beam path by a photomultiplier (PMT). The photomultiplier signal (channel 2 in Fig. 1) then controls the brightness of a monitor via a video amplifier and the variable tandem amplifier connected after a scan generator effects the x and y assignment of the mirrors and the video monitor. In this way, the scanning procedure produces an incident light image of the specimen. Transmitted-light operation is also possible, but is not relevant to the circuit analyses discussed below and therefore not included in Fig. 1.

In addition to the incident image channel a second video channel is also present, into which any signal - conveniently one correlated with the x and y deflection - can be inserted. The photocurrent induced by the laser probe, known as the OBIC (Optical Beam Induced Current) for short, lends itself to the analysis of ICs. The dual-channel video amplifier can thus input the incident light image into the red channel and the OBIC image into the green channel of an RGB monitor to produce a combined image. (At the editor's request, however, the combined images shown here are merely superimposed black & white images.)

The scanning laser microscope can naturally also be used to advantage for pure surface imaging and inspection in the semiconductor industry, since the SLM stands out from conventional optical microscopes due to its superior contrast. In addition, it allows contrast and brightness adjustment. Fig. 2 shows a site on a p+ doped silicon wafer which was subject to intensive annealing by means of excimer laser bombardment. This incident-light image was recorded by a Zeiss LD Epiplan 40 objective and a differential interference contrast equipment of the Nomarski type.

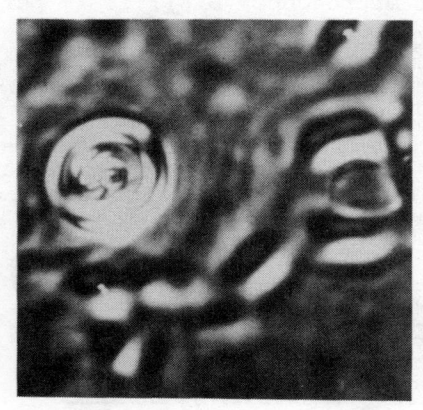

Figure 2. Differential interference contrast image of the intensively annealed point of a p+ doped silicon wafer in the scanning laser microscope.

10 µm

Scanning laser microscopes are available from companies other than Zeiss, e. g.: LASER-SCAN MICROSCOPE 8510 by ICT GmbH, LASCAN 2 by Elektronik & Technik GmbH and SOM 100 by BIORAD/Laser Sharp Ltd. In the last-named of these instruments, the specimen stage is scanned instead of the laser probe.

3. OBIC METHOD

If a laser beam with photon energy E strikes a semiconductor with bandgap energy E_G, the photons are absorbed with the formation of electron-hole pairs when $E > E_G$; see Fig. 3. For silicon, E_G = 1.1 eV and the penetration depth is about 2.5 µm for red laser light of 632.8 nm wavelength whose photon energy E = 1.96 eV. The electron-hole pairs diffuse away from the generation center until they completely recombine, for instance in a homogeneous semiconductor. But if charge carriers are generated close to an inversely-biased pn-junction, as shown in Fig. 3, the electron-hole pairs are separated at the junction.

The minority carriers, in this case electrons, are drawn off from the p-region to the n-region. An external photocurrent can then be measured in the circuit of the pn-junction. The size of this photocurrent depends on the distance of the laser probe from the pn-junction, i. e. on the proportion of the charge carriers which have not yet recombined there.

The OBIC effect described above forms the basis for various methods of circuit analysis /4-8/. Fig. 3, right, shows a combined SLM image of a part of a CMOS microcomputer, the OBIC image being superimposed onto the incident-light image. The OBIC image can also be used to localize ESD (electrostatic discharge) damage in LSI and VLSI devices. This is because pn-junctions destroyed by electrostatic discharge no longer exhibit any blocking behavior, and thus no OBIC effect either.

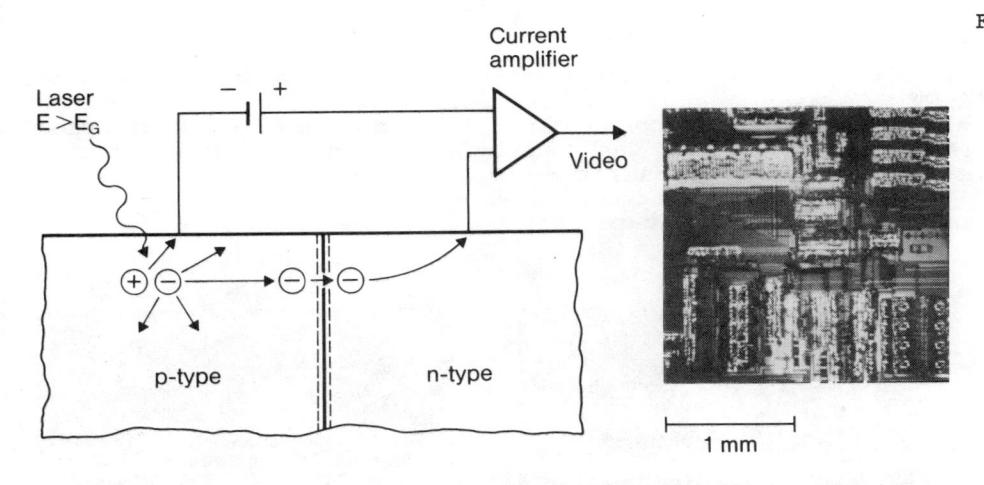

Figure 3. Left: principle of the OBIC (Optical Beam Induced Current) effect.
Right: a combined SLM image of part of a CMOS microcomputer, the OBIC image (bright regions) being superimposed onto the incident-light image.

4. LATCH-UP LOCALIZATION

A CMOS inverter always contains a parasitic bipolar thyristor structure whose undesired activation is known as the latch-up effect /9,10/. This effect leads to defective behavior, in some circumstances even to destruction of the integrated circuit. It can be triggered by an excessive supply voltage V_{DD}, by overshoots in the CMOS inverter output switched to V_{DD}, by undershoots in the output switched to V_{SS} (ground) as well as undershoots at V_{DD} itself or overshoots at V_{SS}. Measures taken to suppress the latch-up effect include adding an epitaxial layer between the substrate and the well or providing additional substrate and well contacts. The first solution is costly, the second one takes up space. In principle, the latch-up effect in a VLSI circuit can be detected by the sudden rise and the self-sustainance of the supply current I_{DD} but cannot be localized.

If the OBIC effect is utilized in the SLM, sites susceptible to latch-up can be localized and the geometrical dependencies of the latch-up determined /11-15/. For this purpose, the IC is operated in the SLM under marginal conditions. For every pixel, the laser intensity is ramped with the aid of the acousto-optical modulator, see Fig. 4. The laser intensity required for triggering the latch-up is an (inverse) measure of the latch-up sensitivity. Fig. 4, right, shows this sensitivity for a section of a CMOS microcomputer. The recording of such an image consisting of 512 x 512 pixels takes about 10 seconds. In this case, the latch-up sensitivity has a maximum close to the well edge. These kinds of analyses permit circuit designers to improve the design at a site localized in this way specifically to preclude the occurence of the latch-up effect there.

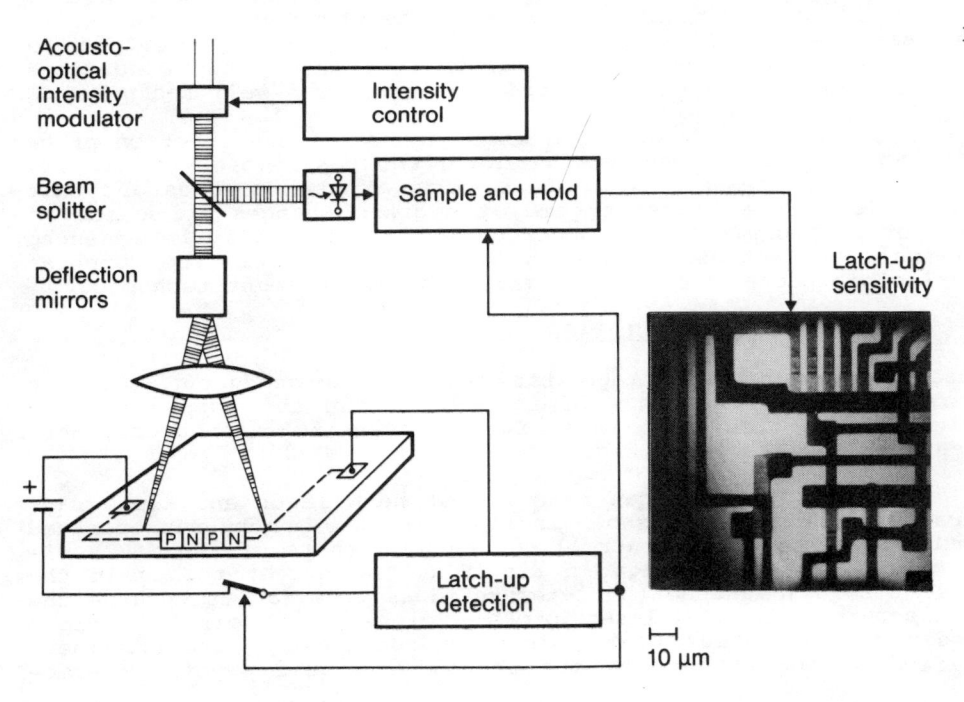

Figure 4. Quantitative analysis of the latch-up sensitivity with spatial resolution using the SLM. Principle of the experimental setup and image of the latch-up sensitivity of a section of a CMOS microcomputer.

5. LOGIC ANALYSIS

The laser probe can also be used to determine logic-level time diagrams at the internal nodes of VLSI circuits /16/. This measuring principle is used in the data probe by Mitsui Comtec Corp. We were also able to implement such a logic analysis with the SLM described above at inverter outputs inside CMOS ICs. This was done by focussing the SLM laser probe in spot mode onto the drain of the p- or n-channel transistor. Depending on the switching state of the inverter, the photocurrent thus induced flows either between V_{DD} and V_{SS} or internally in the circuit and so in the second case can no longer be measured at the terminals.

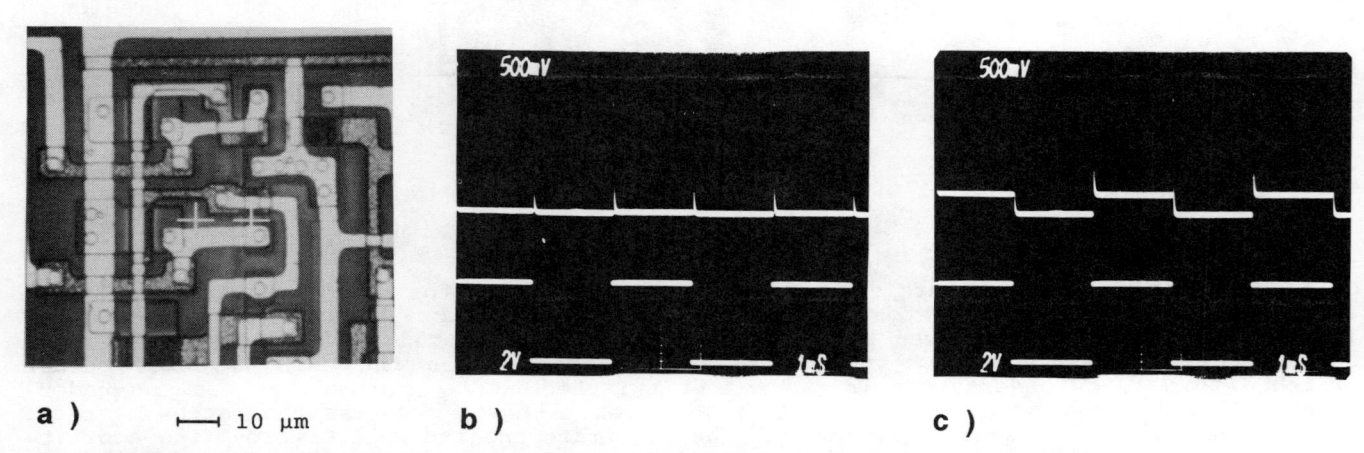

a) ⊢⊣ 10 μm   b)   c)

Figure 5. Logic analysis with the laser probe of the SLM in the interior of a CMOS bus decoder.
    a) Section with two measuring points.
    b) Photocurrent (above) outside the drain, i. e. at the left measuring point in a), and device clock signal (below) for comparison.
    c) Photocurrent at the n-drain of the well transistor, i. e. the right measuring point in a), and clock signal for comparison.

Fig. 5 a shows the inverter of a CMOS bus decoder with two measuring points (white crosses). Fig. 5 b and 5 c show, for both these points, the photocurrent time curves (above) measured between V_{DD} and V_{SS} compared with the input clock signal at the device. In Fig. 5 b the laser probe is positioned outside the drain (left measuring point in Fig. 5 a). Here no voltage dependence of the photocurrent can be detected. In Fig. 5 c the laser beam strikes the n-drain in the p-well (right measuring point in Fig. 5 a). In this example, a voltage dependence of the photocurrent can be seen, the photocurrent and inverter output behaving in the same sense, i. e. when photocurrent flows the output is switched to high. In the event of irradiation at the p-drain (not shown) the conditions are reversed, irrespective of whether the inverter is designed in p- or n-well technology.

Logic analysis with the laser probe has, compared with the use of the electron probe /17,18/, the advantage of obviating a vacuum and not removing the passivation on the IC. However, it has the drawback that measurements can only be made at static nodes and that these must lie in the drain region. At a transfer gate, the drain changes to source so that logic analysis is no longer possible. The photocurrent must additionally be measured between the V_{DD} and V_{SS} terminals, between which a high capacitance exists. The temporal resolution is therefore in principle poorer than in electron-beam measurement technology.

6. FAULT STIMULATION

The laser probe can also be used to advantage for testing semiconductor memories (RAMs). If a memory is operated at its tolerance limits, e. g. for the supply voltage /19/, then faults can be triggered by stimulating them with the scanning laser probe, thus allowing possible circuit weaknesses to be localized.

This fault stimulation (soft error testing) can make use of both laser and alpha rays /20/, whose effects are mutually correlated. As shown in Fig. 6 c, the memory was operated with a tester whose fault output was used for controlling the brightness of the SLM monitor. This was then used to obtain the fault image in Fig. 6 a. The bright points in the memory cell field (Fig. 6 a left below) must not be interpreted as true faults because the cell field capacitors are discharged by the laser probe. However, the circled bright points in the address decoder indicate true weak points in the memory circuit. These points could also be identified as weak points by the method of liquid crystal thermography /21/.

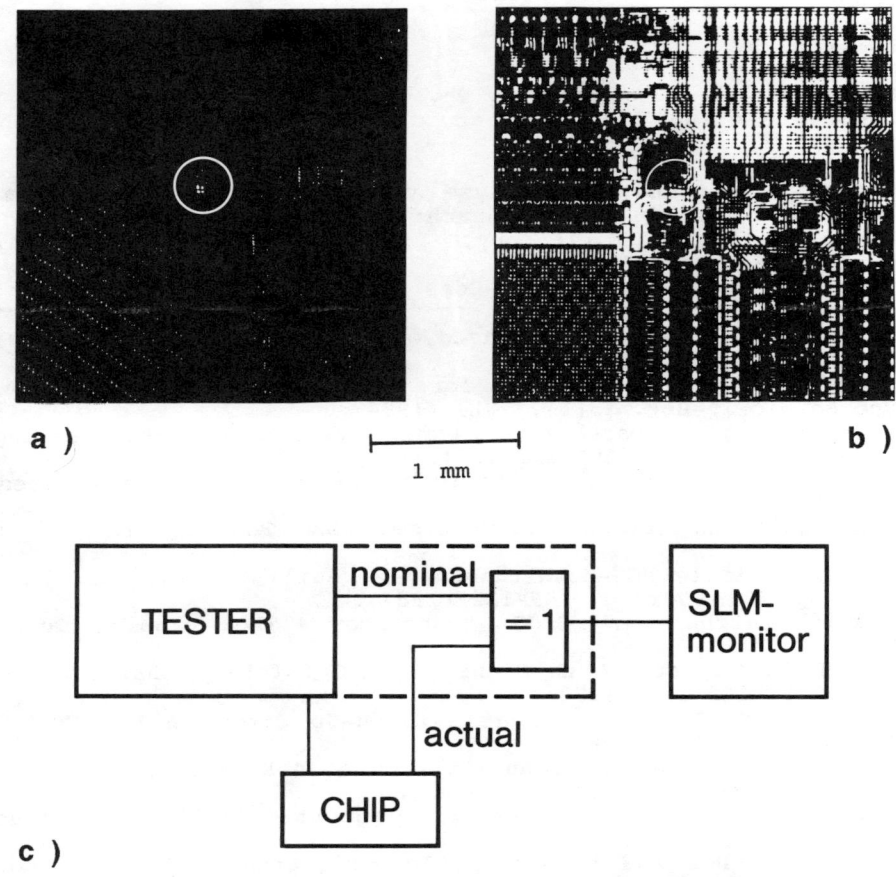

Figure 6. Fault stimulation with the laser probe.
a) Fault image of a 64K memory operated at the upper limit of its supply voltage.
b) Incident-light image of the same memory area.
c) Principle of the experimental setup.

7. OUTLOOK

Today, laser pulses can be generated in the picosecond region and even down to a few tens of femtoseconds. By exploiting electro-optical effects, such picosecond lasers are used to measure very fast signals in discrete semiconductors or integrated bipolar circuits /22,23/. A combination of these methods with a scanning laser microscope appears to be very effective for analyzing the interior of future fast highly-integrated ICs.

8. ACKNOWLEDGEMENTS

We would like to thank Messrs. M. Bischoff, U. Goedecke, J. Luber and Dr. V. Wilke of the Zeiss company for fruitful discussions and valuable suggestions for making modifications to the scanning laser microscope. We also thank Dr. E. Bayer for making Fig. 2 available to us and Prof. K. Goser, Prof. H-J. Pfleiderer and Dr. E. Wolfgang for their constant support. Our thanks are also due to Mr. R. Michell for providing the English translation.

Parts of this work were supported by the Federal Ministry of Research and Technology (Project No. NT2696). The authors alone are responsible for the contents.

9. REFERENCES

Abbreviations used:

IEEE/IRPS = IEEE International Reliability Physics Symposium
SMECS/EA87-1 = Spring Meeting of the Electrochemical Society, Philadelphia, Extended Abstracts, Volume 87-1

1. J. Otto and E. Plies, "Schaltungsanalyse von ICs durch Messungen im Bausteininneren," NTG-Fachberichte 87, Berlin und Offenbach: VDE-Verlag, 110-115 (1985).
2. V. Wilke, "Laser Scanning in Microscopy," Proc. of SPIE - The International Society for Optical Engineering, 396, 164-172 (1983).
3. V. Wilke, "Optische Rastermikroskopie - das Laser-Scan-Mikroskop," Physik in unserer Zeit 15(1), 13-18 (1984).
4. M. Nagase, "A Device Analysis System Based on Laser Scanning Techniques," Microelectronics and Reliability 20(5), 717-735 (1980).
5. R. Müller, "Scanning Laser Microscope for Inspection of Microelectronic Devices," Siemens Forsch.- und Entwickl.-Ber. 13(1), 9-14 (1984).
6. G. R. Woolhouse et al., "Research into Advanced Methods of Laser Testing of Integrated Circuit Chips," SMECS/EA87-1, 175-176 (1987).
7. E. Ziegler and H.-P. Feuerbaum, "Failure Analysis of Integrated Circuits," SMECS/EA87-1, 179 (1987).
8. E. Plies and J. Quincke, "Electron and Laser Beam Testing of ICs," SMECS/EA87-1, 163-164 (1987).
9. K. Horninger, "Integrierte MOS-Schaltungen," 2. Aufl. Berlin, Heidelberg, New York, London, Paris, Tokyo: Springer-Verlag, 109-112 (1987).
10. R. R. Troutman," Latchup in CMOS technology; The problem and its cure," Boston:Kluwer (1986).
11. F. J. Henley, M. H. Chi and W. G. Oldham, "CMOS Latch-Up Characterization Using a Laser Scanner," IEEE/IRPS, 122-129 (1983).
12. D. J. Burns and J. M. Kendall, "Imaging Latch-Up Sites in LSI CMOS with a Laser Photoscanner," IEEE/IRPS, 122-129 (1983).
13. T. Shiragasawa et al., " Latch-Up Analysis on a 64 KBit Full CMOS Static RAM Using a Laser Scanner," , IEEE/IRPS, 63-68 (1984).
14. A. H. Johnston and M. P. Baze, "Lasers as Diagnostic Tools for Latchup in Integrated Circuits," SMECS/EA 87-1, 177 (1987).
15. J. Quincke, F. Dielacher and K. Goser, "Investigation of Surface Induced Latch-up in VLSI CMOS Using the Laser Probe," Microelectronic Engineering 7, 371-375 (1987).
16. F. J. Henley, "An Automated Laser Prober to Determine VLSI Internal Node Logic States," IEEE/International Test Conference, 536-542 (1984).
17. W. Argyo et al., "Analyse elektrischer Funktionen im Inneren von integrierten Schaltungen mit der Elektronensonde," Siemens Forsch.- u. Entwickl.-Ber. 14(4), 216-222 (1985).
18. E. Wolfgang, "Electron Beam Testing," Microelectronic Engineering 4, 77-106 (1986).
19. D. J. Ager, G. F. Cornwell and I. W. Stanley, "The Application of Marginal Voltage Measurements to Detect and Locate Defects in Digital Microcircuits," Microelectronics and Reliability 22 (2), 241-264 (1982).
20. F. J. Henley and W. G. Oldham, "Soft Error Studies Using a Scanning Source," IEEE/IRPS, 88-91 (1982).
21. R. Weyl et al., "Thermographie an integrierten Schaltungen," NTG-Fachberichte 87, Berlin und Offenbach:VDE-Verlag, 116-121 (1985).
22. J. A. Valdmanis and G. Mourou, "Subpicosecond Electrooptic Sampling:Principles and Applications," IEEE J. of Quantum Electronics 22(1), 69-78 (1986).
23. B. H. Kolner and D. M. Bloom, "Electrooptic Sampling in GaAs Integrated Circuits," IEEE J. of Quantum Electronics 22(1), 79-93 (1986).

An Automated Latch-up Measurement System Using a Laser Scanning Microscope

J. Fritz, R. Lackmann, B. Rix

Fraunhofer Institute of Microelectronic Circuits and Systems
Finkenstr. 61, D-4100 Duisburg 1

ABSTRACT

A laser scanning microscope offers a non-destructive technique to locate and analyze latch-up in an IC. We have developed an advanced, fully automated latch-up analyzer coupled with the CAD system used for IC design. It consists of a laser scanning microscope, a x-y table for chip scanning, a monitor TV and a microprocessor based system for control of the test sequence and data analysis. Latch-up sensitivity is measured by stepwise increase of laser beam power using an acousto-optical modulator. The monitoring of the beam position and the modulator voltage while scanning the laser spot over the IC surface and the resulting current changes in the device's power supply locate the latch-up zone and its sensitivity. The sensitive regions found are overlaid graphically over the IC layout data to provide a redesign posibility. As an application example we consider a CMOS A/D converter IC and explain the system performance.

1. INTRODUCTION

Because of its low power consumptions, better α-particle immunity, high noise margin and high speed of operation, CMOS is the preferred technology for VLSI circuit design. A good cost/performance improvement is derived from device scaling. Although an improved technology allows to scale down the n- and p-channel transistors to submicrometer range, the space between them cannot be arbitrarily scaled without inducing the latch-up effect. It results from coupled parasitic bipolar transistors due to the integration of P-N-P-N structures in the layout when complementary MOS transistors come into proximity. While the physical mechanism causing latch-up is well understood [1,2], the multiplicity of triggering mechanisms like current or voltage overshoot and undershoot, temperature and radiation makes sensitivity predictions difficult. Voltage transients which induce latch-up in many cases destroy the IC. Thus it becomes crucial to identify the latch-up locations on the chip and to redesign the layout in order to improve the latchup margin. One easy way to evaluate latch-up hardness is to induce latch-up by a focused laser or electron beam as a localized current generator [3,4]. Our application bases on a laser scanning microscope and the optical beam induced current, the OBIC effect.

2. BASIC PRINCIPLES OF LASER SCANNING MICROSCOPY

Focusing the beam of light from a laser on the surface of an IC leads to a splitting of the total intensity. Depending on the illuminated surface one part of the total intensity is reflected while the other is absorbed by the semiconductor generating electron-hole pairs inside. The carrier-generation rate due to the light beam is taken as

$$I_{(z)} = I_{(z=0)} \exp(-\alpha z) \tag{1}$$

where α is the intensity-attenuation coefficient and z the distance beneath the surface. The inverse of α is the penetration depth which depends on the optical wavelength for which we further require

$$\lambda \geq hc / E_g \tag{2}$$

to ensure the excitation of electron-hole pairs. Using a red HeNe laser at 633 nm leads to a photon energy of 1.96 eV (E_g of Si is 1.12 eV at 300 K). These photo-generated carriers will either be collected by surrounding junctions giving rise to a photo current or recombine.

In contrast to an ordinary optical microscope where an object is illuminated by a wider area light source the image in a laser scanning microscope is taken point by point as the scan takes place. The light reflected by the specimen is directed by a mirror onto a photodetector. The amplified photomultiplier signal is applied to a TV monitor in synchronism with the x-y deflection done by galvanometer mirrors driven by a scan generator. Fig. 1 shows the scheme of the optical system[5]. For the described application we used the spot mode of the system where a fixed point on the specimen is illuminated by the laser beam. The scanning is achieved by moving the device under test (DUT) mounted on a x-y table under computer control. Tab. 1. summarizes the technical data of the used laser scanning microscope.

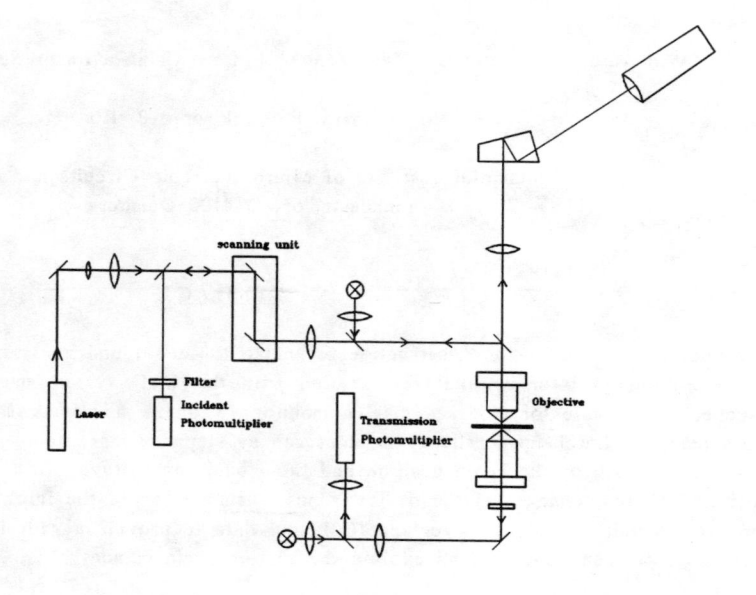

Figure 1. Optical scheme of the laser scanning microscope

Wavelength λ	633 nm
Penetration depth (Si)	2.5 μm
max. laser power on probe surface	≈ 0.5 mW
Spot diameter on probe surface (40x)	≈ 1 μm
x-y table position accuracy	1 μm

Table 1. Technical data of the used scanning system

3. LATCH-UP EFFECT

Because the CMOS technology necessarily involves the fabrication of complementary devices, PNPN structures are localized in many parts of the layout. These structures may be bistable, they can either be off in a high impedance state or on in a low impedance state. Fig. 2 shows the cross section of a CMOS inverter with its typical parasitic elements. These are two lateral NPN- and two vertical PNP bipolar vertical transistors connected together via the substrate and well resistance. Important parameters for latch-up triggering are the base-emitter resistance and the transistor current gains. The cross section indicates that all parasitic transistors are "tied" across the power supply so that a turn on may destroy the circuit by excessive heat.

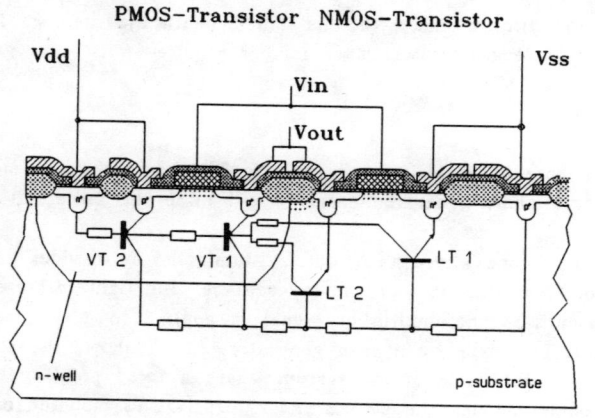

Figure 2. Cross section of a CMOS inverter

In the case of a N-well technology the photo-induced holes produce majority carriers in the substrate. The resulting current tends to increase the base potential of the lateral NPN transistor while the electron current decreases the base potential of the PNP transistor. Which transistor turns on first depends on the bypass resistor values[6]. The increase of the emitter potential by hole current leads to an electron injection into substrate which flow either as a collector current to the well or recombine as base current in the substrate. The electron current in the well on the other hand decreases the emitter potential of the vertical PNP-transistor and leads to hole injection into the well which flow into substrate or recombine in the well. In summary a sufficient condition for latch-up is met once either of the above transistors switches on.

4. LATCH-UP DETECTION SYSTEM

The intensity of the laser beam penetrating into silicon is controlled with an acousto-optical modulator (AOM). Fig. 3 shows the measured laser power on the probe surface versus the DC control voltage at AOM input (40x, N.A. 0.5). An increase in laser power leads to a higher rate of electron-hole pairs and thus to a higher photoinduced current. As part of a complete measurement scheme modulator voltage and current through the DUT are controlled by a programmable microcomputer (MICEL) developed for this task. Fig. 4 shows a block diagram of the whole system and its basic principle. The microcomputer communicates via IEEE-bus with the VME based host.

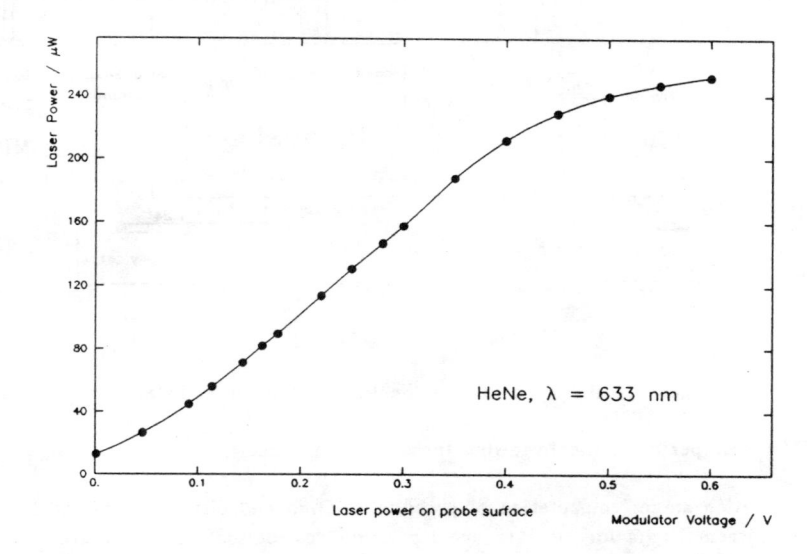

Figure 3. Laser power versus AOM voltage

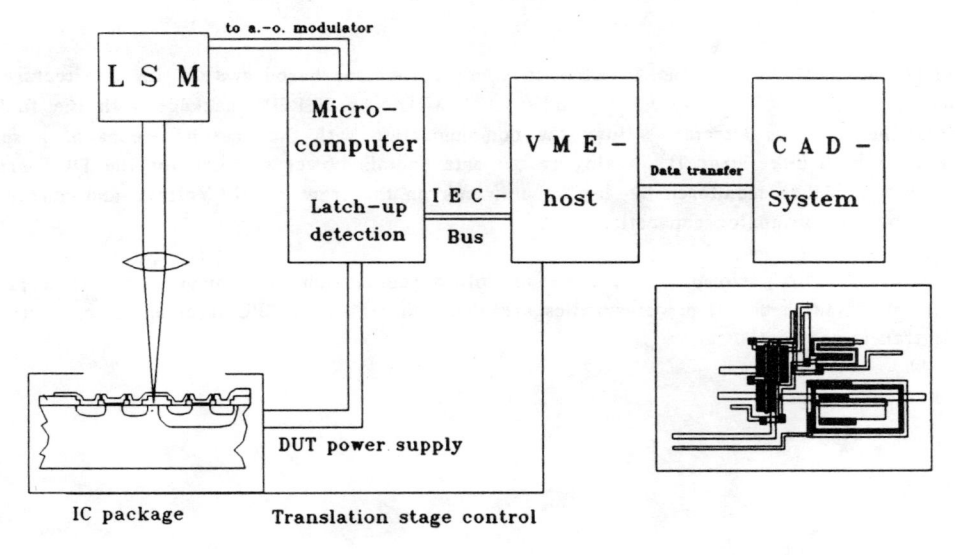

Figure 4. Schematic of the whole detection system

The host controls the IC position by moving a x-y translation stage, stores the measured sensitivity data when latch-up occurs and supports a link to the CAD system via a parallel interface for data transfer and CAD data base access after a complete measurement session. After an initialization the system works fully automated. Fig. 5 shows the block scheme of the complete test equipment used at IMS including the mentioned parts.

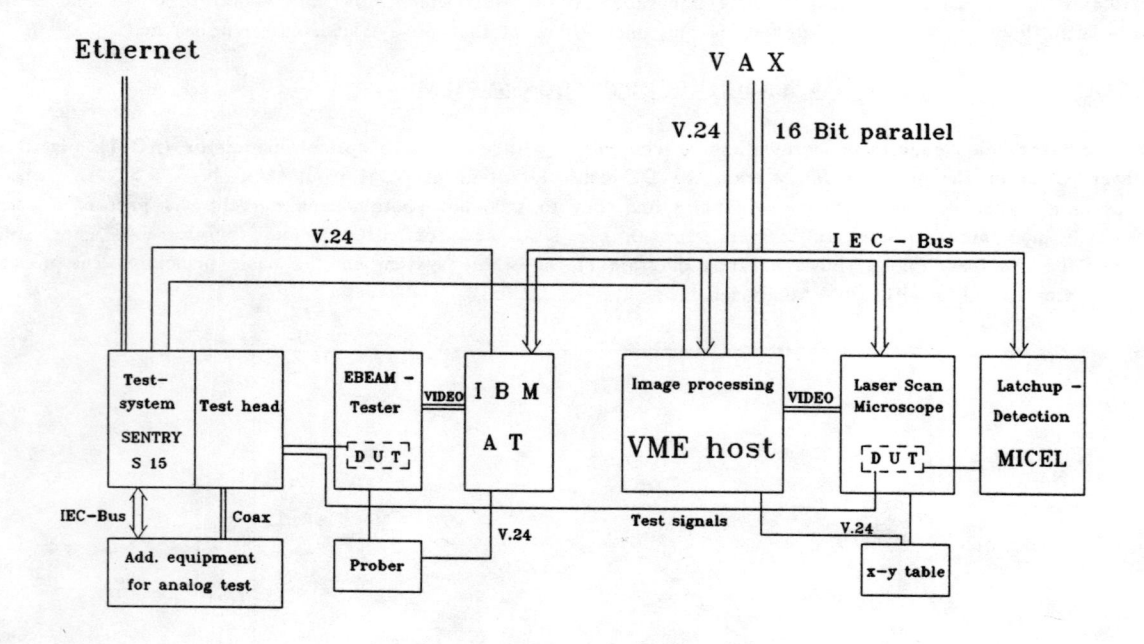

Figure 5. Block scheme of the test equipment at IMS

The main program on the VME host performs the following tasks menu driven:

- calibration of the IC on translation stage, computation of possible rotation and distortions of the IC image
- setup of power supplies and current limits for DUT to avoid possible destruction. All data are transformed to MICEL and switched on when data acquisition begins
- MICEL control for each acquisition step where the voltage at the AOM and thus the OBIC current is sequentially increased.

As already mentioned the heart of the latch-up detection is a microprocessor based system, its architecture is shown in Fig. 6. The CPU is equipped with a 6502 CMOS CPU running at 1 MHz. An EPROM package with the firmware program is allocated on the card. The IEC bus interface allows the communication with the host by means of a specially developed syntax and data transfer in both directions. The analog components contain power supplies for the DUT and the AOM. CPU status and DUT voltages set can be monitored by display elements on the front panel. Voltage and current limits setting is done after D/A conversion via the analog connection lines.

The latch-up detection is done by performing a current to voltage conversion and comparison with a preset value. When DUT current increases due to latch-up all power supplies are switched off and a CPU-interrupt is generated. The corresponding laser intensity is transferred to the host.

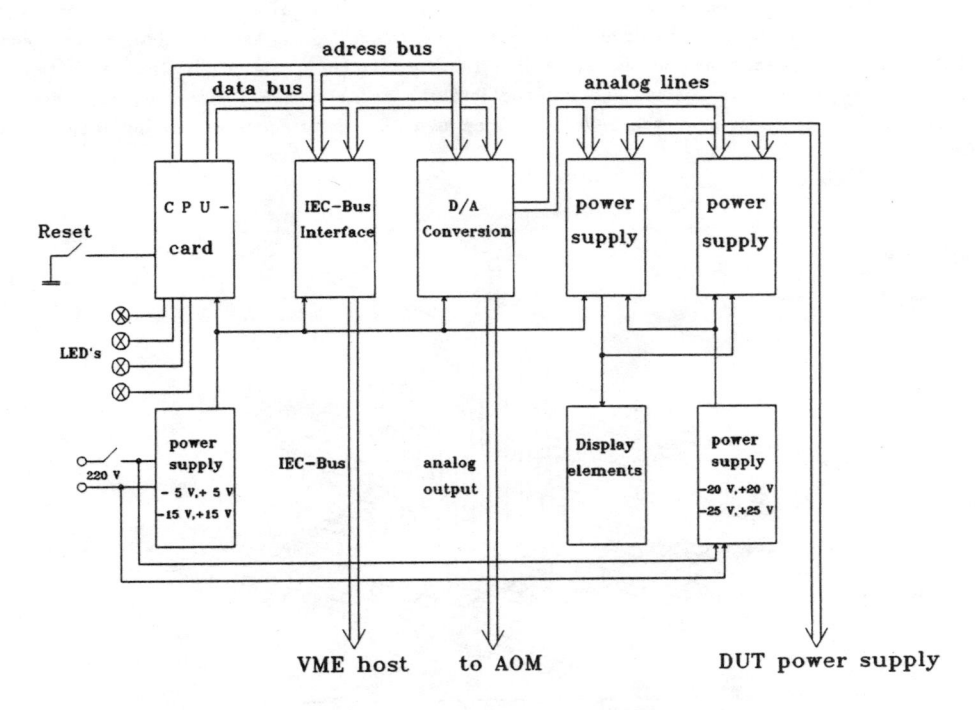

Figure 6. Architecture of detection hardware

To minimize the test time a fast scan is performed first. This means that the IC is scanned with medium resolution under the laser switched to maximum power. One can thus localize the sensitive regions for a second high resoltuion run. Scanning is now performed over the found regions by iterative subdivision of the scan interval regions. This solution has the advantage of providing an overview while scanning takes place and minimizing the acquisition time. Fig. 7 show the complete measurement system with the online graphics display. The MICEL detection system is located on the right together with power and clock supplies for the DUT and connected via IEEE to the VME host not shown here. The progress in data taking is visualized on the color graphics display connected with a hard copy unit below where the bright region is a latch-up sensitive part of the whole layout shown schematically as a rectangle.

Figure 7. Photograph of the complete measurement system

5. DATA BASE COUPLING AND DATA PROCESSING

Because the inverse of the laser intensity is proportional to the latch-up sensitivity at a given position after a complete data acquisition run we obtain a sensitivity map. For further processing the data are transferred via the 16 bit parallel interface to the central computer. After data transfer a conversion program generates a graphic output file, corrects possible IC rotation and scaling distortions and divides measured data into classes, each class connected to a different color. It is now possible to display the measured results as various maps like the one shown in Fig. 8. In the next step the graphic file is transformed to the CAD workstation data format and entered into the user's design library. A menu driven program running on the workstation allows the user to overlay the sensitivity map over his design and switch between a color display of the IC layout and the measured data.

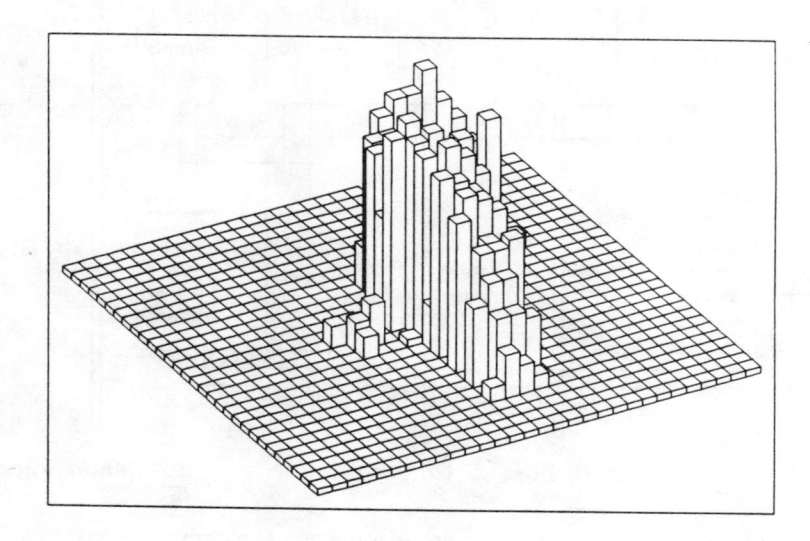

Fig. 8 Latch-up sensitivity map

6. EXPERIMENTAL RESULTS

As an application example we consider a CMOS IC designed by one of us (R.L.) at our institute. It is a 3 $^1/_2$ digit monolithic analog to digital converter for direct driving of LED displays. All needed active elements like LED drivers, decoders, reference voltage and clock generator are integrated on the chip. Part of the manufactured IC's were strongly latch-up sensitive. The localized region is shown in Fig. 9 while the corresponding CAD layout overlaid with the measured data is shown in Fig. 10. This is normally a color display not reproducable here. The shown part of the layout is the magnified region marked in Fig. 9.

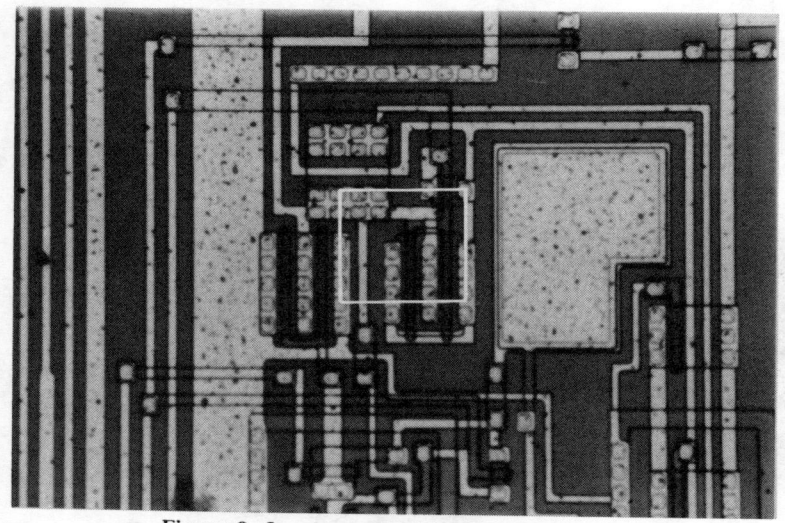

Figure 9. Latch-up sensitive part of the layout

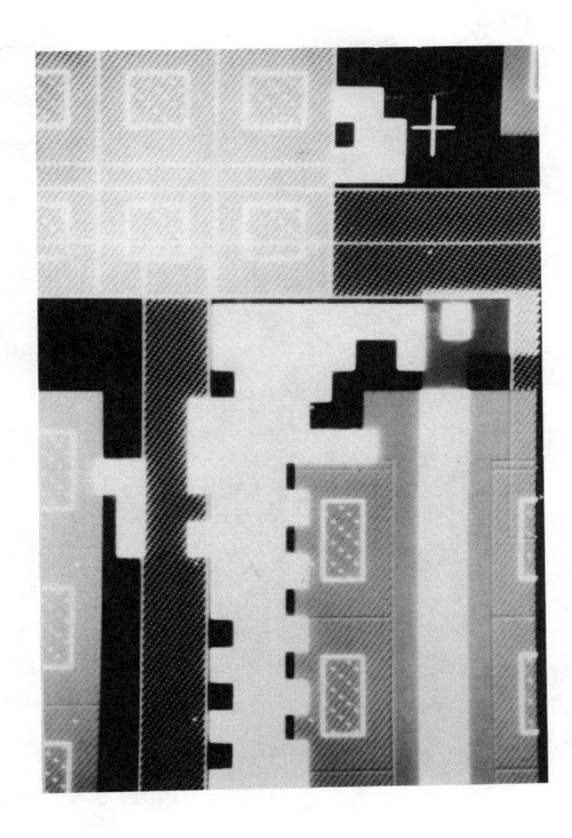

Figure 10. CAD layout of marked region

7. CONCLUSIONS

A laser scanning microscope is a powerful tool for latch-up analysis. An approach to a fully automated system coupled to the CAD data base has been described showing an easy way to redesign the latch-up sensitive regions. An application example proofs the system performance.

8. ACKNOWLEDGMENTS

The authors wish to thank Prof. G. Zimmer for encouragement and W. Budde for helpful discussions.

9. REFERENCES

1. R. Troutman, Latchup in CMOS Technology, Kluwer Academic Publishing (1985)
2. W.M. Coughran, M.R. Pinto, R.K. Smith, "Computation of Steady-State CMOS Latchup Characteristics", IEEE Trans. on CAD, 7(2), 307-323, (1988)
3. F. Shiragasawa, H. Shimura, K. Kagawa, T. Yonezawa, M. Noyori, "Latch-up Analysis on a 64k Bit Full CMOS Static RAM Using a Laser Scanner", Proc. of the IRPS, 63-68, (1984)
4. C. Canali, M. Giannini, E. Zanoni, "Techniques for Latch-up Analysis in CMOS IC's Based on Scanning Electron Microscopy", Microelectr. Reliab., 28 (1), 119-161 (1988)
5. V. Wilke, "Optical Scanning Microscope - The Laser Scan Microscope", Scanning, 7, 88-96, (1985)
6. D. B. Estreich, "The Physics and Modeling of Latch-up in CMOS Integrated Circuits", Tech. Rep. No. 6-201-9, Stanford Electronics Lab., Stanford Univ., (1980)

ECO1
SCANNING IMAGING

Volume 1028

SESSION 6

Scanning Microscopy II

Chair
G. J. Brakenhoff
University of Amsterdam (Netherlands)

Infrared Laser Scan Microscope

Eberhard Ziegler and Hans Peter Feuerbaum

ICT GmbH Klausnerring 1a Postfach 1333, 8011 Heimstetten bei München, FRG

ABSTRACT

The infrared laser scan microscope is especially suitable for applications in material science and in the semiconductor industry. The infrared laser beam is deflected in x and y directions by a mirror system and the scanning beam is focussed with the help of a light microscope onto the sample. Four different IR (infrared) lasers can be used; the semiconductor lasers AlGaAs/GaAs with a wavelength of about 820nm and GaInAsP/InP with a wavelength of about 1.3µm and the helium-neon-lasers emitting at 1.152µm and 1.523µm. The lasers are situated outside the optics and are connected by a special monomode glass fiber. The He-Ne laser, with a wavelength of 1.152µm, is of special significance in the investigation of silicon and ICs because its wavelength exactly corresponds to the energy gap of silicon at room temperature (1.07eV). It is therefore possible to image silicon at different depths, to test devices from the back and, moreover, to produce an OBIC (Optical Beam Induced Current) signal from the back of the device.

1. INTRODUCTION

Using the infrared laser scan microscope it is possible to localize lattice defects and electrical defects on the surface and in the bulk of solid state materials, especially in semiconductors and integrated circuits. The wavelength of the infrared laser used and the properties of the investigated material determine whether this information comes from the surface or the bulk.

Infrared laser scanning is nondestructive and has almost no influence on the specimen. The energy of the infrared photons (about 1eV) is negligible in comparison with x-rays or electron beams, the energy of which is at several kV and, in contrast to electrons, photons have no charge.

Some investigations are only possible using infrared laser scanning. The volume of semiconductors and integrated circuits can not only be imaged by the reflected infrared light, but also the internal electric fields can be shown by the infrared optical beam induced current (OBIC). The production of the infrared OBIC is the most important advantage of the infrared laser scan microscope in comparison with the infrared light microscope.

2. EXPERIMENTAL SET UP OF THE INFRARED LASER SCAN MICROSCOPE (IR LSM)

The experimental set up is similar to that of the LSM in the visible range[1-5], but the optics must be corrected for infrared and an infrared coating is preferable.

The IR LSM allows operation with four different lasers, with two semiconductor lasers and two gas lasers. The two semiconductor lasers GaAlAS / GaAs and GaInAsP / InP emit at a wavelength of about 820 nm and 1.3µm, respectively. The two He-Ne lasers emit at 1.152µm and 1.523µ, respectively.

All lasers are situated away from the other parts of the IR LSM (Fig.1) and connected by a single mode glas fiber to the optical system (left part of Fig.1). The laser beam can be pulsed by an acousto-optical modulator. A system of deflecting mirrors, controlled by a digital scan generator is used to position the laser in x - and y - directions. Thus the laser beam can be directed onto defined points, scan single lines, or scan over areas line by line. The light microscope section focusses the scanning laser beam onto the specimen. It is possible to use commercial light microscopes suitable for visible light or, if a higher infrared transmission is needed, light microscopes with a special coating for infrared.

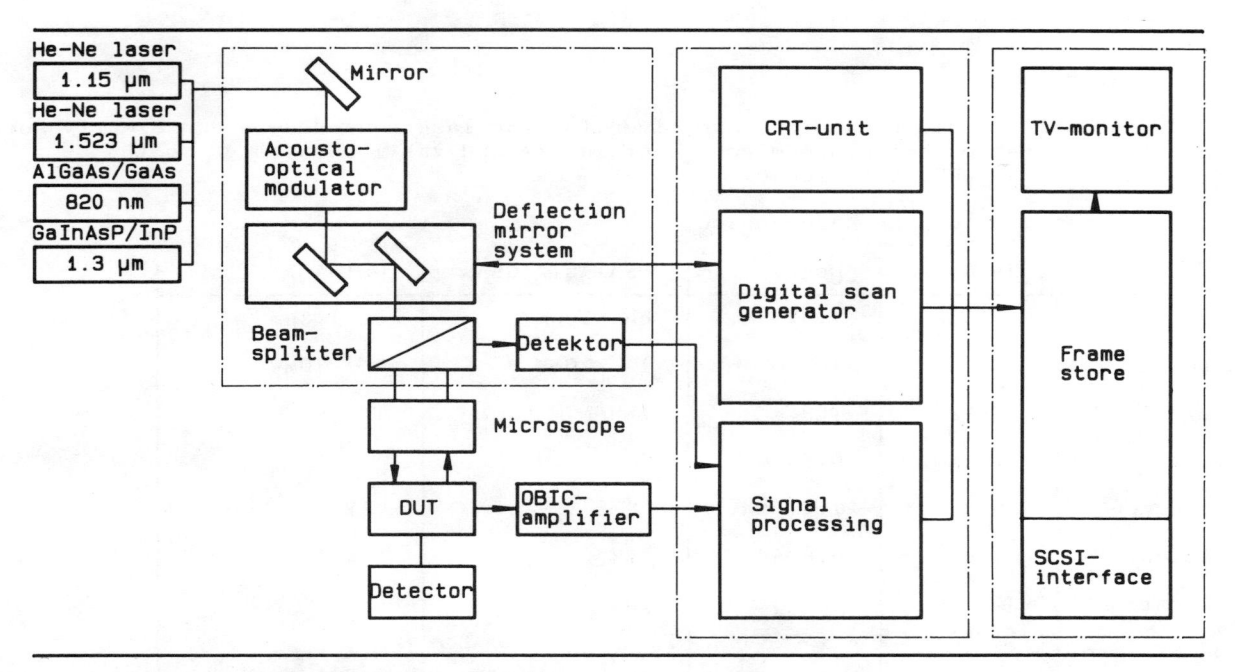

Figure 1. System diagram of infrared laser scan microscope

The reflected and the transmitted infrared light is detected by a Germanium photo diode. The absorbed IR is responsible for producing the OBIC signal. The signals (reflected light, transmitted light, OBIC) can, after signal processing, be directly imaged on the monitors or stored in a framestore. The framestore is not just used for the conversion of the slow scan primary image into a TV image, but also contains important image processing routines.

2.1 Glas fiber

For the transmission of the laser beam from the laser to the optical system, a single mode glas fiber is needed with a suitable optial coupling at each end.

For the He-Ne lasers an objective with a coating for infrared is sufficient, but for semiconductor lasers one has to use a collimator. The single mode fiber ensures that only one mode can be transmitted and that the laser beam keeps its property of coherence. The disadvantage of such a single mode fiber is the small core diameter and the difficulties in adjustment resulting from this. Usually[6], the following relation for the core diameter of a single mode fiber is given as

$$\delta = \frac{\pi d}{\lambda_G} \cdot N_A \leq 2.4 \tag{1}$$

Here, d = core diameter λ_G = wavelength of the laser light
δ = structural parameter N_A = numerical aperture of the fiber

For the numerical aperture, which can be determined directly from the refractive law,

$$N_A = \sqrt{n_K^2 - n_M^2} \tag{2}$$

is valid.
n_K = refractive index of core n_M = refractive index of cladding
With the help of equations (1) and (2) it is easy to determine which core diameter is needed for monomode transmission of laser light of a certain frequency.

If laser light is transmitted with a lower wavelength than that calculated for this core diameter, then the transmission is not monomode. For a wavelength higher than that calculated, cladding modes play an increasingly important role. The wavelength of the transmitted laser light should lie between the calculated wavelength λ_G and about $1.3 \lambda_G$. For the wavelengths used in this equipment (820nm, 1.152µm, 1.3µm and 1.523µm) we use different single mode fibers with core diameters from 6µm to 11µm.
The coupling optics to the laser and to the optical system are integrated into the cables. The transmission efficiency of the cables including optics is, for all four wavelengths, about 70%.

2.2. Infrared lasers

In Table 1 some parameters are given for the lasers used. One can clearly see principle differences between semiconductor lasers and gas lasers, e.g. in the efficiency, or the beam divergence.

Table 1. Parameters of the lasers used

	He-Ne	He-Ne	GaAlAs / GaAs	GaInAsP / InP
wavelength	1.152µm	1.523µm	ca 820nm	ca 1.3µm
output	2mW	1mW	1 - 10mW	1 - 10mW
mode	TEM_{00}	TEM_{00}	TEM_{00}	TEM_{00}
beam diameter	0.72mm	0.83mm		
divergence	2mrad	1.6mrad	∥25° ⊥35°	∥20° ⊥30°
efficiency	0.005%	0.005%	20%	20%
astigmatism	no	no	yes	no
type of laser diode	--	--	gain-guided	index-guided
lifetime	30 000h	20 000h	100 000h	20 000h

Semiconductor lasers have the advantage that their dimensions are very small (some orders of magnitude smaller than that of gas lasers). The total conversion efficiency is higher than for all other lasers. This means that only a small amount of energy is converted into heat and only a metal block is needed for cooling. The light output power can be simply controlled by the current. In the same way, the laser intensity can be modulated and pulsed up to frequencies in the GHz range. The drivers for the laser diodes do not need a high voltage. They work at several volts and require a few milliamps. The coupling of the laser diode with a single mode glas fiber is relatively simple since many laser modules are available with integrated optics and pigtail.

However, the high beam divergence of laser diodes is unwanted. Moreover, the divergence is different parallel and perpendicular to the light emitting p-n junction. The biggest disadvantage for imaging systems is the strong astigmatism of gain-guided semiconductor lasers. The gain-guided diode emitting at 820nm is not used in our equipment for surface imaging but for the determination of the frequence dependence of logic states in integrated circuits. During this test the wide modulation bandwidth and the good triggering ability of the laser diode are exploited.

Because the threshold current and the emitted frequency of a laser diode depend on temperature, it is preferable to regulate the temperature of the semiconductor laser. (The increase of the threshold current with the temperature is caused by the increase of nonradiating recombination processes with temperature; the temperature dependence of the emitted wavelength correlates with the temperature dependence of the gap of the semiconductor material.)

Attractive features of the gas lasers are the high frequence stability, the purity of the modes and the small beam divergence.

Not so advantageous is the small conversion efficiency and the need for a high voltage power supply. Modulation and pulsing is more problematic than with laser diodes. In the IR LSM the gas laser beams are modulated and pulsed, using an external acousto-optical modulator.

3. INVESTIGATIONS ON SEMICONDUCTOR MATERIAL AND ICs

Table 2 shows the lateral resolution using the four different infrared lasers and penetration depths in silicon an GaAs. One can see that the AlGaAs/GaAs laser yields information in silicon and GaAs from a region near the surface, but the lasers with wavelengths of 1.3µm and 1.523µm penetrate Si and GaAs. Only inhomogeneities on the surface, in the bulk, or in the back of the material can be detected. A special case for Si investigations is the He-Ne laser, emitting at 1.152µm, because its energy corresponds exactly with the band gap of Si at room temperature (1.07eV). Thus silicon is transparent for wavelengths greater than or equal to 1.152µm and an IC can be investigated from the underside of the device, through the material. The energy of the photons of this laser is, however, just great enough to produce an OBIC signal. Therefore, not only is reflected light available from the underside of the silicon, but also the OBIC signal with which internal electric fields of the chip can be imaged.

Table 2 Lateral resolution and penetration depth in Si and GaAs

Laser type	wavelength [nm]	[eV]	resolution [µm]	penetration depth [µm] in Silicon	in GaAs
AlGaAs/GaAs	≈820	1.51	0.7	10	1
He - Ne	1152	1.07	0.9	several mm	transparent
GaInAsP/InP	≈1300	0.95	1.1	transparent	transparent
He-Ne	1523	0.81	1.2	transparent	transparent

The possibility of focussing onto different planes in the silicon bulk and, moreover, of producing an OBIC signal is due not only to this laser, but also to the properties of Si. Si is an indirect gap material and such substances have a relatively weak increase of the absorption edge. The absorption edge of direct gap materials exhibits a steep increase.

Figure 2 shows the same area of a device taken in the reflected infrared mode from the back, through the silicon, with the wavelength 1.152µm and 1.3µm, and also the infrared OBIC image from the back of the device. A defect region under the contact can be recognized in the images.

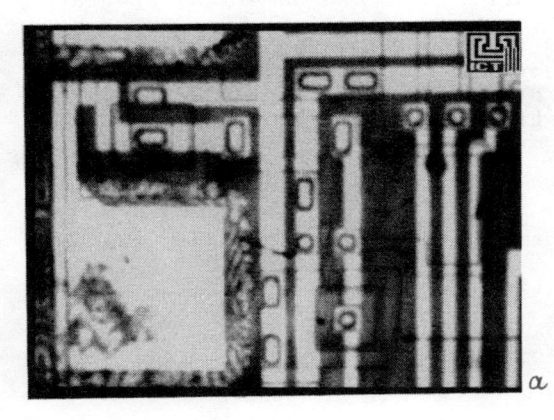

Figure 2:
Damage under a contact made visible by reflection of infrared light (a:1.152µm, b:1.3µm) and infrared OBIC from the back of the device through the substrate (c).
(magnification 600X)

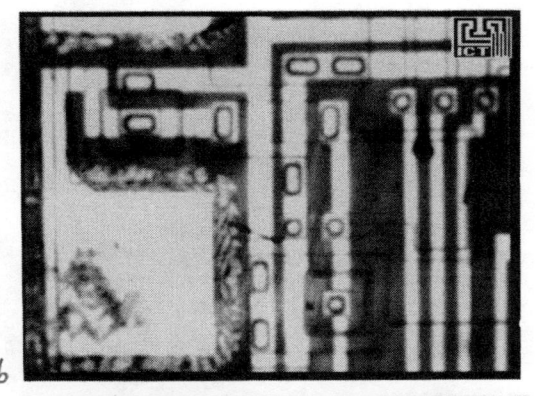

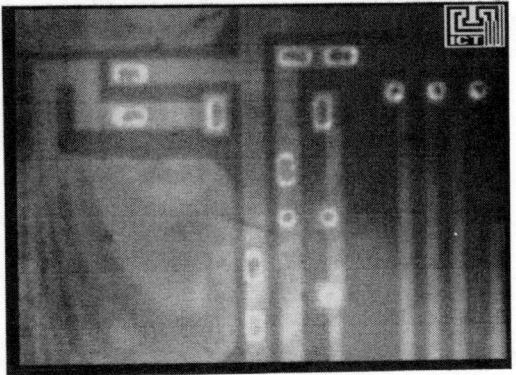

Figure 3 shows a reflected light image taken in the same manner as in Fig.2, but with the He-Ne laser emitting at 1.523µm.

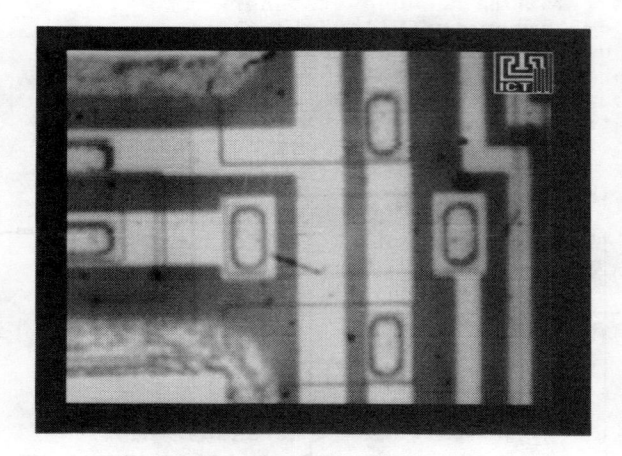

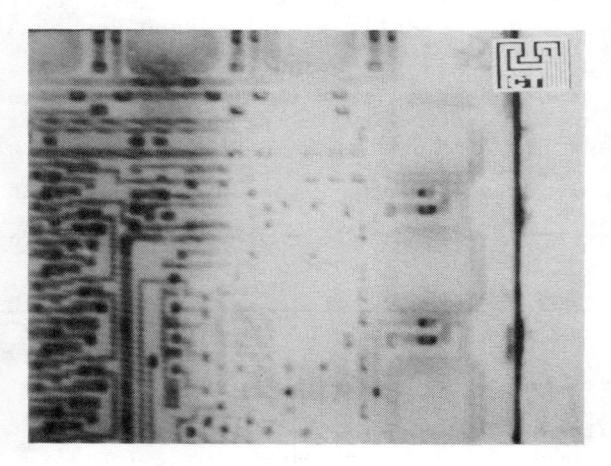

Figure 3. Reflected infrared light image taken with 1.523µm from the back (magnification 1200X)

Figure 4. Infrared OBIC image of a CMOS device from the back. The bright region near the center reveals the p - well (magnification 100X, numerical aperture 0.1)

During such investigations of semiconductor material one has to be sure that the numerical aperture of the objectives used is not too small. Imaging of the reflected light is in practise not possible with objectives of low numerical aperture, because of the high refractive index of the semiconductor material. In contrast to that an imaging of the OBIC signal at low numerical apertures is still possible.

Figure 4 shows an OBIC photograph at small magnification and numerical aperture. The bright region near the center of the image is caused by the well of this CMOS device.

Besides the recognition of mechanical and electrical defects under the metallization of devices, the IR LSM can be applied in the testing of solar cells and silicon wafers at different depths. A special application of the laser diode at 820nm is the determination of the frequence dependence of logic states (on the face of device).

4. REFERENCES

1. C.J.R. Sheppard and A. Cloudhury, Image formation in the scanning microscope, Optica Acta 24, 1051 - 1073 (1977)
2. C.J.R. Sheppard and T. Wilson, Image formation in scanning microscope with partially coherent source and detector, Optica Acta 25, 315 - 325 (1978)
3. G.J. Brakenhoff, P. Blom and P. Barends, Confocal Scanning light microscopy with high aperture immersion lenses, J. Micros. 117, 219 - 232 (1979)
4. V. Wilke, Optical Scanning Microscopy - The Laser Scan Microscope, Scanning 7, 88 - 96 (1985)
5. E. Ziegler and H.P. Feuerbaum, IC Testing using optical beam induced currents generated by a Laser Scan Microscope, Microelectronic Eng 7, 309 - 316 (1987)
6. S. Geckeler, Lichtwellenleiter für die optische Nachrichtenübertragung pp.14ff, Springer Heidelberg (1986)

Computerized Surface Plasmon Microscopy

Eric M. Yeatman and Eric A. Ash

Imperial College of Science and Technology
Department of Electrical Engineering
Exhibition Road, London SW7 2BT, England

ABSTRACT

A method is presented for calculating the reflectivity of a beam of arbitrary profile, in a prism coupler, which is generating plasmons on a non-uniform surface. The particular case of a line discontinuity is considered, and computations for specific systems are presented. Implications for various methods of surface plasmon microscopy are discussed, including the use of image processing in a computerized system.

1. INTRODUCTION

Surface plasmons are electromagnetic guided waves associated with the oscillation of carrier electrons on the surface of a conductor. Optical excitation of surface plasmons using a prism coupler was first reported by Otto[1]; the geometry now most commonly employed is the Kretschmann configuration[2], as illustrated in figure 1. Light polarized in the TM mode is incident on the base of a prism at an angle θ greater than the critical angle, and is thus totally reflected. An evanescent field extends through a thin metallic layer, and when the parallel wave vector of this field matches that of the surface plasmon mode for the far surface of the metal, plasmons are generated. This causes a strong field enhancement at this far surface, and a drop in the reflected signal due to absorbtion in the silver.

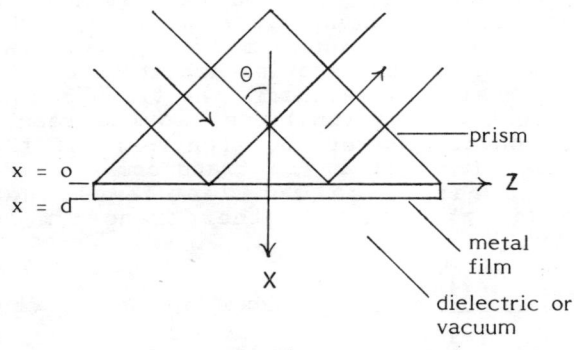

Figure 1. The Kretschmann configuration for optical excitation of surface plasmons.

If the reflection coefficient in the geometry above is measured as a function of angle, a sharp dip occurs where the plasmon phase matching condition is satisfied. The width of the dip is proportional to the decay coefficient of the plasmon mode, and the depth is determined by the coupling strength, set by the metal thickness. Silver is the metal most commonly used, as its high conductivity gives a particularly sharp resonance (a few tenths of a degree). The resonance angle and width are highly sensitive to the surface properties of the metal, due to the strong enhancement of the intensity of the field, and its confinement to a thin region adjacent to the surface. For instance, biological monolayers deposited on the metal are easily detected. For this reason, the use of surface plasmon measurements for sensing has been the subject of much investigation[3,4].

We have been interested in extending this application of plasmons in sensing to the measurement of spatial variations on a surface. We have reported initial experimental results for surface plasmon microscopy (SPM)[5,6], in which images of thin oxide films on silver were produced using both scanned focussed beams and wide beam illumination. Thickness sensitivity of about 3 angstroms was achieved with a lateral resolution of about 20 microns. Similar results have since been independently obtained by another group[7].

2. A DIFFRACTION THEORY FOR SPM

Fundamental considerations indicate that the resolution of a microscopy technique using lossy guided waves will be in some way limited by the decay length of these waves (typically 10 to 50 microns for surface plasmons). This can be looked at in two ways: first, to illuminate a spot smaller than the decay length requires a beam with an angular width greater than that of the resonance, and therefore when the measurement is made the reflectivity dip will broaden and sensitivity will be lost; equivalently, the maximum precision in measuring the plasmon wavevector, and thus the maximum sensitivity, cannot be obtained unless the incident beam interacts with it coherently over at least its decay length.

The following analysis describes a method by which the interaction of incident light and surface plasmons in the presence of spatial variations can be calculated. We begin by formulating the relevant fields; this we will do for convenience in terms of the magnetic field vectors, as for TM polarization they lie entirely in the y direction. Using the geometry of figure 1, the incident and reflected waves are given by:

$$H_i = A \exp[i(\beta_a z + k_{xp} x - wt)] \quad (1)$$

$$H_r = B \exp[i(\beta_a z - k_{xp}x - wt)] \tag{2}$$

The plasmon field is given by:

$$H_{sp} = D \exp[i(\beta_a z - wt) - K_d(x-d)] \quad (x \geq d) \tag{3}$$

where:

$$k_{xp} = (n_p^2 k_o^2 - \beta_a^2)^{\frac{1}{2}} \tag{4}$$

$$K_d = (\beta_a^2 - n_d^2 k_o^2)^{\frac{1}{2}} \tag{5}$$

$$k_o = w/c \tag{6}$$

$$\beta_a = n_p k_o \sin\theta \tag{7}$$

and n_p and n_d are the refractive indices of the prism and dielectric medium respectively.

In the prism coupled geometry, a solution to Maxwell's equations is given by a surface plasmon at the metal/dielectric interface, having a propagation vector:

$$\beta_{sp} = \beta_r + i(\Gamma_i + \Gamma_r) \tag{8}$$

where β_r is the resonant spatial frequency (real part), and Γ_i and Γ_r are the decay coefficients due to internal absorbtion and re-radiation into the prism respectively.

For plane waves incident on a uniform system of infinite extent, the reflectivity can be solved exactly using Fresnel reflection equations for a multi-layered system[8]. A useful explicit approximation for the reflectivity as a function of angle can be obtained using an expansion about the resonance. This gives, for the amplitude reflection coefficient:

$$\frac{B}{A r_{pm}} = 1 - i \frac{2\Gamma_r}{(\beta_a - \beta_r) + i\Gamma_t} \tag{9}$$

where $\Gamma_t = \Gamma_i + \Gamma_r$ \tag{10}

and r_{pm} is the reflectivity for the prism/metal interface for an infinite metal thickness. The reflectivity will reach zero at resonance $(\beta_a = \beta_r)$, if the impedance matching condition $\Gamma_i = \Gamma_r$ is satisfied.

The resonant amplitude can be obtained in a similar way:

$$\frac{D}{A} = \frac{-r_{pm}}{1-r_{pm}} \exp(K_m d) \frac{i2\Gamma_r}{(\beta_a - \beta_r) + i\Gamma_t} \tag{11}$$

In the uniform plane wave case, the amplitude of the plasmon field is constant in the z direction, at the equilibrium value given by (11) above. If we consider this to be instead the local amplitude of the plasmon which is the modal solution of Maxwell's equations in this geometry, then it will have associated with it a re-radiation field in the prism with an amplitude which we shall call B^+, given by:

$$B^+ = \frac{1-r_{pm}}{\exp(K_m d)} D \tag{12}$$

If we write this in terms of the incident amplitude responsible for the plasmon amplitude, we get:

$$B^+ = -r_{pm} A \frac{i2\Gamma_r}{(\beta_a - \beta_r) + i\Gamma_t} \tag{13}$$

Using equation (9) we can now see that the total reflectivity can be written as two terms, one due to the direct reflection of the incident beam, and the other due to the re-radiation of the plasmon field.

$$B = r_{pm} A + B^+ \tag{14}$$

Thus when spatial variation is introduced into the system, we can consider the effect on these two components independently.

2.1 Reflection of a focussed beam

A focussed incident beam can be thought of as an angular spectrum of plane waves, and the reflectivity calculated accordingly. In the following discussion we will assume that the system and the fields are uniform in the y direction, so we will only consider focussing in one dimension. For a beam incident at or near the resonant angle, $r_{pm}(\beta)$ will be a very slowly varying function compared to B^+, so we can approximate it as a constant. Then we can write:

$$\frac{B(z)}{r_{pm}} \exp(i\beta_a z) = \int_{-\infty}^{\infty} A(\beta) \exp(i\beta z) d\beta - \int_{-\infty}^{\infty} \frac{A(\beta) i2\Gamma_r \exp(i\beta z)}{(\beta - \beta_r) + i\Gamma_t} d\beta \tag{15}$$

These integrals are Fourier transforms from spatial frequency into position. The first, being the transform of the spectral amplitude, gives the spatial amplitude of the incident beam. The second is the transform of the product of two functions, so the result is the convolution of the transforms of the individual functions. The transform of $A(\beta)$ gives, again, the incident amplitude $A(z)\exp(i\beta_a z)$, and the other transform is given by:

$$\int_{-\infty}^{\infty} \frac{i2\Gamma_r \exp(i\beta z)}{(\beta - \beta_r) + i\Gamma_t} d\beta = \begin{cases} 2\Gamma_r \exp[(i\beta_r - \Gamma_t)z] & z \geq 0 \\ 0 & z < 0 \end{cases} \tag{16}$$

If we rewrite (15) as:

$$B(z)/r_{pm} = A(z) + B^+(z)/r_{pm} \tag{17}$$

then $B^+(z)$ is the convolution of the incident amplitude function with the decaying exponential of the plasmon mode, according to:

$$B^+(z)/r_{pm} = -2\Gamma_r \exp(-i\beta_a z) \times \int_{-\infty}^{z} \exp[(i\beta_r - \Gamma_t)(z-\sigma)] A(\sigma) \exp(i\beta_a \sigma) d\sigma \quad (18)$$

where β_a is now the centre spatial frequency of the angular spectrum.

Because the plasmons travel in the +z direction, the amplitude of the plasmon field, and thus of B^+, at some position z_1, is due only to the nature of the medium and incident amplitude in the region $z<z_1$. If we take the derivative of (18) we obtain:

$$dB^+(z)/dz = -[\Gamma_t + i(\beta_a - \beta_r)]B^+(z) - 2\Gamma_r A(z) r_{pm} \quad (19)$$

Equations (19) and (17) provide a complete solution for the effective source amplitude distribution of the reflected beam, for an arbitrary amplitude distribution incident on a guided wave supporting structure of arbitrary spatial variation. The intensity distribution at some distance away from the prism base will be given by an integration of this source distribution multiplied by the appropriate phase function; the far-field image will simply be the Fourier transform of $B(z)$.

In the special case of a guide without intrinsic loss, $\Gamma_i = 0$, this formulation reduces to one equivalent to that described by Ulrich[9] in the analysis of a tapered coupler. As in that case, the solution does not include the effects of scattering of the plasmons into radiation modes due to perturbations of very high spatial frequency. This is a subject we are currently investigating.

2.2 Line Features

We can include the influence of scattering in one particular type of feature: that of a line separating two surfaces with different plasmon propagation vectors. Let us consider the geometry shown in figure 2; a finite beam is incident on two uniform media, characterized by propagation vectors $\beta_1 + i(\Gamma_{i1} + \Gamma_{r1})$ and $\beta_2 + i(\Gamma_{i2} + \Gamma_{r2})$ respectively, which meet at $z=0$.

When a plasmon field arrives at the boundary, part of the power will be transmitted as a plasmon into medium 2, while the rest will be partially scattered as free radiation, and partially reflected as a backwards-going plasmon in medium 1. It can be shown that the reflection coefficient will in general be very small, and as the backwards-going waves are not phase matched to the forward-going ones, we need not include them in our calculations of the re-radiated signal; they will re-radiate light in the prism back towards the incident beam. If two line features close together have high plasmon reflection coefficients, the contribution of secondary reflections might become significant, but this is not something we will consider further in this discussion.

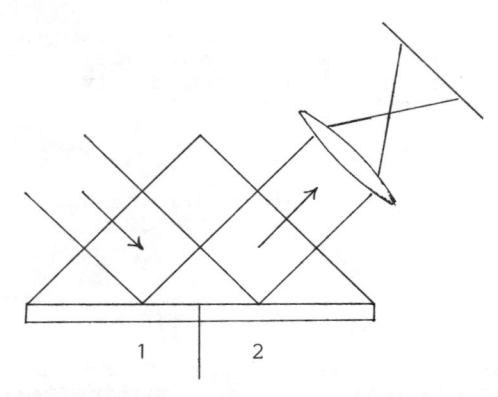

Figure 2. Geometry for analysis of line feature. A lens can be inserted in the reflected beam to create an image of the reflected intensity distribution at the prism base.

The amplitude transmission coefficient for plasmons can be calculated using the method described by Leskova[10]; alternatively, we can use an overlap integral, normalized by the square root of the phase velocities:

$$t_{sp} = (\beta_1/\beta_2)^{\frac{1}{2}} \frac{\int_0^\infty \exp(-K_{d1}x)\exp(-K_{d2}x)dx}{\int_0^\infty \exp(-2K_{d2}x)dx} \quad (20)$$

where the K values are as defined by (5). This gives:

$$t_{sp} = (\beta_1/\beta_2)^{\frac{1}{2}} \frac{2K_{d2}}{K_{d1} + K_{d2}} \quad (21)$$

If the power transmission coefficient is calculated using this expression, the result is equivalent to that given by Barlow and Brown[11] for radio frequency surface waves.

The calculation of $B^+(z)$ using equation (19) must now be supplemented by the multiplication of B^+ by the appropriate value of t_{sp} at any line feature. The subsequent behaviour will continue to be governed by (19). If we assume that $A(z)$ is slowly varying, as with a broad incident beam, then B^+ will have an equilibrium value in each medium given by:

$$B^+_{eq} = \frac{-2\Gamma_r A(z) r_{pm}}{\Gamma_t + i(\beta_a - \beta_r)} \quad (22)$$

Equation (19) can then be rewritten as:

$$dB^+(z)/dz = -[\Gamma_t + i(\beta_a - \beta_r)](B^+(z) - B^+_{eq}) \quad (23)$$

The solution to this differential equation is straightforward:

$$B^+(z) = B^+_{eq} + C\exp[-(\Gamma_t + i(\beta_a - \beta_r))z] \quad (24)$$

where C is a constant to be determined by the boundary conditions.

Let us assume that B^+ is at its equilibrium value in medium 1 when it arrives at the boundary $z=0$. Then its value just after the boundary will be given by:

$$B^+(0^+) = t_{sp}B^+_{eq1} \qquad (25)$$

The constant C can now be determined, giving the complete solution for medium 2:

$$B^+(z>0) = B^+_{eq2} +$$
$$(t_{sp}B^+_{eq1} - B^+_{eq2})\exp[-(\Gamma_t + i(\beta_a - \beta_r))z] \qquad (26)$$

Experimental observation of the transient oscillations indicated by equation(24) have been reported[12]. They allow the determination of both β_r and Γ_t of the second medium, by measurement of the period and decay rate of the oscillations, respectively.

3. COMPUTED RESULTS

The following figures show source intensity distributions, $|B(z)|^2$, computed by the method described above. In figure 3, a wide Gaussian beam is incident on the junction between two regions having resonances separated by 5 resonance widths. The transient oscillations are clearly visible, and have their maximum intensity when the incident beam is phase-matched in medium 1 and mis-matched in medium 2. The reflected intensity reaches its new equilibrium most rapidly on crossing a boundary from a mis-matched to a matched region. If the incident field is poorly matched to both resonances, the presence of the junction is not clearly seen. In figure 4, the resonances are much closer together and have different widths. The line boundary is still clearly visible, particularly when the beam is resonantly absorbed in the first medium.

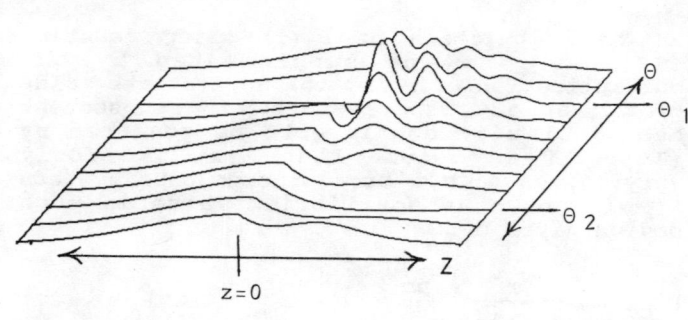

Figure 3. Source intensity distributions, $|B(z)|^2$, for a wide Gaussian beam incident at a series of angles. The values θ_1 and θ_2 are the angles of resonant excitation in the two regions. Relevant parameters are:
Plotted range: $1200/k_o$ Beam width: $1000/k_o$
$\beta_1 = 1.08k_o$, $\beta_2 = 1.03k_o$,
$\Gamma_{i1} = \Gamma_{r1} = \Gamma_{i2} = \Gamma_{r2} = 0.0025k_o$

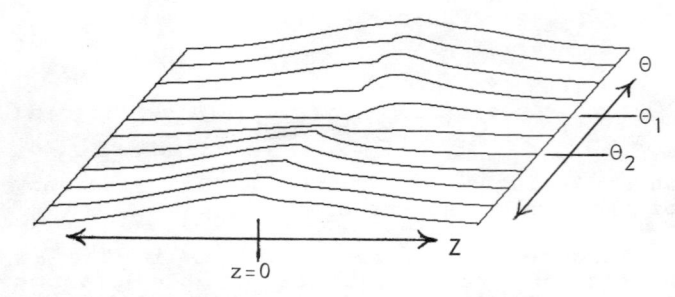

Figure 4. Source intensity distributions, for $\beta_1 = 1.04k_o$, $\Gamma_{i2} = 0.0075k_o$, other parameters as for figure 3.

If a narrow beam is used to excite plasmons, the excitation distribution is the convolution of the incident field with the natural decaying exponential of the plasmon, as we have seen, and the change in the reflected signal will depend on the surface properties over the whole of the excitation range. Only that part of the incident angular spectrum within the plasmon angular resonance range can be absorbed. Figure 4 shows the intensity distribution for a beam of width less than the plasmon decay length, exciting plasmons on a uniform surface, calculated exactly using Fresnel's equations and a superposition of plane waves.

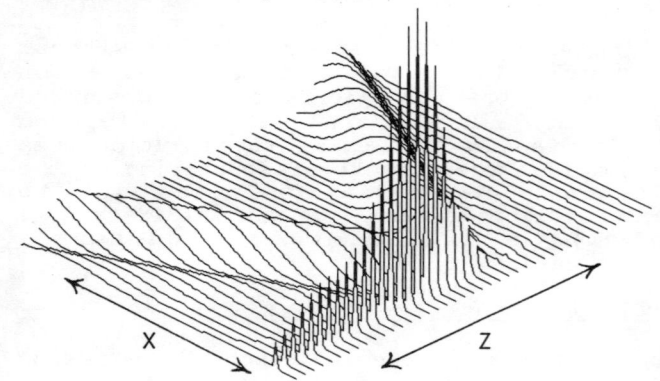

Figure 5. Intensity of the incident and reflected beams and the plasmon excitation, in the plane of incidence, for a narrow Gaussian beam incident on a uniform silver film at the base of a coupling prism. Plotted area is about 100x100 microns.

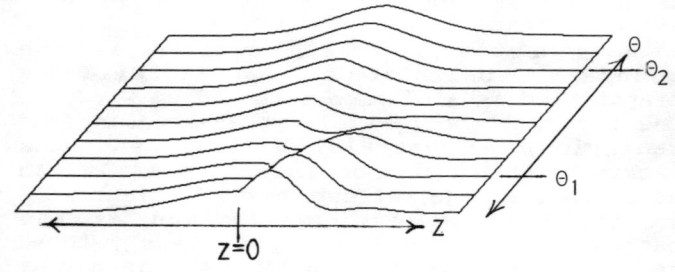

Figure 6. Source intensity distributions for a narrow Gaussian beam incident at a series of angles. Parameters are as in figure 3, except:
Plotted range: $400/k_o$ Beam width: $200/k_o$

Such a narrow beam will not reveal the line feature as clearly, as we can see in figure 6. Here the incident beam width is equal to the plasmon decay length. If it is reduced further, the boundary between the two surfaces becomes rapidly more difficult to resolve.

3.1 Scanned measurements

We can also investigate the results of scanning a focussed beam across the line feature. Figure 7 gives the source distributions for a beam width of half the plasmon decay length, incident at a series of positions on the surface. In each case the center angle of the beam is the resonance angle of the first medium. Changes in source intensity can be seen at the line boundary even when the incident beam has not reached it, but the effects are not as strong as for the wide beam. Figure 8 gives the far-field radiation pattern, which is equivalent to the angular spectrum of the reflected beam. The total power increase gradually as the beam approaches the boundary, over a range corresponding to the plasmon decay length. The shape of the field, however, varies in a complex way due to the influence of two different resonance curves.

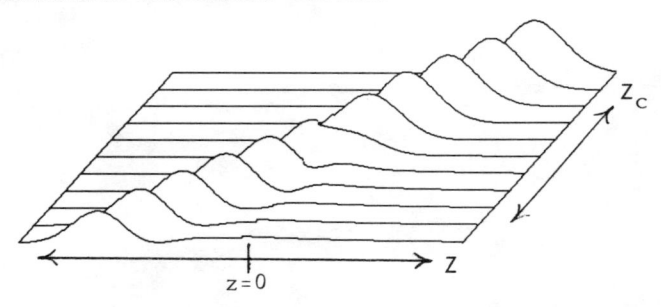

Figure 7. Source intensity distributions for a focussed beam, for various positions of incidence Z_c. Beam width is $100/k_o$, other parameters are as for figure 6.

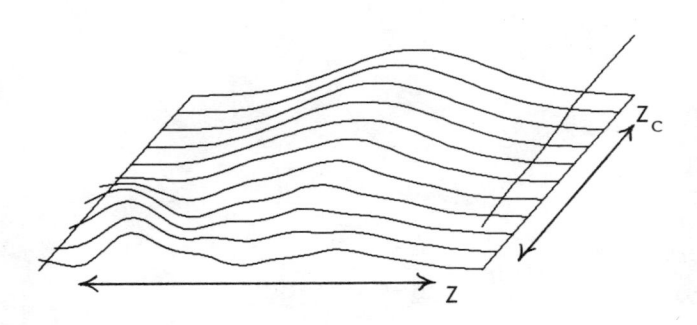

Figure 8. Far-field intensity distributions for the source distributions of figure 7. The additional line shows the variation of the integrated area of the plotted distributions, corresponding to the total power.

4. CONCLUSIONS

The analysis we have presented allows the prediction of the images to be obtained by surface plasmon microscopy from specific surface features, and gives a quantitative explanation for the transient oscillations resulting from plasmon interference at a line boundary. Thus it should be useful in the general interpretation of SPM images. We are investigating the further application of the analytic results by developing a computerized surface plasmon microscope.

This device will digitize and store a series of images of a surface in one of two ways; wide beam source images at a series of incident angles, or far-field angular spectrum images for a focussed beam at a series of positions. The results described above seem to indicate that the analysis is more straightforward for the first method. The object is to develop digital image processing routines that can recover wavelength-limited resolution without loss of sensitivity, possibly using deconvolution routines developed from the analysis in section 2.

5. ACKNOWLEDGEMENTS

The authors are grateful to Professor M. Green for his assistance and support, and to the Science and Engineering Research Council for its financial support. E. Yeatman also gratefully acknowledges a Commonwealth scholarship provided by the Association of Commonwealth Universities.

6. REFERENCES

1. A. Otto, "Excitation of nonradiative surface plasma waves by the method of frustrated total reflection," Z. Phys. 216, 398-410 (1968).

2. E. Kretschmann, "The determination of the optical constants of metals by the excitation of surface plasmons," Z. Phys. 241, 313-324 (1971).

3. M.T. Flanagan and R.H. Pantell, "Surface plasmon resonance and immunosensors," Electron. Lett. 20, 968-970 (1984).

4. C. Nylander, B. Liedberg and T. Lind, "Gas detection by means of surface plasmon resonance," Sens. and Actuators 3, 79-88 (1982/83).

5. E. Yeatman and E. Ash, "Surface plasmon microscopy," Electron. Lett. 23(20), 1091-1092 (1987).

6. E. Yeatman and E. Ash, "Surface plasmon scanning microscopy," in Proc. SPIE Vol. 897, _Scanning Microscopy Technologies and Applications_, 100-107 (1988).

7. B. Rothenhausler and W. Knoll, "Surface plasmon microscopy," Nature 332, 615-616 (1988).

8. W. Hansen, "Electric fields produced by the propagation of plane coherent electromagnetic radiation in a stratified medium," J. Opt. Soc. Am. 58, 380-390 (1968).

9. R. Ulrich, "Optimum excitation of optical surface waves," J. Opt. Soc. Am. 61, 1467-1477 (1971).

10. T.A. Leskova, "Theory of a Fabry-Perot type interferometer for surface polaritons," Solid State Comm. 50, 869-873 (1984).

11. H.M. Barlow and J. Brown, <u>Radio Surface Waves</u>, (Clarendon Press, Oxford, 1962), p. 140.

12. R. Rothenhausler and W. Knoll, "Interferometric determination of the complex wave vector of plasmon surfcae polaritons," J. Opt. Soc. Am. B 5, 1401-1405 (1988).

Computerized analysis of high resolution images by scanning acoustic microscopy

Daniele D. Giusto, Bruno Bianco, Andrea Cambiaso, Massimo Grattarola, and Mariateresa Tedesco

Dept. of Biophysical and Electronic Engineering, Univ. of Genoa
Via all'Opera Pia 11A, I-16145 Genoa, Italy

ABSTRACT

High resolution scanning reflection acoustic microscopy is a new scanning technique which provides information about the local elastic properties (both at the surface and in depth) of various kinds of objects. In the present work, two applications of the scanning system (Leitz ELSAM microscope, frequency range: 0.8-2.0 GHz) are considered. The first is the characterization of the adhesion of mouse neuroblastoma cells to a silicon substratum: rings of alternate intensities, originated by acoustic interference fringes, and shown in the cell image, are utilized to obtain information about cell morphology. In the second application, instead, acoustic microscopy is proposed as a non-destructive, inexpensive, and fast technique for characterizing semiconductor devices. The work is focused on the low-level processing (filtering, segmentation, and feature extraction) of the resulting acoustic images, to restore the original information and to measure several features useful in characterizing and understanding an object. The final goal is to determine the acoustic impedance and the acoustic attenuation of the object considered, and, in the case of living cells, to monitor them in time.

1. INTRODUCTION

Though acoustic imaging is adopted by a group of well established disciplines related to macroscopic objects (e.g., geological prospection, sonar techniques, and medical echography), acoustic microscopy is a relatively unknown technique. At resolution levels comparable with those of standard optical microscopy, this novel technique has been developed from prototypes (thanks to the efforts of several researchers, in particular, Quate and coworkers [1]) to commercial apparatus only in this decade. As a consequence, the actual capabilities of acoustic microscopes are not yet completely clear. Certainly, in comparison with optic apparatus, the acoustic microscope exhibits a number of unvaluable advantages. It is conceptually very simple: in fact, a single lens can accomplish the major imaging task; this simplicity is due to the focusing properties of acoustic lenses, which are an order of magnitude more accurate than optic ones. As a consequence, a single acoustic lens can be equivalent to an optic objective made up of about 10 lenses (equivalence refers to resolution, field aperture, and magnification). Another advantage of acoustic microscopes, maybe the major one, lies in their ability to "see" under the surface of an (optically) opaque sample.

The main drawbacks of acoustic microscopes are probably the following:

- the need for a highly sophisticated (hence expensive) mechanic and electronic control;
- the low speed of image acquisition;
- the need for a coupling liquid between sample and lens;
- from the point of view of parameter extraction, the great complexity of the relation between the local properties of a sample and the image acquired.

In the present paper, we present two kinds of applications: analysis of the adhesion of mouse neuroblastoma cells to a silicon substratum, and a technique for characterizing semiconductor devices by means of some image processing algorithms, among which fractal analysis.

2. MICROSCOPE MODEL AND IMAGE FORMATION

In this section, reference is made to the ELSAM (Ernst Leitz Scanning Acoustic Microscope of Leitz Gmbh, FRG) apparatus available in the authors' laboratory (Figure 2.1) and used for the image acquisition, as described later on. A strongly simplified block diagram of ELSAM is shown in Figure 2.2. The acoustic lens (Figure 2.3) is mechanically scanned under the control of the X-Y scanning module, that drives two magnetic vibrators in X and Y direction. For each pixel scanned, the sequence of operation is the following: an RF signal from the oscillator reaches, via the solid state switch (SW), the piezoelectric transducer (actually, the frequencies allowed can be either 100 MHz or a frequency in the 0.8-2 GHz range; the transducer is a sheet of zinc oxide). This signal is converted to a longitudinal elastic wave which propagates inside the body of the lens. The latter is made of sapphire, and the lens cavity is spherical. A coupling liquid (water) is interposed between the lens and the sample: this is needed in order to allow sufficient acoustic power to reach the sample (if the coupling fluid were air or another gas, the incident wave would be almost completely reflected by the lens-gas interface). The spherical lens transforms the plane wave into a spherical one, centered at some point F, the (second) focus of the lens. This is actually an approximation, but the caustic is much more spread than in the optic case, owing to the fact that the spread of the caustic (i.e., the focusing performances) is controlled by the relative refractive index between the media on either side of the spherical surface. This index is about 1.5 for an optic lens (glass-air), and around ten times higher for an acoustic one (sapphire-water). In general, if the focus coincides with a point of the sample surface, the wave reflected by the surface toward the lens is no longer spherical. This reflected wavefront is more or less different from a sphere depending on the reflectance properties of the sample (which, in turn, depend on its elastic parameters, e.g., the Lame' coefficients). The lens transforms the wavefront into longitudinal waves generally not plane, and the transducer converts them to an RF voltage distributed, in general non uniformly, on its surface. During this "reflection" period, the solid state switch is turned in such a way that the oscillator signal is prevented from reaching the transducer, whereas the transducer signal, related to the reflected wavefront, can reach the demodulator section, whose output is a continuous level proportional to the average amplitude of the voltage on the transducer. An "image" is a set of such levels (one for each scanned point), or its representation in analog form on a TV screen. The image is spatially uneven in intensity, in accordance with the local reflectance of each point. Similar considerations hold true in the case where the focus does not coincide with a surface point.

The image resolution, defined according to Rayleigh's criterion, is given by the classical equation:

$$Dx = [0.6 \ast wl] / \sin(a)$$

where wl is the wavelength in water and a the lens aperture. At 2 GHz, for example, wl in water is about 0.75 um; for a=70 degrees, one has Dx=0.5 um, a resolution of the same order as in optics. It must be stressed that, in an optic microscope, the resolution value can be doubled through the proper use of a matched condenser; this is not the case with the ELSAM system.

3. CELL ANALYSIS

This application is part of a long-range project dealing with the functional coupling of an array of ISFET-like devices to a network of living cells characterized by neuronal behaviour [2].

This goal has far-reaching implications for several research areas, including those of biosensors, electrophysiology, and clinical rehabilitation.

We are presently characterizing the adhesion of mouse neuroblastoma cells to wafers covered with Si_3N_4, by utilizing the techniques of fluorescence image analysis and scanning reflection acoustic microscopy. This paper is only concerned with the use of the second technique.

We used cells of the line NB41A3 (mouse neuroblastoma) purchased from the American Type Cell Colture: they were grown under standard conditions. The Si_3 wafers were covered with laminin in order to improve cell adhesion. Square 20 mm x 20 mm wafers were supplied by SGS-Microelettronica SpA: they were covered with a thin layer of SiO_2 (about 200 angstroms) plus a thin layer of Si_3N_4 (about 300 angstroms). The double layer of insulators was chosen in accordance with an approach commonly adopted for ISFETs, in order to obtain a greater stability of the devices [3].

In Figure 3.1 one can see a neuroblastoma cell acquired at 1.6 GHz and 35'C, utilizing the culture medium as the coupling liquid between the lens and the cell. The focal plane is near the surface of the substratum. Small dark spots are visible near the cell periphery: they could be interpreted as sites of stronger adhesion [4].

This image was processed with two low-level processing techniques in order to extract these points. In particular, we used an edge-preserving smoothing to reduce the high-frequency noise that affected the image: this technique replaces the grey level of a pixel with the medium grey level of a mask with the minimum variance among a set of masks in the neighbourhood of that pixel [5]. Then we applied a region-growing segmentation technique which split the image into a set of elementary regions characterized by homogeneous grey levels and by contours above a predefined threshold (automatically determined by analyzing the local histograms of the differential image) [5]. Then, the regions interpretable as sites of adhesion (with dark grey levels and circular shapes) were extracted, as one can see in Figure 3.2.

Figure 3.3 depicts a cell acquired under the same conditions as for Figure 3.1, except for the focal plane, which was near the top of the cell (i.e., above the surface of the substratum). In this way, interference fringes (equi-depth contours) were generated, which were related to the cell structure [6]. This image was processed in order to extract the profile of the grey level values along an interference fringe (Figure 3.4). First, we applied a low-pass filter (unweighted spatial mean on a square of 3*3 pixels) to the original image, and then we subtracted the processed image from the original one. In this way, the "offset" values of the alternate rings corresponding to the interference fringes were eliminated. Subsequently, a two-level segmentation was performed to detect rings. By this procedure, we can extract the variations in the grey levels along "pieces" of four rings (calculated on the original picture). Once all the artifacts have been eliminated from the images, the variations in the grey levels along the rings (which should be equi-depth contours) can be interpreted in terms of local elastic properties of the cell structures.

4. MICROELECTRONIC DEVICE ANALYSIS

The relation between the output voltage of the piezoelectric transducer, and a defocusing in the z direction, is one of the most quantitative measurements allowed by acoustic microscopy. This parameter is usually denoted by $V(z)$: V stands for the output of the transducer, and z indicates the related position between lens and sample. Theoretical [7,8,9] and experimental [10] works show that the shape of the $V(z)$ curve is very characteristic for the material under investigation, hence this curve can be regarded as the acoustic signature of the material. The trend of the $V(z)$ curve is normally an oscillating function that decreases when the lens is moved toward the sample. The period of each oscillation depends on the material being measured, but it is generally very small (e.g., 3 um for aluminium and 7 um for silicon). For this reason, if the sample consists of zones characterized by slight differences in thickness, it may be possible to find a focus position that gives prominence only to zones with the same thickness (in Figure 4.1 one can see a portion of a thyristor: zones at different zones are visible). In order to detect the different zones, we recall that no edges are present, and that there are only differences in the texture. Then we can extract the texture parameter for each pixel and segment an image by analyzing this feature.

The fractal-based technique is one of the most recent approaches in texture analysis [11]. To extract the fractal dimension "D" [12] of each pixel, we used an estimation technique based on an adaptive mask selection: different masks (of different sizes, shapes and positions) are considered for each pixel, and the most homogeneous neighbourhood (in terms of fractal dimension) is selected by using a focusing mechanism and a minimum-variance criterion [13]. The image so obtained is then processed (by means of the same edge-preserving smoothing and region-growing segmentation technique) and the zones at different foci are separated (Figure 4.2).

5. CONCLUSIONS

We have presented some application of high-resolution scanning reflection acoustic microscopy, in particular, the characterization of the adhesion of mouse neuroblastoma cells to a silicon substratum, and the analysis of semiconductor devices. By using some low-level image processing techniques we have extracted informations and measured several features useful in characterizing and understanding an object.

Future work will be focused on the determination of the acoustic impedance and of the acoustic attenuation of the object considered, and, in the case of living cells, the monitoring of these quantities in time.

6. REFERENCES

[1] C.F.Quate, A.Atalar, and H.K.Wickramasinghe, "Acoustic microscopy with mechanical scanning: a review", Proc. IEEE 67/8, pp.1092-1114 (1979).

[2] M.Grattarola, M.Tedesco, A.Cambiaso, G.Perlo, and A.Sanguineti, "Cell adhesion to silicon substrata: characterization by means of optical and acoustic cytometric techniques", Biomaterials 9, pp.101-107 (1988).

[3] A.Sibbald, "Recent advances in field-effect chemical microsensors", J. Molec. Electronics 2, pp.51-83 (1986).

[4] J.Hildebrand, D.Rugar, R.Johnston, and C.Quate, "Acoustic microscopy of living cells", Proc. Natl. Acad. Sci. USA 78, pp.1656-1660 (1981).

[5] M.Nagao and T.Matsuyama, _A structural analysis of complex aerial photographs_, Plenum Press, New York (1980).

[6] J.Hildebrand and D.Rugar, "Measurements of cellular elastic properties by acoustic microscopy", J. Microscopy 134, pp.245-260 (1984).

[7] A.Atalar, "An angular spectrum approach to contrast in reflection acoustic microscopy", J. Appl. Phys. 49, pp.5130-5139 (1978).

[8] A.Atalar, "A physical model for acoustic signatures", J. Appl. Phys. 50, pp.8237-8239 (1979).

[9] H.L.Bertoni, "Ray-optical evaluation of $V(z)$ in the reflection acoustic microscope", IEEE Trans. on Sonics and Ultrasonics 31, pp.105-116 (1984).

[10] M.Hoppe, A.Atalar, W.J.Patzelt, and A.Thaer, "The ELSAM acoustic microscope: Application in material science - first results", Int. Rep., E.Leitz Wetzlar Gmbh (1983).

[11] S.Peleg, J.Naor, R.Hartley, D.Avnir, "Multiple resolution texture analysis and classification", IEEE Trans. on Pattern Analysis and Machine Intelligence 6, pp.518-523 (1984).

[12] B.B.Mandelbrot, _The fractal geometry of nature_, Freeman, San Francisco (1982).

[13] F.Arduini, C.Dambra, S.Dellepiane, S.B.Serpico, G.Vernazza, and R.Viviani, "Fractal dimension estimation by adaptive mask selection", Proc. of ICASSP-88, pp.1116-1119 (1988).

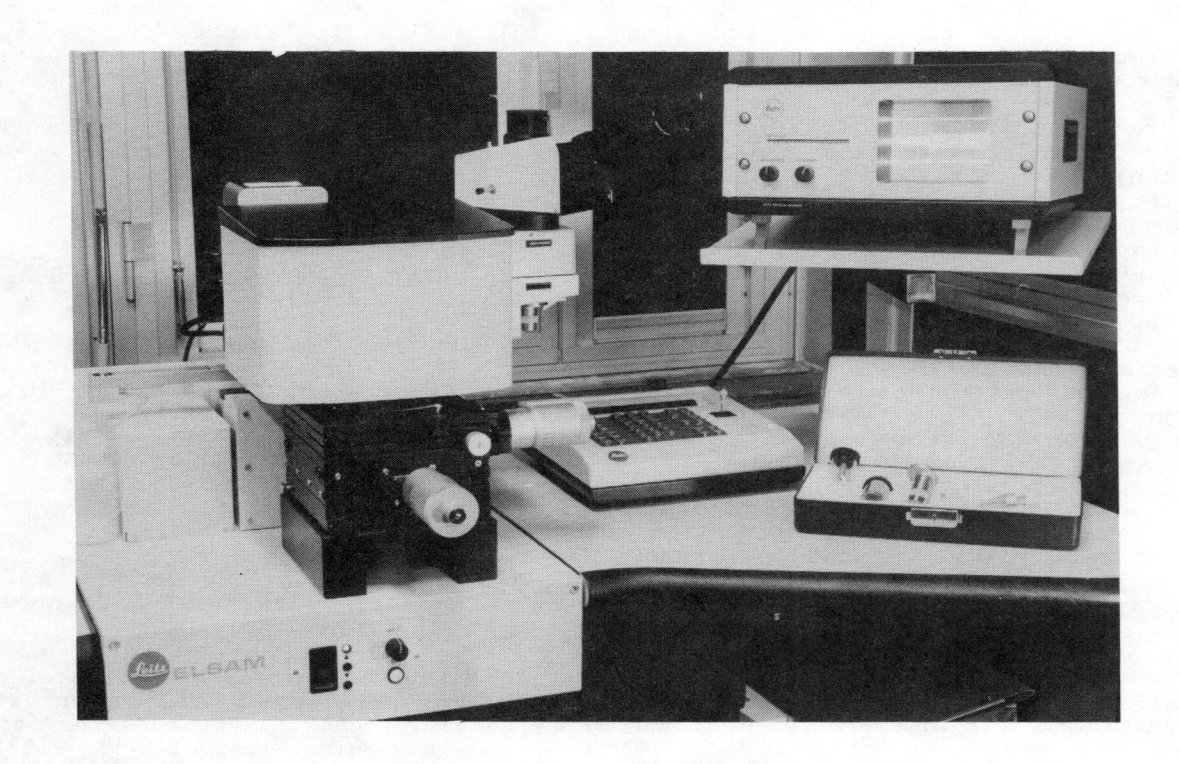

Figure 2.1. The ELSAM acoustic microscope.

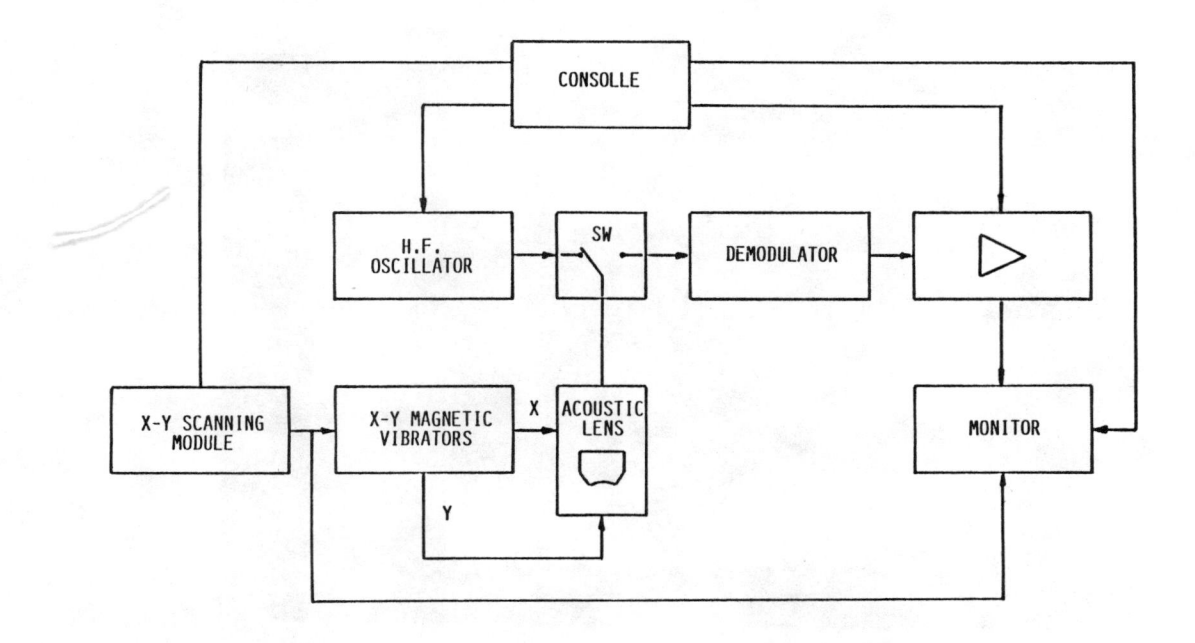

Figure 2.2. A simplified block diagram of ELSAM acoustic microscope.

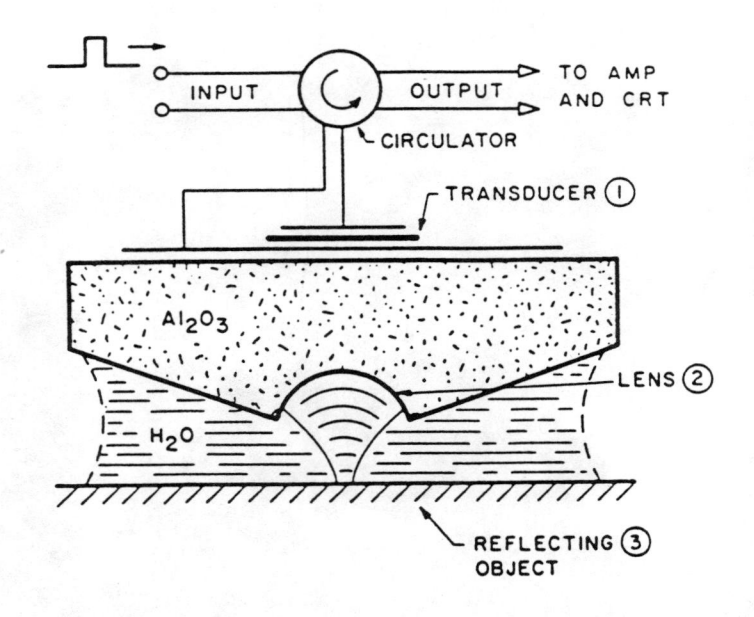

Figure 2.3. A sketch of the lens-sample interface.

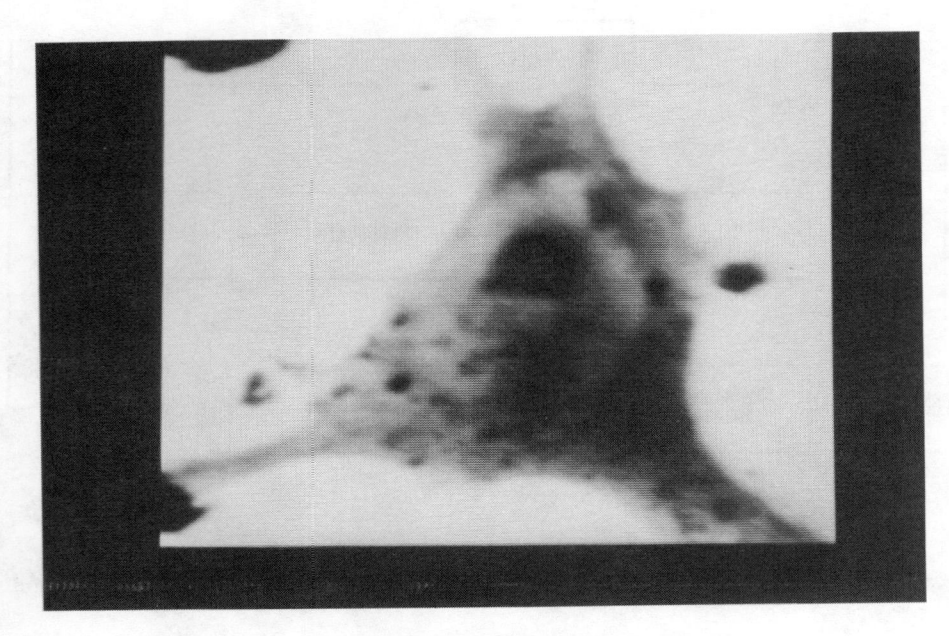

Figure 3.1. A neuroblastoma cell acquired at 1.6 GHz and 35'C, utilizing the culture medium as the coupling liquid between the lens and the cell. The focal plane is near the surface of the substratum. Small dark spots are visible near the cell periphery: they could be interpreted as sited of stronger adhesion.

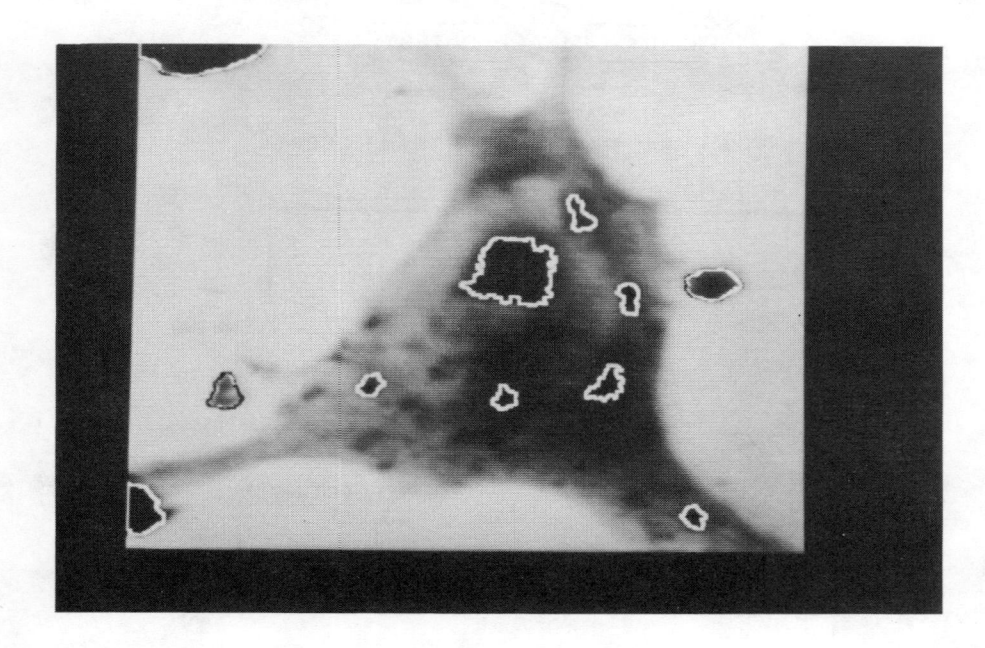

Figure 3.2. After the low-level processing, the sites of stronger adhesion (together with other structures characterized by large areas) are extracted and their contours are superimposed on the original image.

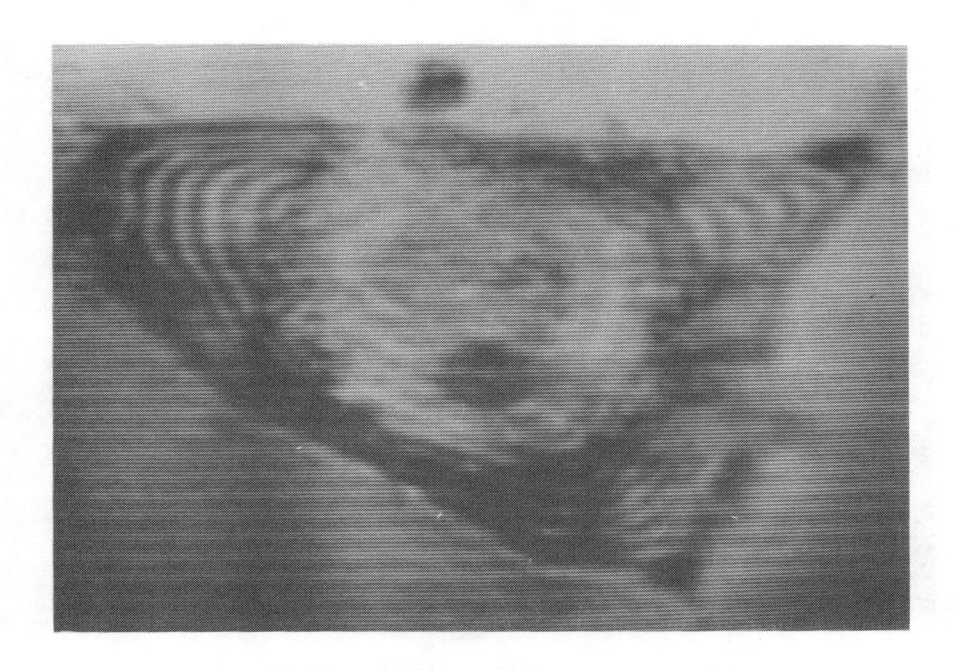

Figure 3.3. A cell acquired under the same conditions as for Figure 3.1, except for the focal plane, which was near the top of the cell (i.e., above the surface of the substratum). In this way, interference fringes (equi-depth contours) were generated, which were related to the cell structure.

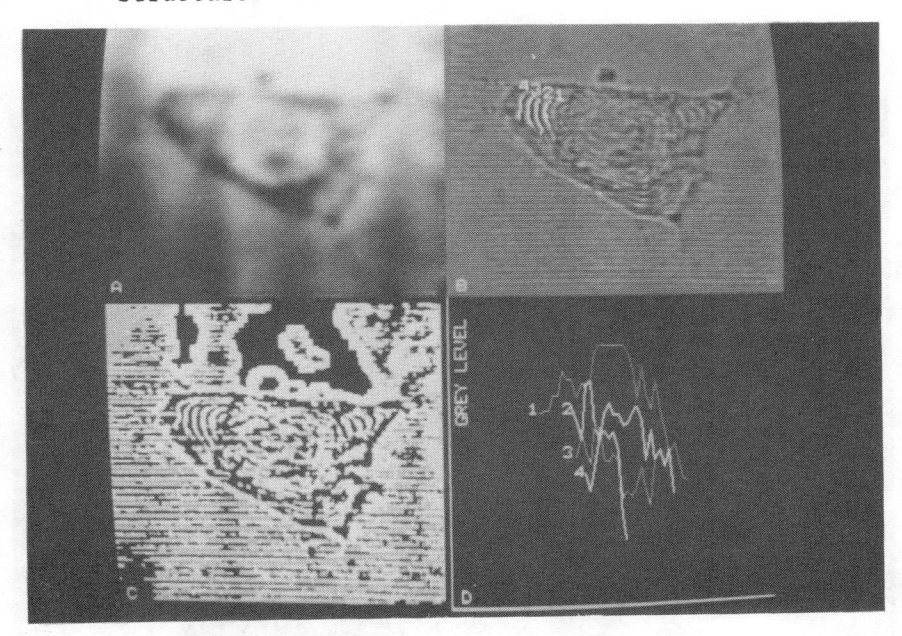

Figure 3.4. Collage of software elaborations of Figure 3.3. Panel A shows the result of the application of a low-pass filter to the original image. Panel B shows the image resulting after a point by point subtraction of image A from the original one: in this way, the "offset" values of the alternate rings, corresponding to the interference fringes were eliminated. Panel C shows a two-level segmentation of the image, useful for detecting the rings. The graphs of panel D represent the variations in the grey levels along "pieces" of four rings, indicated in panel B (and calculated on the original picture). Once all the artifacts have been eliminated from the images, the variations in the grey levels along the rings (which should be equi-depth contours) can be interpreted in terms of local elastic properties of the cell structures.

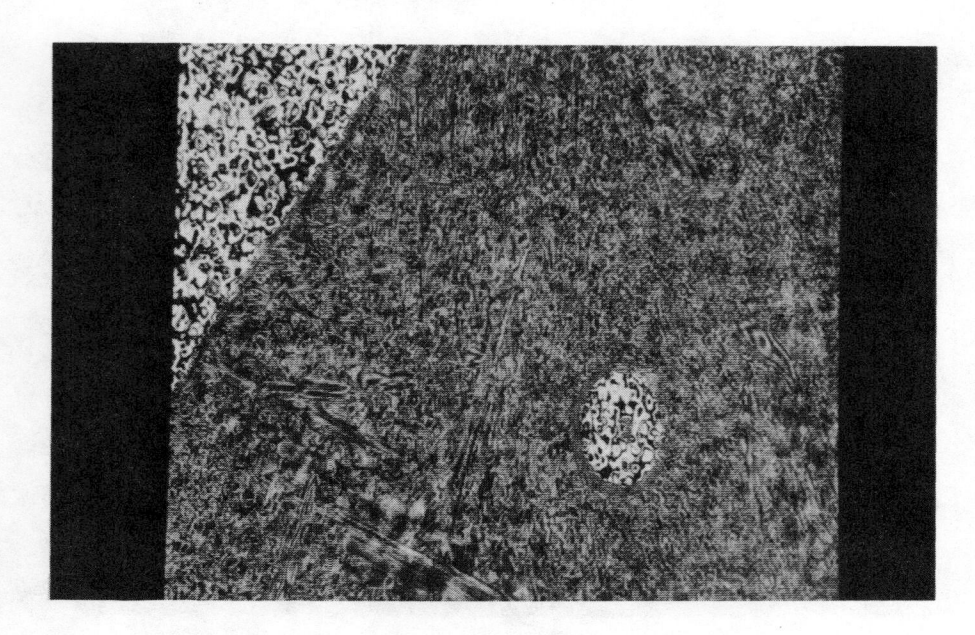

Figure 4.1. An image of the surface of a semiconductor device (thyristor): zones at different foci (characterized by different texture) are clearly visible.

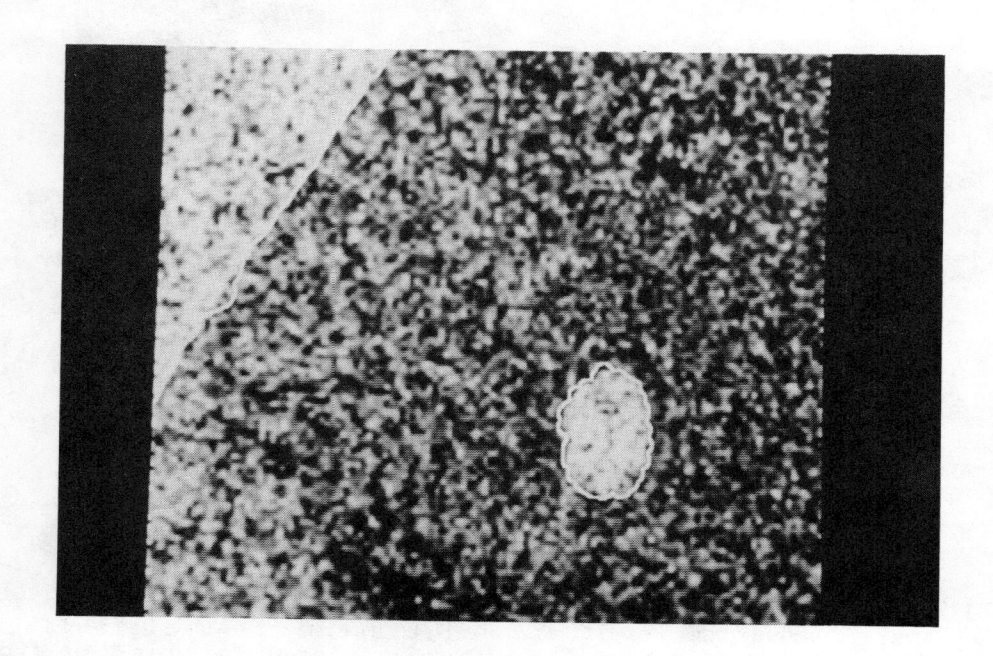

Figure 4.2. The zone at different foci are separated. The contours regions are superimposed on the texture (fractal) image.

Semiconductor Laser Digital Scanner

H.Sekii, A.Fujimoto, T.Takagi, K.Imanaka, and M.Shimura

Central R&D Laboratory, OMRON Tateisi Electronics Co.,
Shimokaiinji, Nagaokakyo, Kyoto, 617, Japan

ABSTRACT

A novel semiconductor laser digital scanner composed of a monolithic semiconductor laser array and a conventional plastic lens is proposed. A high speed scanning, a simultaneous light emission, and a wide angle scanning are achieved. The semiconductor laser array consists of monolithically integrated ten index-guided 780nm emitters spaced by 0.3mm. The lens is an aspheric plastic lens which is widely used in the compact disc system with 4.5mm focal length and 0.45 N.A.. The maximum scanning angle is about 43 degrees and the beam diameter (FWHM) on a screen varies from 0.10mm to 0.53mm. Computer simulation using the ray tracing method is performed. We have found that the calculated results on the spheric lens case fit well with experimental ones using the aspheric lens.

1. INTRODUCTION

Recently, a laser beam scanner has been widely applied to various optical systems such as a laser beam printer and an optical bar code reader. There are three types of scanning methods; a mirror reflection type, a hologram type, and a surface acoustic wave(SAW) type. A typical application of the mirror reflection type scanner is a laser beam printer. A high speed scanning can be performed with a many-sided mirror(polygon mirror) rotating by electrical motor. However, the polygon mirror is so big that the scanner is difficult to be small. The scanning speed of the beam is limited by the mirror rotating speed and the number of mirrors. Therefore, the scanner becomes complex to obtain the high scanning speed. The hologram type scanner is used in a conventional optical bar code reader. In this type, the laser beam is diffracted by hologram rotated by electrical motor.[1] The scanning speed of the beam is limited by the hologram rotating speed. In addition, it is difficult to obtain the large diffraction angle and the high scanning speed. The SAW deflection type scanner is based on the diffraction of the light beam by SAW. This type scanner can be operated statically without mechanical noises and vibrations. The scanning speed is very high(a few nsec), however, the beam deflection angle is very small and the deflection efficiency is also very low.[2] In this paper, we describe the structure and the operating characteristics of a novel semiconductor laser digital scanner which can be operated statically with a extremely high speed and a high repeatability. The comparison with experimentally obtained deflected beam intensity profiles and calculated ones is also described.

2. FABRICATION OF THE SCANNER

Figure 1 shows a semiconductor laser digital scanner composed of a monolithic semiconductor laser array and an aspheric plastic lens. The semiconductor laser array consists of ten index-guided 780 nm emitters spaced by 0.3mm as shown in Fig.2. Each emitter is electrically isolated by etching process. Therefore, emitters in a semiconductor laser array can emit the light independently. The laser array is mounted with the junction-up configuration on a Cu heat sink by Au-Sn solder. Threshold current of each laser is around 60 mA. The lens is an aspheric plastic lens for the compact disc system with 4.5mm focal length, 3.2mm thickness, 1.49 refractive index and 0.45 N.A.. The semiconductor laser array is normal to the optical axis of the lens, and the distance between the lens and the semiconductor laser array is slightly larger than the focal length so as to obtain an image formation length of 105 mm from the lens. The light emitting from the laser diode is refracted at two surfaces of the lens. The light going out of the lens reaches a screen. The observation of beam patterns and beam intensity profiles on the screen is performed by high precession TV camera C-1000 (HAMAMATSU PHOTONICS Co.). The beam intensity profiles are measured in the tangential image surface plane. Y-axis shown in Fig.1 is coincident with the direction of the laser diode arrangement. The base point of Y-axis is the intersection of the optical axis of the lens and the semiconductor laser array. It corresponds to the middle point of the 5th and the 6th laser diode.

Merits of the present semiconductor laser digital scanner are as follows:
(1) It does not need mechanical actions. Therefore, it has no mechanical noises and vibrations. It can operate statically with the extremely high speed and repeatability.
(2) It has many light sources which can be operated independently.
(3) The beam position on the screen depends on the distance between the laser diodes and the lens.

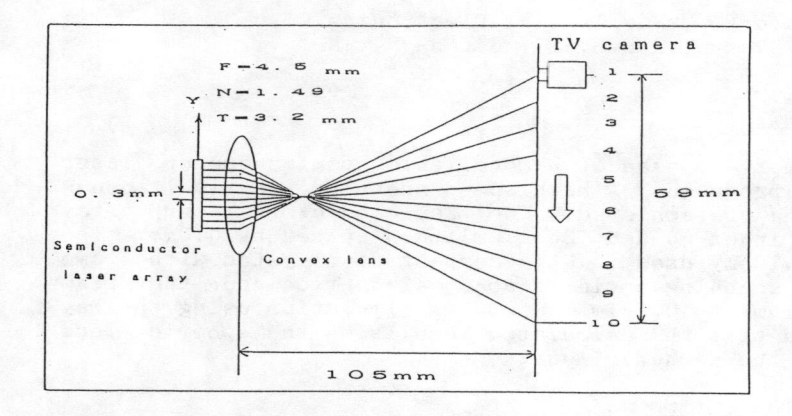

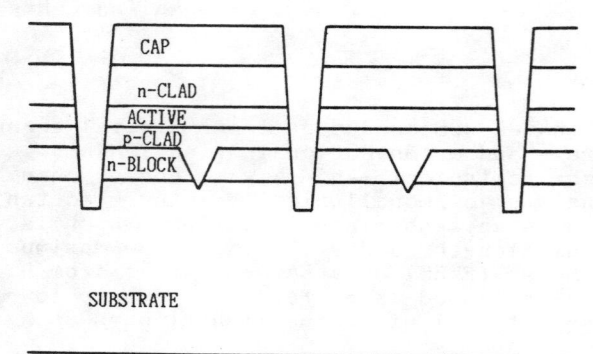

Figure 1. Construction of semiconductor laser digital scanner.

Figure 2. Cross-section of a monolithic semiconductor laser array.

3. EXPERIMENTAL RESULTS

Figure 3 shows the relation between the positions of laser diodes and the scanning angles of the beams. The maximum deflection angle is about 43 degrees as shown in Fig.3. The average deflection rate is about 15 degrees/mm. These values depend on the distance between the laser diode and the lens. Some nonlinearlity is observed in Fig.3. It could be caused by the aberration of the aspheric plastic lens. Figure 4 shows the relation between the positions of laser diodes and the beam patterns on the screen. The beam patterns have nearly a circular shape, and the beam size changes a little when the position Y is less than 0.75 mm (corresponding to the deflection angle of 13 degrees). While, they are quite different from a circular shape and the beam size becomes large for the Y beyond 1.05 mm (deflection angle of 17 degrees). We suppose that these results are originated from the off-axis aberration of the lens and so on.

Figure 3. Experimental laser beam deflection angles as a function of a laser diode position.

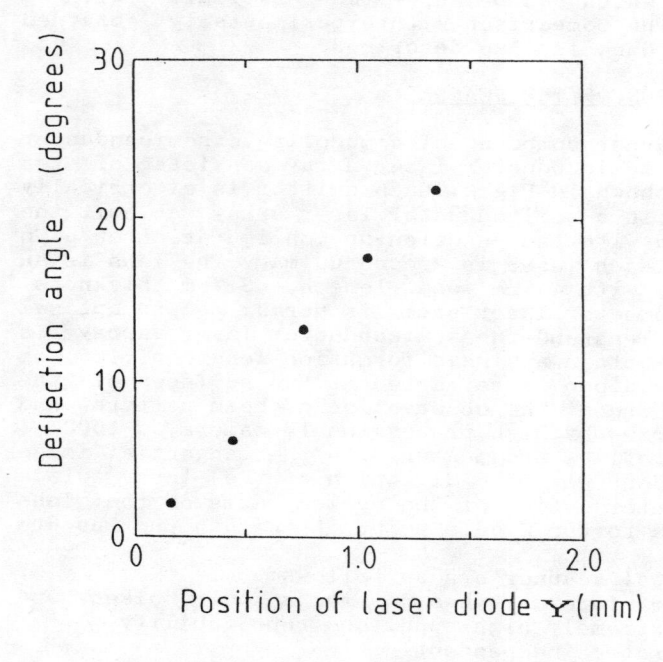

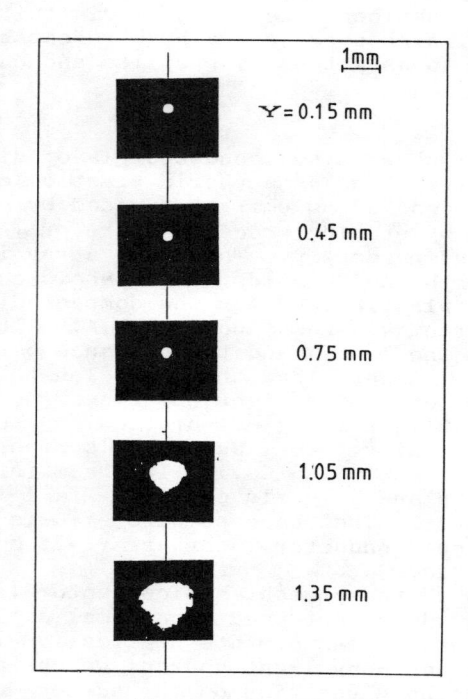

Figure 4. Deflected laser beam patterns with a different laser diode position Y

4. CALCULATION METHOD

We calculate the beam intensity profiles on the screen using the ray tracing method. In this calculation, we have made following assumptions:
(1) The lens has 4.5mm focal length, 3.2 mm thickness, 1.49 refractive index and 0.45 N.A.. they are the same values as those of an aspheric plastic lens.
(2) The effective aperture of the lens is 3 mm.
(3) The diffraction and the interference of the light can be neglected.
(4) The curvature of an aspheric plastic lens is unknown, therefore, a convex shape is assumed.
(5) The distance between the lens and the screen is 105 mm which coincides with the experimental condition.
(6) The size of light sources can be neglected and the intensity profile of a laser beam has a Gaussian distribution.
(7) Only rays in meridional plane contribute to the light intensity.

According to the assumption (1), undefined values of the lens are the radius of curvature. We define the radius of curvature on the light source side as R1 and on the screen side as R2. Under the condition that the focal length is constant, R1 and R2 must satisfy the relation expressed by

$$R2 = \frac{N \cdot (N-1) \cdot R1 - (N-1)^2 \cdot T}{N \cdot R1/F - N \cdot (N-1)} \quad (1)$$

where N, T, and F represent the refractive index, the thickness, and the focal length of the lens, respectively. The object principal plane of the lens depends on R1 and the image principal plane of the lens depends on R2. Consequently, we adjust the distance between the lens and the semiconductor laser array under the assumption (5) whenever the value of R1 changes. Parameter values adopted in this calculation are as follows:

Table 1. Parameter values

R1	R2
2.23	103.87
2.55	9.57
2.88	5.97

The lens form resembles to a plano-convex lens when R1 becomes smaller. In computer simulation using the ray tracing method, the screen and the lens are divided by an interval of 0.01 mm along the direction of Y-axis, respectively. To obtain the beam position on the screen, we calculate the angle between the incident vector of the light and the normal vector at the lens surface, and the refraction angle by Snell's law. The calculation is performed for all of divided points of the lens. To get the outline of a beam profile, calculation results are converted according to the value of the beam position on the screen such as a histogram distribution. Finally, they are weighted with the Gaussian distribution of the light source.

5. CALCULATION RESULTS

Figure 5 shows experimental results of the beam intensity profiles and calculated results when R1 is 2.23 mm. The calculated results agree well with experrimentally obtained ones. The calculated results are already weighted with the Gaussian distribution of the light source. Figure 6 shows experimental results of a beam size (FWHM) and calculated results when R1 is 2.23, 2.55 or 2.88 mm. Experimental results are smaller than calculated ones for the position Y less than 0.75 mm. In addition, experimentally obtained change in the beam size is larger than that of calculated results. These results imply that the aberration of the aspheric plastic lens is smaller than that of the convex lens. We suppose that off-axis aberration of the aspheric plastic lens is especially small. Figures 5 and 6 show that the calculated results using the spheric lens fit well with experimental results using the asperic lens when R1 is 2.23 mm. Figure 7 shows the peak intensity of the beam profiles shown in Fig.4 as a function of laser diode positions. We have found that the peak intensity is independent of R1 for the position Y beyond 1.05 mm. However, the peak intensity strongly depends on R1 for the position Y less than 0.45 mm. In the case of an usual lens, it is designed to satisfy that the beam intensity becomes a maximum value at Y=0. We suppose that small changes of the beam size and the beam intensity profile are better for a beam scanner. Figures 6 and 7 show that the beam size change becomes larger when the change in the beam intensity profile is large. In the case of a spheric lens, it is suitable for the beam scanner when R1 is small and the lens form resembles to a plano-convex lens. Figure 8 shows the calculated results of spherical aberration. In this calculation, we assume that the rays parallel to the optical axis of the lens are incident to the lens. We calculate the ray's intersection with the optical axis of the lens. The lateral axis in Fig.8 represents the distance of the axial focal point from the axial intercept of a peripheral ray. The spherical aberration becomes larger when the lens form resembles to a plano-convex lens. This result means that the lens suitable for the beam scanner is inferior to converge rays parallel to the optical axis on a focal point. In addition, it is coincident with the result in Fig.7.

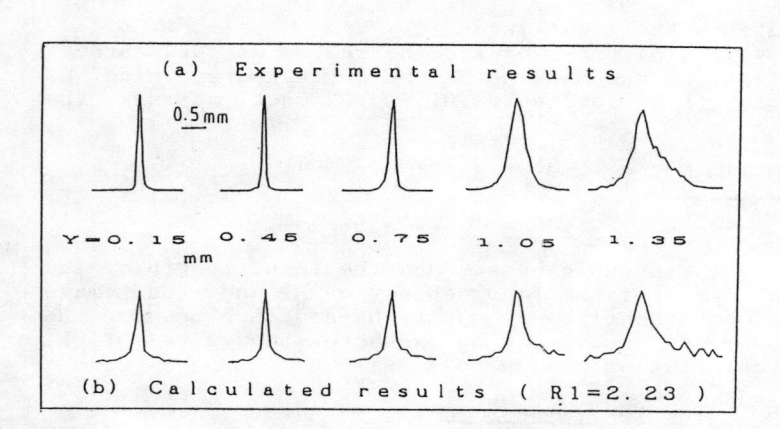

Figure 5. Deflected beam intensity profiles as a function of laser diode positions:
(a) experimental results
(b) calculated results

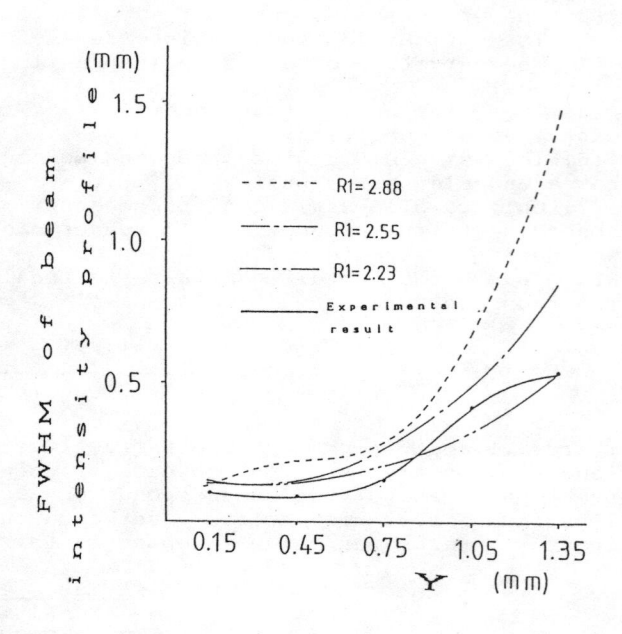

Figure 6. Calculated FWHM of the deflected beam intensity profiles as a function of laser diode positions

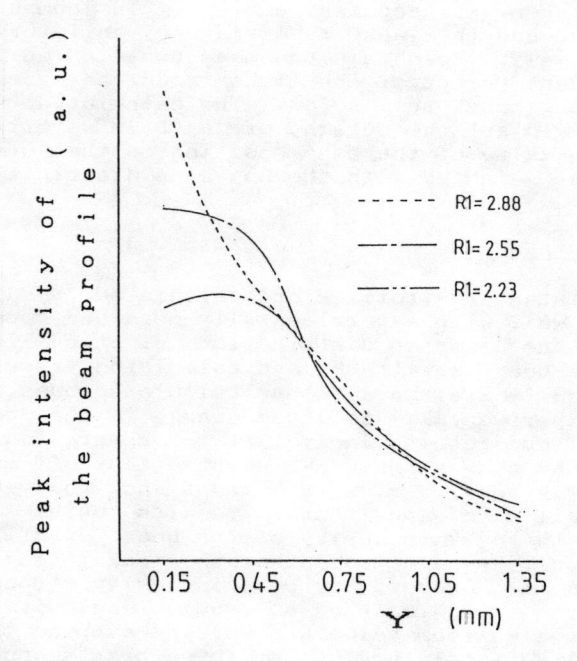

Figure 7. Peak intensity of the beam profiles as a function of laser diode positions.

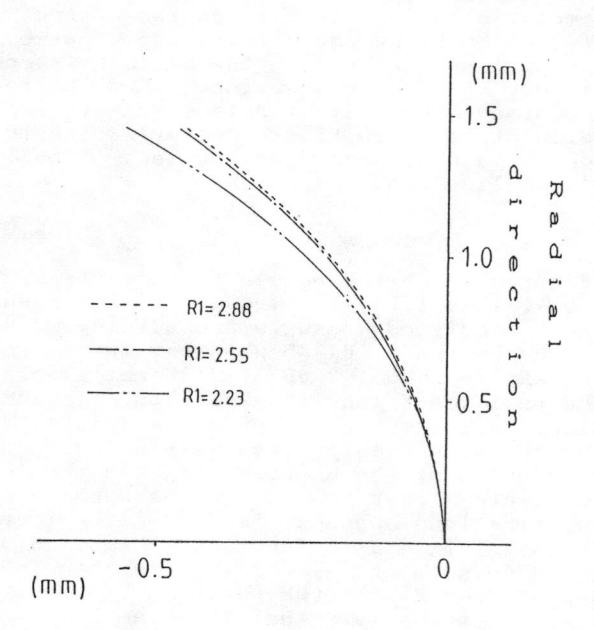

Figure 8. Spherical aberration. R1 represents the radius of curvature on the light source side.

6. CONCLUSIONS

A novel semiconductor laser digital scanner composed of a monolithic semiconductor laser arrry and an aspheric plastic lens is proposed. It can be operated without any mechanical actions. The scanning speed is expected to be near the semiconductor laser switching time. When the scanner consists of 0.3 mm period 10 elements laser array and 4.5 mm focal length aspherical surface convex lens, the maximum deflection angle is about 43 degrees and the average deflection rate is about 15 degrees/mm. In addition, the beam patterns have nearly a circular shape and the beam size change is a little for the laser position less than 0.75 mm, however, they are quite different for the laser position beyond 1.05 mm. The beam size at 1.35 mm laser position is about 5 times larger than that at 0 mm. Computer simulation using the ray tracing method is performed. The calculated results using the spheric lens fit well with experimental results using the aspheric lens which has 2.23 mm curvature radius on the light source side. The changes in the beam size and the beam intensity become smaller when the lens form resembles to a plano-convex lens. These results indicate that the present beam scanner is suitable for use as various optical systems, for example, a profile sensor, a laser beam printer, an optical bar code reader, and so on.

7. REFERENCES

1. S.K.Case and V.Gerbig, "Laser beam scanners constructed from volume holograms," Opt.Eng. 19 (5), 711-715 (1980).
2. T.Yamashita, M.Matano, N.Inoue, and M.Katoh, "Efficient Light Deflection of a Narrow Guided Beam in LiNbO Using Two SAW Pulses," Jpn. J. Appl. Phys. 24 (1), 108-109 (1985).

Measurements of optical waveguides by a near-field scanning technique

Jerzy Helsztyński, Tadeusz W. Kozek

Warsaw University of Technology, Institute of Electronics Fundamentals
Nowowiejska 15/19, 00-665 Warsaw, Poland

ABSTRACT

The transmission near-field scanning technique can be used for determining the refractive index profile, the maximum core-cladding refractive index difference and/or numerical aperture, the geometrical characteristics and the mode-field dimensions in single-mode waveguides.

In the experimental apparatus presented in the paper, the near-field pattern under test is magnified onto the faceplate of a vidicon camera, the video signal is sent to a computer-controlled video digitizer with single frame buffer memory organized as 512x512x8 bits. The incoming signal is digitized in real time to 8 bit resolution. Algorithms developed to optimize the scanning routine and to derive waveguide parameters from the measured data are described. As measured signals are usually noisy, techniques which reduce the noice level are also presented. Futhermore, techniques applied to compensate the vidicon sensitivity nonuniformity and nonlinearity are reported. Some examples of the results of measurements of the refractive index profiles and geometrical parameters like: core and cladding diameters, core and cladding non-circularities and core/cladding concentricity errors are presented.

1. INTRODUCTION

Among different measurement methods of optical waveguides, transmission near-field scanning technique plays a significant role. This relatively simple method can be used for determining the refractive index profile, the maximum core-cladding refractive index difference and/or numerical aperture, the geometrical characteristics and the mode-field dimensions in single mode waveguides.

CCITT issued recommendations concerning characterization and measurement techniques of circular optical waveguides i.e. optical fibres used in telecommunication networks (which can be adapted for other applications), namely: "Recommendation G.651 - Characteristics of 50/125 μm multimode graded index optical fibre cables" (1984) and "Recommendation G.652 - Characteristics of single-mode optical fibre cable" (1984). CCITT has established for each measurement parameter the Reference Test Method which is related to the definition of the parameter, and the Alternative Test Method - more simple in technical realization but not directly related to the definition. This paper deals with the measurement of optical fibres which comply with CCITT recommendations. The generalization of measurement techniques from conventional telecommunication fibres to other circular or non-circular fibres (e.g. polarization - maintaining fibres or three-dimensional waveguides) is possible.

The transmission characteristics of optical fibres depend on dielectric materials from which they are fabricated, in particular it is the shape of the refractive index profile that is responsible for bandwidth response of optical fibres. E.g. an optimum index profile of the multimode fibre is the near parabolic profile. Such a profile provides a minimum modal time dispersion. In single-mode fibre the material dispersion is compensated by the waveguide dispersion which may be described by the index profile. By proper profiling chromatic dispersion controlled fibres can be fabricated, namely: dispersion optimized at 1.3 μm, dispersion shifted fibres (1.5 μm) and dispersion flattened fibres.

Geometrical characteristics are also very significant. Connection-related losses caused by intrinsic properties of the fibres are due to: different cladding diameters, different core or mode-profile (in single-mode fibres) diameters, core concentricity errors, cladding non-circularities and core non-circularities.

2. THE NEAR-FIELD SCANNING TECHNIQUE

The measurement is based on the scanning of the magnified image of the end-face of the fibre under test. A microscope objective magnifies the output near-field and focuses it onto the plane of the scanning photodetector. One of the following techniques can be used:
a) mechanical scanning: scanning photodetector controlled by stepping motors,
b) electronic scanning: scanning vidicon or CCD device.

In the case of mechanical scanning system, modulated light source and lock-in amplifier for photocurrent signal processing are used. The single pixel analysis time should provide desirable signal to noise ratio. The light source has to be very stable in its position, intensity and wavelength over a time-period sufficiently long to complete the measurement procedure of the whole image.

In the electronic scanning system a vidicon or a CCD device can be used. Vidicons have poor linearity (i.e. output signal current versus irradiation). Visible radiation vidicon has a power law response ($\gamma = 0.5 \div 0.6$), and infrared vidicon-rather a power series response. Moreover, in a vidicon camera it is very difficult to attain a good scanning linearity and the stability of image dimensions. One more problem characteristic of both types of devices is nonuniformity of the sensitivity over the photoconductive area. These imperfections can be corrected by the computer. In TV standard each complete frame is scanned in 1/25 sec, so there is no problem of the light source stability, but signal to noise ratio is rather poor. Averaging by means of the summing the video signals of many frames improves signal to noise ratio.

2.1. Multimode fibre refractive index profiling

In multimode fibres, geometrical optics interpretation to the light acceptance which follows from the behaviour of the rays, can be implemented. According to such an approach Gloge and Marcatili[1] showed, that if the light source uniform over full front-face of the fibre core emits light according to the Lambertian law, then irradiance accepted by the core (taking into account meridional rays) as a function of the radius r is given by:

$$\frac{H(r)}{H(o)} = \frac{NA^2(r)}{NA^2(o)} = \frac{n^2(r) - n_2^2}{n^2(o) - n_2^2} \approx \frac{n(r) - n_2}{n(o) - n_2} \tag{1}$$

where: $NA(r)$ is a local numerical aperture, $n(r)$ - index profile to be determined and n_2 - cladding index. In a short length of the fibre (~1m) attenuation and mode-coupling can be ignored, and then emitance $W_{NF}(r)$ distribution over the end-face of the fibre follows irradiance distribution $H(r)$. The principle of the measurement method is shown in Fig.1. $W_{NF}(r)$ is measured across the diameter of the fibre. It follows from (1), that:

$$W_{NF}(0) = H(0) \doteq NA^2(0) \approx 2n(n(o) - n_2) \tag{2}$$

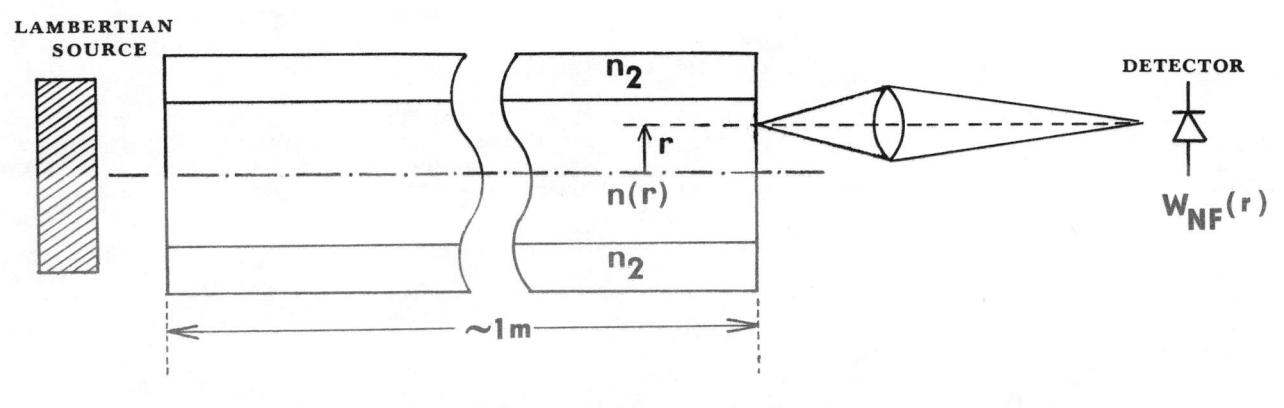

Figure 1. The near-field method (idea).

We see, that near-field emitance in the vicinity of the core centre is proportional to the maximum refractive index difference of the core and of the cladding or to the squared numerical aperture, so we can estimate the values of these parameters by the calibration of the known fibre.

Some doubts arise in connection with the problem of the possible propagation of the leaky rays (those rays that are skew to the circular fibre axis), not taken into consideration by Gloge and Marcatili. However, according to experimental data of Petermann[2] and Costa with Sordo[3], in the case of parabolic index profile fibres, skew rays disappear after one metre of length of such fibres. Petermann suggests that this effect is caused by non--circularity of the core.

According to G.651 document, the near-field technique of multimode fibre index profiling is the Alternative Test Method. The light should be incoherent, the launch optics should be arranged to overfill the fibre (NA greater than 0.3, light focused on the front-face of the fibre to the spot greater that 70 μm). In view of our experiments a diffuzer before the fibre must be used.

2.2. Single-mode fibres refractive index profiling

In G.652 document transmission parameters of single-mode fibres have mainly been normalized disregarding the index profile shape and its measurement techniques. Nevertheless the importance of proper profiling of single-mode fibres, during the stages of designing and fabrication, is obvious. Geometrical approach to the fibres propagating one mode or even few modes is not valid. Coppa et al.[4] proposed simple method of determining the index profile $n(r)$ from the near-field amplitude $E_{NF}(r)$ distribution of the fundamental propagation mode, with the help of the scalar wave equation:

$$E''_{NF}(r) + E'_{NF}(r)/r + [k^2 n^2(r) - \beta^2] E_{NF}(r) = 0 \tag{3}$$

where: $k = 2\pi/\lambda$, β - mode propagation constant and $E_{NF}(r) \doteq \sqrt{W_{NF}(r)}$ where $W_{NF}(r)$ is the measured emitance distribution across diameter of the end-face of the fibre. The measurement should be done at the wavelength longer than a cut-off wavelength of the second mode of the fibre, i.e. in the case of conventional telecommunication fibre at 1.3 μm or 1.5 μm. The evaluation of the refractive index profile is independent from launching conditions.

2.3. Fundamental mode-field diameter measurement

Presently, there is no agreement on the choice of the fundamental mode-field diameter definition. At first this diameter has been determined as equal to the width at the 1/e points of the optical near-field amplitude $E_{NF}(r)$ distribution. Now the view is held that this definition is adequate only for fibres with near-gaussian mode-field distribution. Different definitions have been proposed, among them e.g. Petermann's mathematical definitions [5,6]:

$$d_m = 2 \left(2 \int_0^\infty E_{NF}^2(r) \, r^3 \, dr \bigg/ \int_0^\infty E_{NF}^2(r) \, r \, dr \right)^{1/2} \tag{4}$$

$$d_j = 2 \left(2 \int_0^\infty E_{NF}^2(r) \, dr \bigg/ \int_0^\infty (dE_{NF}/dr)^2 \, r \, dr \right)^{1/2} \tag{5}$$

2.4. Geometrical characteristics measurements

According to CCITT Recommendations G.651 and G.652, near-field measurement technique of geometrical characteristics is an Alternative Test Method - for multimode fibres, and a Reference Test Method - for single-mode fibres. In the following table geometrical parameters of multimode and single-mode fibres with their nominal values and tolerances are given.

Table 1.

Fiber Parameter	multimode acc.to G.651	single-mode acc.to G.652
core diameter	50μm ± 6%	-
mode-field diameter	-	9-10μm ±10%*
cladding diameter	125μm ± 2%	125μm ± 2,4%
core non-circularity	6%	-
mode-field non-circularity	-	6%
cladding non-circularity	2%	2%
core concentricity error	6%	-
mode-field concentricity error	-	0.5-3μm**

* in fibers optimized for 1.3μm

** depending on the connection technique and required connection loss

It has to be pointed out that in single-mode fibres fundamental mode-field diameter (which is not a strict geometrical parameter) is normalized instead of the core diameter.

The measurement of geometrical parameters imposes the necessity of scanning the whole end-face of the fibre, i.e. in the area of core and cladding, so the cladding modes have to be also propagated. Modern fibres characterize themselves by very high attenuation of cladding modes, so the specimen shall be a short length (few cm) of the optical fibre.

3. TEST APPARATUS FOR FIBRES PARAMETERS MEASUREMENT

Two different computer-controlled measurement systems for fibers parameters evaluation by near-field technique have been realized: one based on mechanical scanning, the other on electronic scanning principle. Only the last one will be described here.

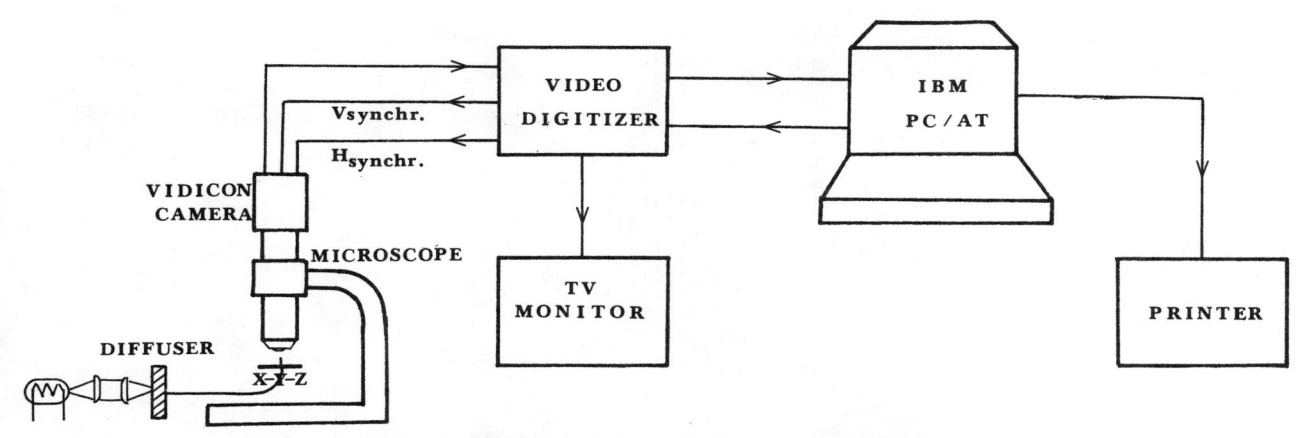

Figure 2. Test apparatus for fibres parameters measurement.

The basic components of the measurement system are shown in Fig.2. The front-face of the short length of the fibre under test is illuminated from tungsten light source or from pigtailed 1.3 μm LED (or 1.3 μm laser diode). The near-field pattern magnified by a microscope objective (x40,0.65 NA or x100,0.87 NA) localized in Biolar microscope (PZO, Poland), is focused onto the photoconductive layer of one of the following vidicons: visible radiation vidicon P831, (EEV Co.Ltd., England) or infrared vidicon TV2201 (Teltron, USA). The camera TP-K16 (WZT, Poland) contains a FET low noise preamplifier. The video signal is sent to a video digitizer VFG-512 Frame Grabber (Visionetics Inc., Taiwan), which has a form of a plug-in card for the IBM PC/AT. It has 256 K bytes of frame memory organized as 512x512x8 bits. The VFG Frame Grabber digitizes in real time the incoming video signal to 8 bit resolution (i.e. video signal is quantized to 256 possible gray levels) and stores into the frame memory. A 10 MHz pixel clock is generated for digitizing and displaying image. The VFG Frame Grabber is used for the synchronization of an external camera. Video output permits the observation of the image being encoded and processed on a TV-monitor.

4. PROCESSING OF MEASUREMENT DATA

Computer-controlled video digitizer offers a means for fast evaluation of imaging device characteristics. These results can be used to improve a metrological performance of the measurement system by appropriate processing of measurement data.

In general, the output signal of the vidicon imaging devices versus the input irradiation can be expressed as a power series response. For visible radiation vidicons, this relation is usually approximated by the power law function

$$S = K E^{\gamma} + S_0 \tag{6}$$

where: S is the output signal, E - the input irradiation, S_0 - the dark current, K - proportionality constant and γ - the measure of the linearity of the device. Both the dark current and vidicon sensitivity vary on the photosensitive area of the device. The vidicon nonlinearity, the nonuniformity of the dark current and of the sensitivity can substantially influence measurements. The nonuniformity of sensitivity of the visible radiation vidicon was determined by measurements performed for uniform irradiation of the photosensitive area. To eliminate the accompanying noise the average of 16 frames was taken. The standard deviation of pixels values calculated for the averaged calibration image was about 15% of the mean value. The histogram of the calibration image is shown in Fig.3.

Measurements of the dark current have indicated that its influence can be neglected in our measurements as its value was less than resolution limit of the video digitizer (disregarding a few defects in the corners of the photosensitive area that cause "dark current spikes"). We have not measured the nonlinearity coefficient γ of the vidicon. The results of the measurements are corrected by using gamma value given by the manufacturer.

The correction of the nonlinearity is based on the algorithm:

$$S_{dc} = A\, S_d^{1/\gamma} \tag{7}$$

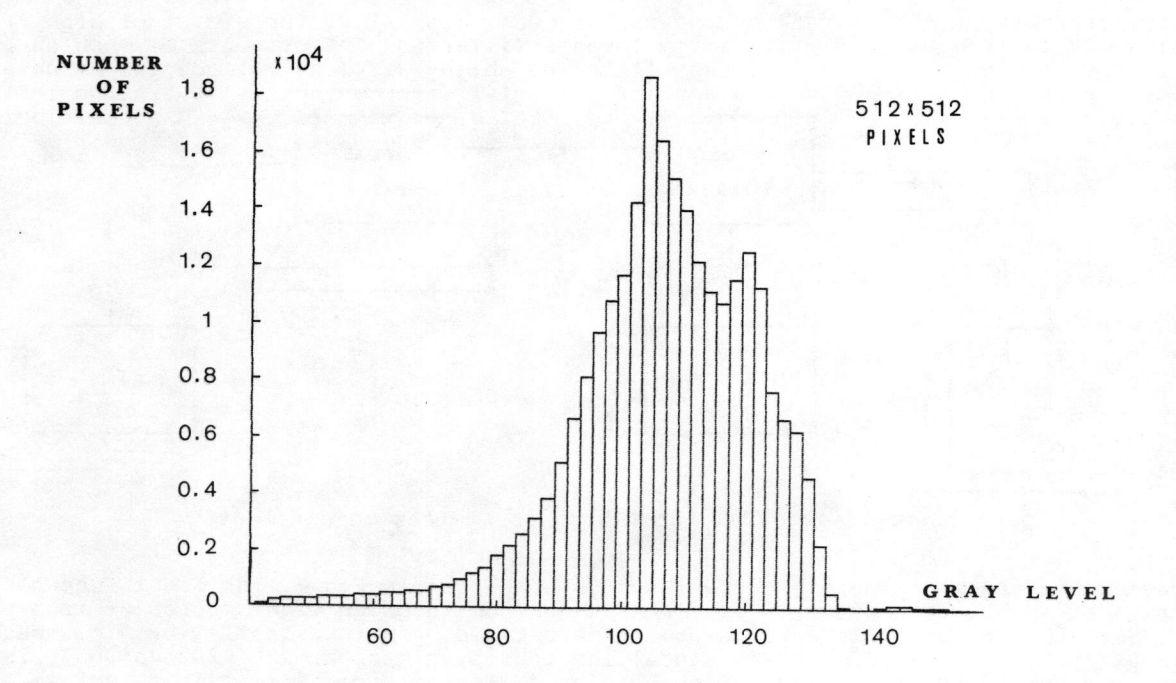

Figure 3. The histogram of the averaged calibration image for the visible radiation vidicon

where: S_d is the digital output value from the video digitizer, S_{dc} - the corrected value and A is a normalization constant.

Pixel-to-pixel sensitivity nonuniformities are corrected by using the calibration image for the uniform irradiation. The corrected value of a given pixel is evaluated from the expression:

$$S_{dc} = \bar{S}_{du} \frac{S_d}{S_{du}} \tag{8}$$

where: S_d is the uncorrected value, S_{du} - the value of a given pixel of the calibration image and $\bar{S}_{du}$ - the mean value of the calibration image. Fig.4 shows the effect of the nonlinearity and sensitivity nonuniformity correction on a diameter cross-section of emitance distribution for the multimode fibre.

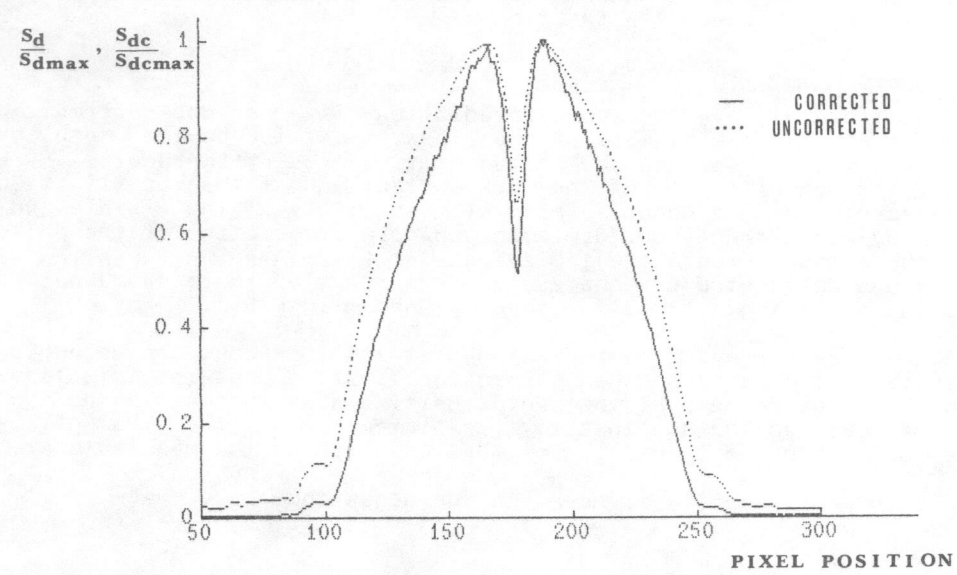

Figure 4. Corrected diameter cross-section of the emitance distribution for the multimode fibre i.e. index profile.

The noise accompanying the signal is another problem. Basically, the effect of high frequency noise can be reduced by digital image lowpass filtering.[7] The drawback of such an approach is a smoothing property of lowpass filtering which may cause elimination of details of an image. Instead of applying a more sophisticated filtering we applied averaging in the temporal domain as the alternative technique. Fig.5 shows the measured emitance distribution for a multimode fibre obtained from a single frame and as the average of 16 frames. The effect of noise reduction is distinctly visible.

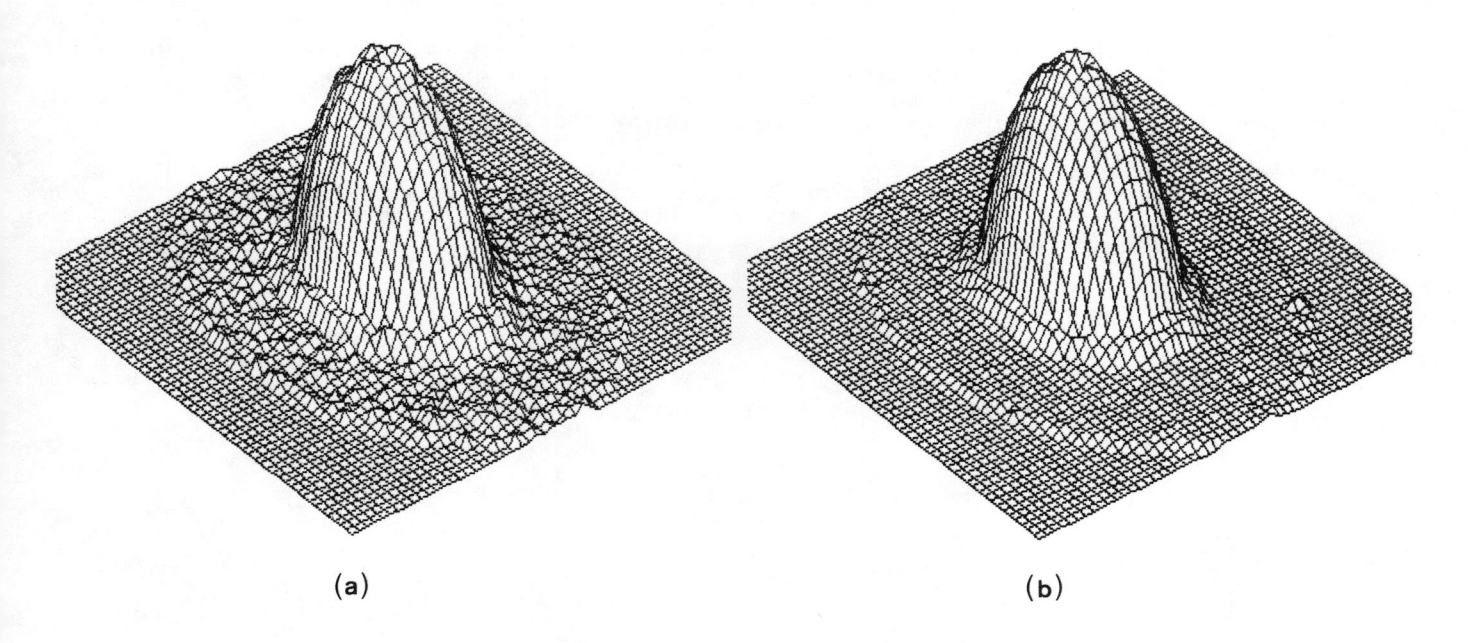

Figure 5. The near-field patterns of a multimode fibre obtained from: (a) a single frame (b) the average of 16 frames.

Fig.6 shows cross-sections of the emitance distribution for a multimode fibre. The effect of the lowpass filtering is compared with the averaging in the temporal domain. The lowpass filtering caused considerable reduction of the depth of the fibre's central dip.

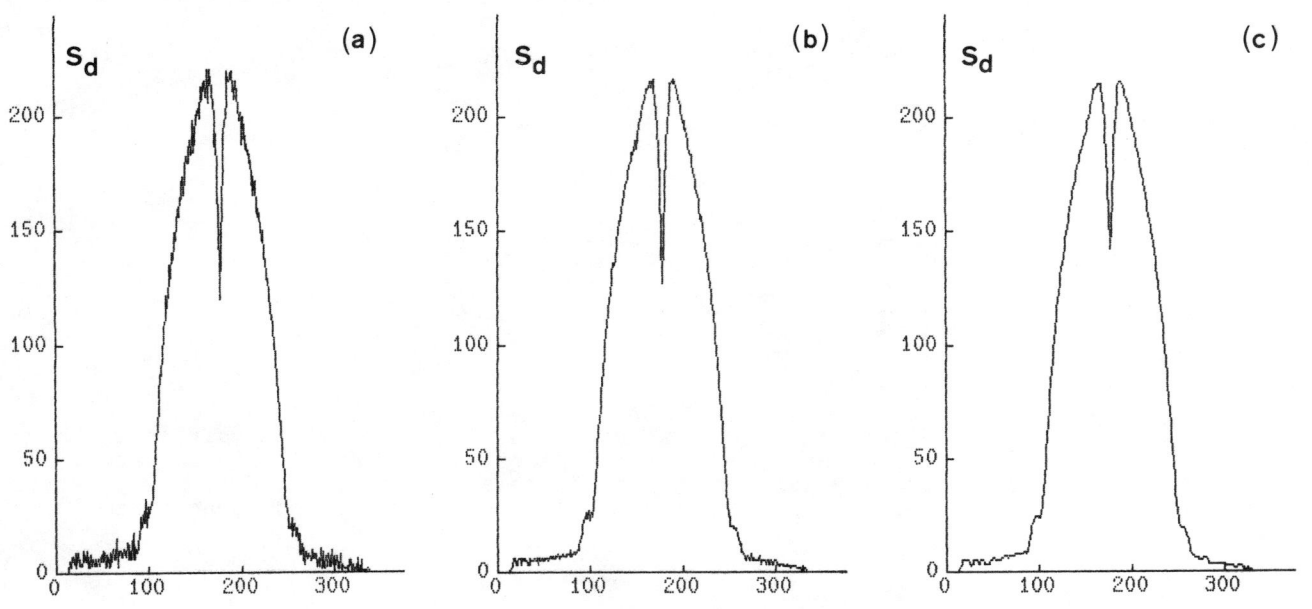

Figure 6. Diameter cross-sections of the emitance distribution obtained: (a) from a single frame (b) from the average of 16 frames (c) 2x lowpass filtering.

4.1. Geometrical characteristics evaluation

The basic part of the software which was developed for optical fibres measurements constitutes the algorithm for evaluation of the geometrical characteristics. The algorithm starts with finding of the geometrical centres of the core and of the cladding. The routine begins with searching for points at a given gray level in a fixed direction (a row or a column). The first approximation of the geometrical centres position is the midpoint of the segment between the detected two points at the same gray level. Next step is to find the midpoint of the analogous segment on the bisector of the previous one. The above procedure is iterated until succesive passes produce no significant change of the centre co-ordinates. Points on the interfaces (core/cladding or outside cladding) are detected by analysis in radial directions, equally spaced around the circumference of a circle with the found centre. The basic test for the interfaces detection is based on the rate of change of intensity in a given direction. Two points on the interface are obtained at each radial scan.

The parameters characterizing geometry of the fibre are obtained by finding the elipses which best approximate, in the sense of the least squares, the experimental points determining each interface. The calculation algorithm consists of solving the overdetermined system of linear equations with respect to parameters a_{ij} (one of these parameters should be fixed:

$$a_{11}x_i^2 + 2a_{12}x_iy_i + a_{22}y_i^2 + 2a_{13}x_i + 2a_{23}y_i + a_{33} = 0 \tag{9}$$

where x_i, y_i are the Cartesian co-ordinates of the i-th point on the interface ($i=1,\ldots,N$; $N \geq 5$). The geometrical centre co-ordinates are given by

$$x_c = -\frac{1}{D}\begin{vmatrix} a_{13} & a_{12} \\ a_{23} & a_{22} \end{vmatrix} \qquad y_c = -\frac{1}{D}\begin{vmatrix} a_{11} & a_{13} \\ a_{13} & a_{23} \end{vmatrix} \tag{10}$$

where: $D = \begin{vmatrix} a_{11} & a_{12} \\ a_{12} & a_{22} \end{vmatrix}$

Then the axes of the ellipse are:

$$a^2 = -\frac{A}{\lambda_1 \lambda_2^2} \quad , \qquad b^2 = -\frac{A}{\lambda_1^2 \lambda_2} \tag{11}$$

where λ_1, λ_2 denote the roots of the characteristic equation:

$$\begin{vmatrix} a_{11}-\lambda & a_{12} \\ a_{12} & a_{22}-\lambda \end{vmatrix} = 0$$

and $A = \begin{vmatrix} a_{11} & a_{12} & a_{13} \\ a_{12} & a_{22} & a_{23} \\ a_{13} & a_{23} & a_{33} \end{vmatrix}$

Assuming: $a_{11}=a_{22}=1$ and $a_{12}=0$ and solving (9) we can find parameters of the circle which approximates the interface. In our software both elliptical and circular fits are used. All basic parameters characterizing the geometry of the fibre can be determined from the parameters of the ellipses which approximate the core/cladding and the cladding boundaries.

```
time:  15:08:03              date: 08-16-88      time:  14:53:39              date: 08-15-88
       MONOMODE FIBER GEOMETRY MEASUREMENTS             MULTIMODE FIBER GEOMETRY MEASUREMENTS

Experiment number: 12/08                         Experiment number: 11/08
Fibre name: sm10/08                              Fibre name: mm06/08
Comment: test - TK                               Comment: test - TK

Unit = micrometer                                Unit = micrometer

CORE                                             CORE
Circular fit:                                    Circular fit:
     diameter =  9.39                                 diameter = 58.63
     center X= 97.98 Y= 71.47                         center X= 103.68 Y= 64.47

Elliptic fit:                                    Elliptic fit:
     large axis = 9.83  small  axis = 9.59            large axis =58.86  small  axis =57.89
     center X= 98.00 Y= 71.58                         center X= 103.65 Y= 64.40
     non-circularity =  2.48 %                        non-circularity =  1.65 %

CLADDING                                         CLADDING
Circular fit:                                    Circular fit:
     diameter = 110.85                                diameter = 126.50
     center X=  98.21 Y=  71.48                       center X= 105.10 Y=  65.78

Elliptic fit:                                    Elliptic fit:
     large axis =111.20  small  axis =110.69          large axis =127.53  small  axis =124.90
     center X=  98.31 Y=  71.58                       center X= 105.16 Y=  65.81
     non-circularity =  .45 %                         non-circularity =  2.06 %

CORE/CLADDING concentricity error                CORE/CLADDING concentricity error
     = 3.14 %  - for elliptic fit                     = 3.50 %  - for elliptic fit
     = 2.44 %  - for circular fit                     = 3.30 %  - for circular fit
```

Figure 7. Examples of results of geometrical characteristics measurements of optical fibres.

5. CONCLUSIONS

The refractive index profile, the numerical aperture and/or maximum refractive index difference of core/cladding and the geometrical parameters of multimode and single-mode fibres as well as mode-field diameter of single-mode fibres are measured by means of a simple transmission near-field scanning technique. In computer-controlled measurement system electronic scanning by vidicon cameras is implemented. Nonlinearity of transfer characteristics and nonuniformity of sensitivies of vidicon imaging devices are corrected by the computer. The additional problem arizes when we apply this technique to power series response correction of the infrared vidicon, as we need to know the maximum irradiation value of the processed image. To reduce the effect of the noise, the near-field pattern is measured many times.

6. REFERENCES

1. D. Gloge and E.A.J. Marcatili, "Multimode theory of graded-core fibers", Bell Syst. Techn.J. 52,, 1563-1578 (1973).
2. K. Petermann, "Uncertainties of the leaky mode correction for near square law optical fibres", Electron.Lett. 13(17), 513-514 (1976).
3. B. Costa and B. Sordo, "Measurements of the refractive index profile in optical fibres: comparison between different techniques", Second Europ. Conf. on Optical Fibre Comm., Paris, 1976, 81-86.
4. G. Coppa et al., "Characterisation of single-mode fibres by near-field measurement", Electron.Lett. 19(8), 293-294 (1983).
5. K. Petermann, "Fundamental mode microbending loss in graded-index and W fibres", Opt.Quant.Electr. 9, 167-175 (1977).
6. K. Petermann, "Constraints for fundamental-mode spot size for broadband dispersion--compensated single-mode fibres", Electron.Lett. 19(18), 712-714 (1983).
7. T. Pavlidis, Algorithms for Graphics and Image Processing, pp. 56-69, Computer Science Press, Rockville (1982).

Reflectance and Optical Contrast of Old Manuscripts: Wavelength Dependence

Julián Bescós
Fundación Ramón Areces, Madrid

Francisco Jaque
Dept. de Fisica Aplicada, Universidad Autonoma de Madrid

Luis Montoto
IBM, Scientific Center
P. de la Castellana 4, 28046 Madrid, Spain.

ABSTRACT

The reflectance of Spanish manuscripts (paper and ink) dating from the XV to XIX centuries has been studied in the spectral range 350-1000 nm, specially when poor definition and/or manuscript degradation are present. The results have been applied to optimize the readability of the corresponding digital images of manuscript by choosing appropriately the spectral distribution of the source and the spectral sensitivity of the detector in a given digitizing device. For those manuscripts with severe degradation, the method gives enhanced original digital images that could be further processed resulting in optimal final images.

1. INTRODUCTION

The need of historical manuscript conservation and the availability of high density optical storage, makes feasible the storage of those old documents in digital form, and the creation of a pictorial data base for document retrieval in historical archives environments with a big number of manuscripts. In addition, the use of digital image techniques gives the most convenient method to compress, transmit and display those documents.

The readability of texts depends on physical parameters concerning the visual environment as well as on the psychophysical ones depending of the observer. Among the physical parameters such as definition, ilumination, contrast, etc., contrast is the most significant one. In this sense, the CIE Commission has recomended its use in visibility measurements to evaluate visual performance and the difficulty of working on a task [1]. In this work, the contrast of a writen text is defined as

$$C = \frac{R_p - R_i}{R_p + R_i} \quad (1)$$

where R_p and R_i are the reflectance of the paper and of the ink, respectively, and where the illumination component, supposed to be constant, has been droped.

Given the degradation suffered by manuscripts along its life, and in order to preserve, or even enhance, the readability of the corresponding digital images, the digitization of those documents should be done using the optimum parameters that later will affect the resulting digital images. So, it seems clear that knowledge of the optical characteristics of the paper as well as the ink are important in order to optimize their contrast, by an appropriate choice of lamps and detectors.

In this work, the physical parameters that will enhance the readability of Spanish documents from the XV to XIX centuries, while the process of digitization, have been studied. In those cases of severe document degradation, it is shown how this "analog" enhancement is complemented with digital processing, with the result of legible digital images.

2. MEASUREMENT OF MANUSCRIPT REFLECTANCE

The reflectance of manuscripts dating from the XVI to XIX centuries has been studied in the spectral range 350-1000 nm. Different reflectance measurements have been carried on samples of paper and ink using two distinct instruments. The first instrument, a Perkin Elmer spectrophotometer, takes diffuse reflectance measurements on samples of 5 mm2 minimum size, and therefore it requires to build mosaics for the ink reflectance measurements. The second instrument (available to us later) was a PR-710 Spot Spectrascan Radiometer System, that allows direct reflectance measurements of the manuscript strokes using an interposed telescope. The reflectance measurements done with both instruments are in good agreement each other and are summarized in the following.

Figure 1 shows the typical wavelength reflectance dependence of the paper, Rp, direct ink, Ri, and ink bleeding through the paper, Rib, of a document dated on 1800. It is observed that the paper has a reflectance of 70% in the range 500 - 1000 nm, decreasing below 60% at lower wavelengths. This last behaviour accounts for the yellowish colour in old manuscripts. At variance with the paper, the ink reflectance presents a monotonous descent below 1000 nm.

The presence of ink bleeding through the paper is a very usual degradation in old manuscripts. As it was expected, the reflectance given by the bleeding ink is much higher than that of the direct ink, as is shown by the intermediate curve in Fig. 1. The difference is greater in the red spectral region, which accounts for the usual brown colour in manuscripts with severe degradation produced by this bleeding.

Figure 2 shows the difference of reflectances for the two following cases: a) paper and direct ink, Rp-Ri, and b) bleeding ink and direct ink, Rib-Ri. The first difference Rp-Ri represents the main factor concerning contrast for normal manuscripts as stated by expression (1), being its maximum in the center of the visible spectrum (around 550 nm). The second difference Rib-Ri accounts for ink bleeding manuscripts, where, to assure readability, direct ink should be separated as much as possible from bleeding ink. From the figure, Rib-Ri has its maximum around 700 nm.

The computed contrast for normal manuscripts (no ink bleeding present) given by expression (1) is plotted in Figure 3 versus the wavelenght. The contrast reaches the maximum value 0.8 in the blue region.

Similar results have been obtained from measurements of other documents from the XV-XVIII centuries. This fact was expected as the same chemical components were used in the manufacturing of paper and ink along such time period.

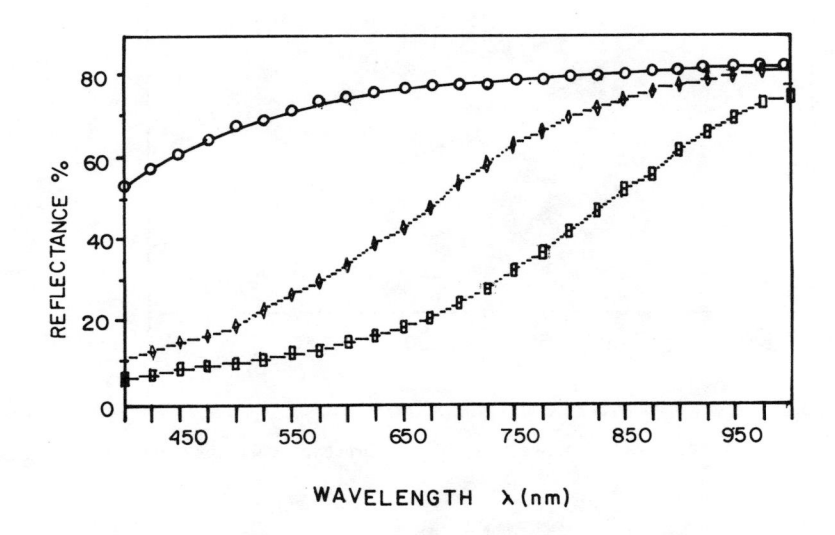

Figure 1. Reflectance measurements of paper (o), direct ink (□) and bleeding ink (◊) versus wavelenght.

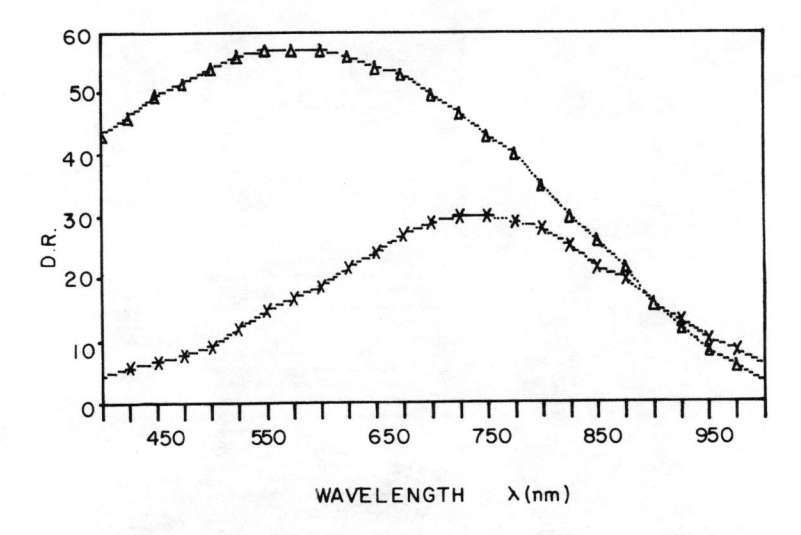

Figure 2. Difference of reflectances between paper and direct ink (△), and bleeding and direct ink (X) versus wavelenght.

3. A REAL CASE: APPLICATION TO MANUSCRIPT DIGITIZATION

In order to evaluate the improvement obtained in manuscript readability when the digitization is done in the spectral regions selected in the previous section, some experimentation has been performed with original manuscripts. Figure 4 shows a mosaic of four digital images obtained with a commercial scanner using their own illumination and detector and 8 bits per pixel resolution. The images recorded from a CRT display, gives illustrative examples of the usual degradations found in manuscripts.

The selection of the adecuate spectral regions has been carried out by introducing optical filters before the objetive lens of a commercially available CCD digitizing camera. From the previous reflectance results, the two following spectral regions have

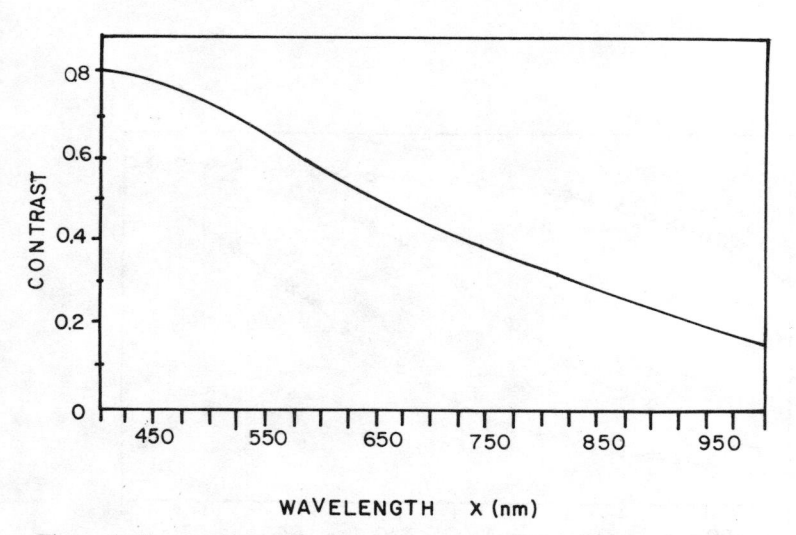

Figure 3. Computed contrast between paper and direct ink versus wavelenght.

Figure 4. Mosaic of four digital images showing typical degradations found in manuscripts. Top left: space variant low contrast and spots of varying size. Top right: low contrast and non-uniform background. Bottom: moderate (left) and severe (right) ink bleeding through the paper.

been considered: (a) visible spectrum for normal manuscripts, and (b) high band of the visible spectrum (red wavelenghts) for degraded manuscripts with ink bleeding. In case (a), the restriction to the visible region has been obtained by the introduction of an IR filter whose spectral trasmittance is represented in Figure 5. In case (b), a red Kodak Wratten 29 filter has been added to the IR filter for selection of the red band. In both cases it has been required to decrease the F number of the objetive lens to the minimum value, as well as to increase the illumination to reach appropiate image acquisition levels.

The restriction to the visible region (case (a) above) has been applied to the original manuscript shown in digital form in the top left of Figure 4. The result is shown in Figure 6 (where only a portion corresponding to the middle-upper-right part of the image is displayed). Figure 6 shows on the left the output of the digitizing camera without any filter, and on the right, the same image obtained using the IR mentioned filter. It can be observed that the constrast as well as the definition are significantly enhanced (right image) just by cutting-off the IR radiance seen by the CCD. In other words, typical CCD's with with high quantum efficiency in the IR region and used in commercially available scanners, should be used with caution in this type of applications.

For degraded manuscripts with ink bleeding (case (b) above), the result of selecting a red band to increase readability can be seen on Figure 7. On the top left of this figure it is shown a digitized image of a portion of this type of manuscripts, and using only the IR mentioned filter. In this image, most of the text corresponds to bleeding ink from the other side, and only a text line placed at the low part of the image corresponds to direct ink. On the top right of Figure 7, it is shown the same image digitized with both the IR and red filters. As expected from the reflectance curves in Fig. 2, it is observed how the selection of a red band produces a neat separation (in gray level sense) among direct and bleeding ink in the digital image, being

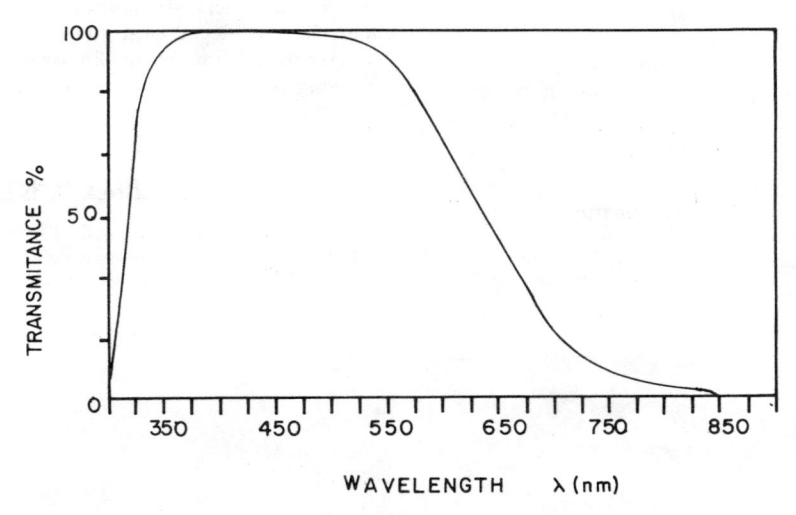

Figure 5. Spectral transmitance of the used IR filter.

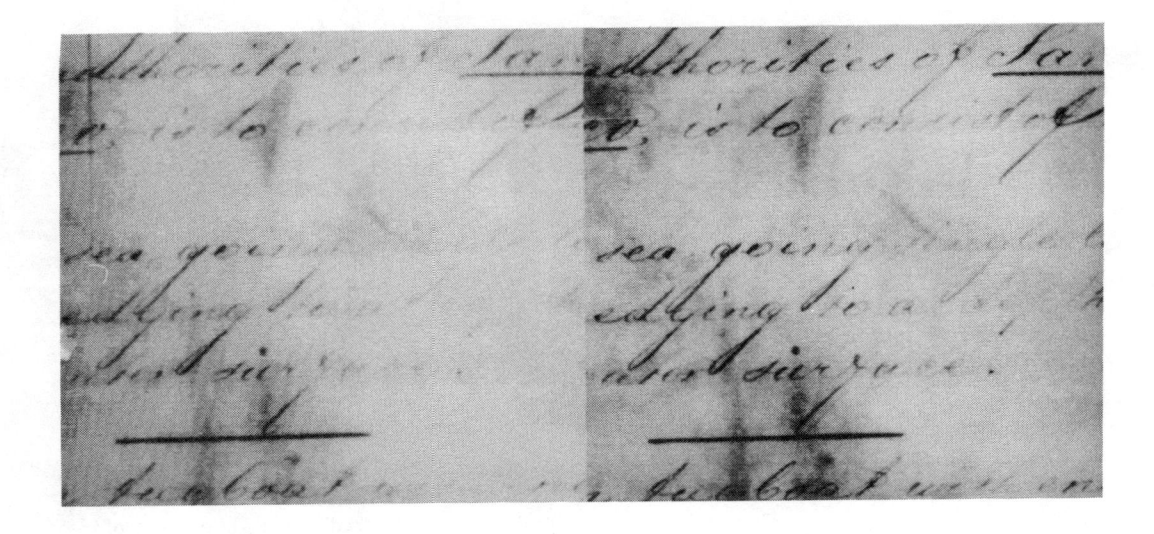

Figure 6. Digital images obtained without (left) and with (right) the IR filter of Figure 5.

the bleeding ink gray levels more similar to those of the background.

The use of the proposed red window in the digitization of manuscripts with ink bleeding problems, gives digital images with separate gray levels for direct and bleeding inks. This fact allows the effective application of standard digital processing techniques to suppress the ink bleeding contribution. In this sense, Figure 7 shows at its bottom the result of application of the same digital transform algorithm to the corresponding two digital images at the top of the Figure. The result of this futher processing applied to the image digitized with the red filter plus the IR filter (image on the top right) is definitive in terms of readability: only the direct ink is present on the final image (bottom right). It should be noted that this final image is not a binary one: the transform used is linear below a threshold, and above it assigns pixels values to the background level. This transform results in a gray level image, which gives the best quality in terms of readability.

4. CONCLUSIONS

Reflectivity measurements of paper, ink and bleeding ink parts of old manuscripts have been performed, in order to properly select the illumination sources and detectors to optimize the readability of the corresponding digital images.

It can be concluded from the reflectance curves, that the digitization should be performed with illumination and detectors of high response in the lowest part of the visible region. Consequently, optical detectors with high quantum efficiency in the visible range must be used. In this respect, amorphous silicon or AlGaAs heterojunction based detectors, which present a maximum efficiency between 500-600 nm, are good candidates. In thoses cases where standard CCD detectors are used, IR filter above 700 nm must be considered.

For degraded manuscripts with ink bleeding, much better contrast should be obtained in the high part of the visible region. The use of a red pass-band gives enhanced digital images in term of readability.

Moreover, images of manuscripts digitized in this way with selected optical components, turn out to be the best input for further digital processing, that are presently being developed for different kinds of manuscript degradations.

5. ACKNOWLEDGEMENTS

This work is done in relation to a joint research project between the Spanish Ministry of Culture, the Ramón Areces Foundation and IBM Spain to create an information system in the Archivo General de Indias (Seville). The authors acknowledges the significant contribution of P. González, from the Ministry of Culture, for his discussions on the problem and for supplying manuscripts. One of us (F.J.) wants to thank the support of those institutions.

6. REFERENCES

1. Commission International de l'Eclairage. "A unified framework of methods for evaluating visual performance aspects of lighting". CIE Publication No. 19 (1972).

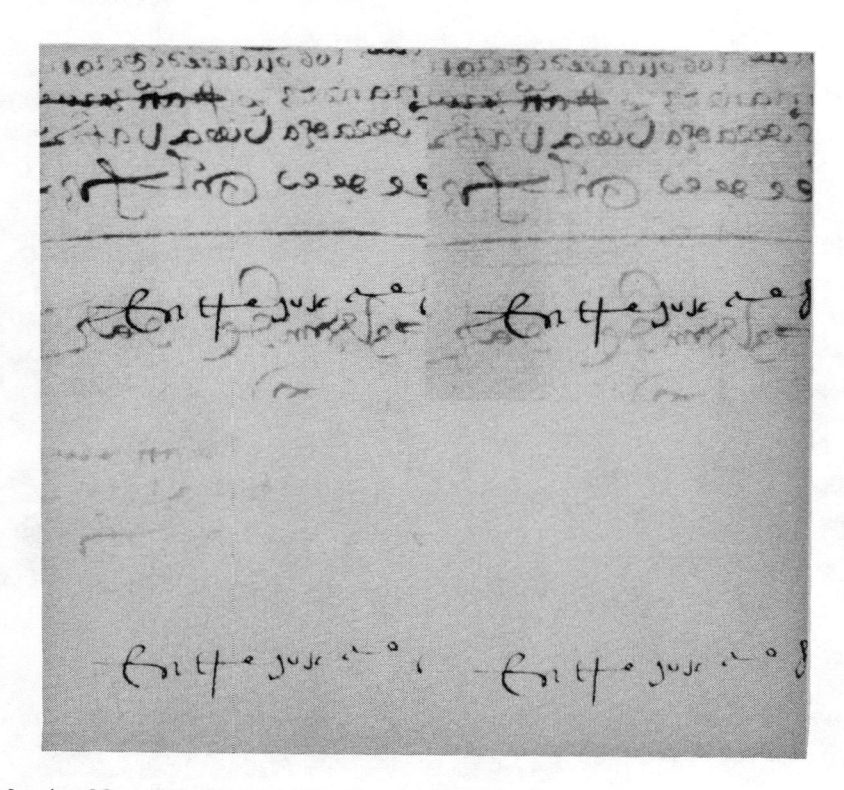

Figure 7. Mosaic of four digital images of a manuscript degraded with bleeding ink. Top: result of using the IR filter (left) and a red pass-band filter (right). Bottom: result of application of a digital transform to the above images. Only the direct ink is enhanced.

A Novel Opto-electronic Method of Position Measurement.

Stephen Gergely, Andrew J. Syson

Coventry Polytechnic, Priory St., Coventry, United Kingdom.

ABSTRACT

The paper contains a description of the principle of operation of a system to measure the position of marked points in a scene. The main application envisaged is in the study of human movement and biomechanics. This system eliminates the need for the time consuming computation involved in tracking particular points from scene to scene. Therefore many more points of interest may be studied than in existing video based systems. The implementation described uses a rotating disc that contains the scanning and sychronising pattern, an optical projection system and light sensors attached to the points of interest and connected to a desktop computer via an interface unit.

1. INTRODUCTION

In the analysis of human movement and in many other applications it is required to determine the positions of points of interest in space and follow their path over a period of time. The position information enables the calculation of absolute and relative velocities, accelerations and relative angles. Traditionally this task was accomplished by recording the scene on film, marking the points of interest on each frame, and making the necessary measurements manually.[1] This method is slow and laborious and so its usefulness is severely limited. More recently the availability of video cameras and recorders led to the introduction of several techniques ranging from the manual to the fully automated. In the latter the points of interest in the scene are marked by a bright spot. The higher light intensity is detected within the video waveform, and the timing information derived from the frame and line synchronising pulses enables the calculation of the position of each of the points within the scene.[2] This is a simple and effective way of measuring position. However for the computation of velocities and other variables each point must be identified in each frame, i.e. the system must be able to follow the time course of the position of each point and relate the series of position coordinates as belonging to a particular point. Existing systems are able to do this for a few points in a reasonable time. As the number of points is increased, the time required for the execution of the identification algorithm is increased to such an extent that the system is only useful in specialist research applications. In some cases processing times can exceed several hours.

The subject of this paper is a system invented by the authors and developed in cooperation with two groups of final year students from the Ecole Nationale Superieure de l'Electronique et de ses Applications in Cergy France. It overcomes the problem of point identification and thus eliminates the need for the tracking calculations, at the expense of having the subject carry a small electronic interface package connected to the processing computer by a cable or a wireless link.

2. PRINCIPLE OF OPERATION.

The scene in which the subject moves is scanned by a pattern of light. A light sensor is attached to each of the points of interest and is connected to the electronics package carried by the subject. The times at which the sensors are illuminated by the pattern are recorded for each scan, and the positions of the points are calculated from it. The need for time consuming computation is removed since the identity of each point is recorded directly, together with the timing and position information. Thus the limit is removed on the number of points of interest imposed by the tracking or identification computation.

3. THE CHOICE OF SCANNING PATTERN.

The first pattern to come to mind when considering this system in relation to the existing video based methods is a point source scanning in a conventional TV. type raster. The generation of such a raster requires a laser source and a two axis deflection system. The speed of scan is limited by the risetime of the output of the sensors (approximately 80 nsec would be required at the normal TV. resolution and scan rate).

The sensor risetime requirement can be relaxed to approximately 30 μsec if the scanning pattern is changed to a pair of lines. The most obvious choice to establish the two position coordinates of the point marked by the sensor in question is one vertical and one horizontal line (Fig.1). Unfortunately the cost of the two axis deflection system could not be met from the resources available for this project, so a new approach was required.

Figure 1. X-Y scan lines, perpendicular motion.

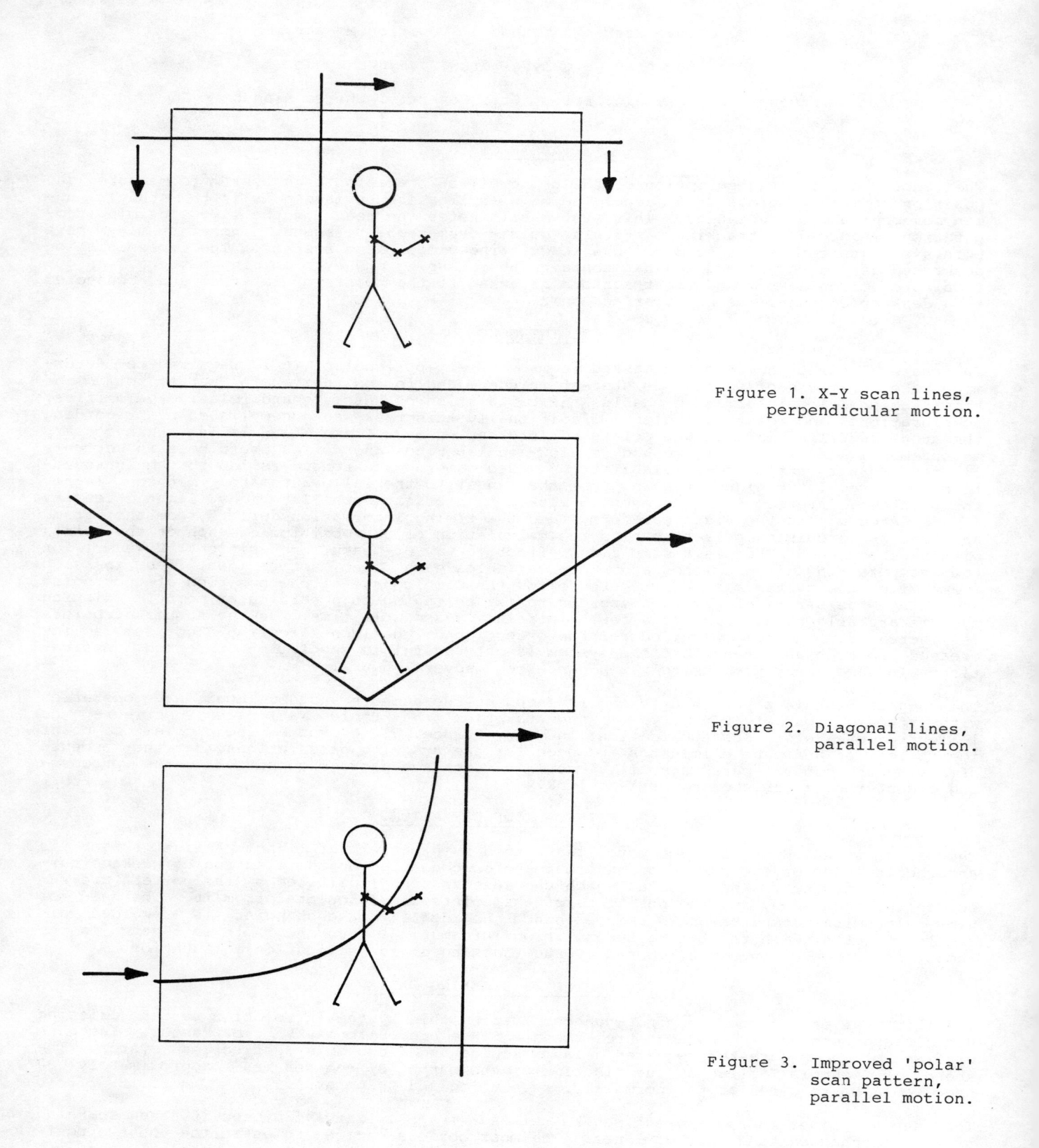

Figure 2. Diagonal lines, parallel motion.

Figure 3. Improved 'polar' scan pattern, parallel motion.

It can be seen from consideration of Fig.2 that the required information can be obtained using a pattern consisting of two perpendicular lines inclined at 45° to a common direction of scan. The most important property of this pattern is that only one direction of scan is required to obtain the two items of information which describe the position of the point in the plane of interest. The one directional scan pattern was implemented by a disk rotating in the image plane of a 35mm slide projector. The cost of this system is far less than the ones considered previously.

It has recently come to the attention of the authors that the modified pattern shown in Fig.3 would also permit a one directional scan and would have advantages compared to the one shown in Fig.2. This pattern consists of a radial line followed by an Archimedian spiral. The elapsed time between the start of the scan and the first intercept of the sensor (by the radial line) is directly proportional to the argument of the polar coordinate of the point with the projected centre of rotation of the disc as the origin. The spiral is designed to intercept the point in such a way that the time difference between the two intercepts is directly proportional to the magnitude of the polar coordinate (the radial distance from the origin). Thus the amount of computation needed to establish the rectangular coordinates within the scan frame is considerably reduced compared to the method actually implemented and described in this paper. An additional advantage is that the pattern occupies approximately half the space required for the two 45° lines (one and two frame widths respectively). Therefore a higher scan rate can be achieved for a given disk speed and diameter.

4. CALCULATION OF THE POSITION COORDINATES.

The rectangular coordinate system (x,y) used to represent the window through which the scene is viewed and the polar coordinate (r,θ) system used to describe the rotating disc are shown in Fig. 4. The two origins are separated by a radial distance R.

Therefore: $x = R \cdot \sin\theta$ and $y = R \cdot (1 - \cos\theta)$

The equations of the two scanning lines expressed in the form $y = a \cdot x + b$ are:

$y = \tan(\theta_1 + 45°) \cdot x + (R \cdot ((1 - \cos\theta_1) - \tan(\theta_1 + 45°) \cdot \sin\theta_1))$ and

$y = \tan(\theta_2 - 45°) \cdot x + (R \cdot ((1 - \cos\theta_2) - \tan(\theta_2 - 45°) \cdot \sin\theta_2))$

For any given point the two X and Y coordinates must be the same thus:

$X = (b_1 - b_2) / (a_2 - a_1)$ and $Y = (a_2 \cdot b_1 - a_1 \cdot b_2) / (a_2 - a_1)$

After substitution and rearranging one obtains:

$X = R \cdot (\cos\theta_1 - \cos\theta_2 + \tan(\theta_2 - 45°) \cdot \sin\theta_2 - \tan(\theta_1 + 45°) \cdot \sin\theta_1) / K$

$Y = (R \cdot \tan(\theta_2 - 45°) \cdot ((1 - \cos\theta_1) - \tan(\theta_1 + 45°) \cdot \sin\theta_1) / K) -$
$\quad - (R \cdot \tan(\theta_1 + 45°) \cdot ((1 - \cos\theta_2) - \tan(\theta_2 - 45°) \cdot \sin\theta_2) / K)$

where $K = (a_2 - a_1) = \tan(\theta_1 + 45°) - \tan(\theta_2 - 45°)$

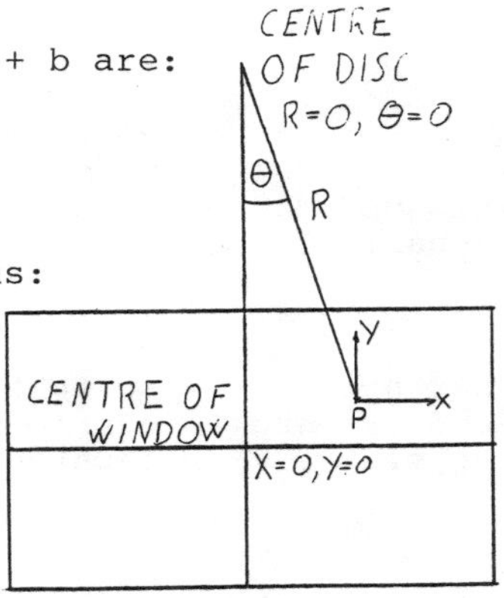

Figure 4. The coordinate systems.

5. SYSTEM DESIGN.

The operation of the system is best understood by reference to Fig.5 which shows the scanning disc. In addition to the track containing the four pairs of scanning lines the disc also carries three additional tracks. Two of these are used to provide start of frame and start of line synchronisation. The third one provides a clock signal that is used to measure the angular displacement of the pattern between the relevant start pulse and the pulse received from the sensor i.e. to measure θ_1 and θ_2 for use in the calculations of X and Y as outlined above. An electronic counter clocked by these pulses is started by one of the start pulses. Each sensor is associated with two latches. The numbers at the output of the counter when the each of the lines illuminates the sensor are held by these latches. These are transferred to a buffer and downloaded to the computer used to carry out the calculations and to store and display the information.

Note that the use of a clock signal derived directly from the disc makes the measurements independent of the speed of rotation of the disc. Therefore no particular measures are required to control the speed of rotation of the disc. A rotational speed of 1500 rpm., of a disc containing four patterns, corresponds to a scan rate of 100 frames per second. This is considered to be sufficient to follow most forms of human movement. The required number of clock pulses may be obtained directly from a Moire fringe type displacement transducer on the edge of the disc. Alternatively a phase locked loop (PLL) can be used to synchronise a higher speed voltage controlled oscillator to the pulse rate obtained from the disc. The inertia of the disc provides sufficient short term speed stability.

The measurement resolution of the system is determined by the number of clock pulses used. In the implementation described here 960 pulses were obtained from the disc per revolution. After multiplication by a PLL this resulted in 946 pulses per scan line corresponding to approximately 2 mm in a 2 m square frame. The resolution is also a function of the rate of change of light intensity (sharpness) at the edge of the scan line. This is dependent on the quality of the optical system used for the projection assuming that the sytem used to print the pattern on the disc is of much higher quality.

The frame to frame repeatability of the measurement requires that the scan patterns on the disc (in this case four) coincide with one an other i.e. produce the same result for a stationary target point. Assuming a 36 mm square window on the projector and a projected pattern size of 2m square any error on the disc is magnified by 2000/36=55.6. Thus if the required repeatability of measurement is 2 mm, the patterns must be positioned with an accuracy of 0.036 mm. The patterns drawn on the disc shown in Fig 5. were generated by software and drawn by an A O size plotter at twice the final size. This was sufficiently accurate. However great care and a special mechanical design of the mounting was required to ensure the correct centring of the disc and its balanced, vibration free rotation. The distortion of the projected image caused by the optical system determines the overall accuracy of measurement.

The maximum number of points a system can be designed to accommodate is governed by the rate of data transfer between the interface unit and the computer, and also by the time taken for computation if the position coordinates are to be calculated on line.
The measurement system was implemented using an IBM PC as the control and storage device. Software was produced in Pascal and Assembler to control the process, make the necessary calculations and to provide an interface to enable the user to carry out the initialisation, data acquisition analysis and storage functions without any specialist technical knowledge. A resolution of 3.5 mm was obtained using a commercial slide projector.

6. CONCLUSIONS.

A system is described which eliminates the need for time consuming computations for tracking points in existing systems in gait analysis. An experimental system was constructed. This validated the proposed principle. It highlighted the importance of the quality of the mechanical and optical parts of the system. It had an accuracy of 0.4% or 4mm in 2 m.

7. ACKNOWLEDGEMENTS.

The authors wish to thank Mlle. Dominique Jacquemin and Messres Eric Pasquier, Thierry Houdayer and Abdelkrim Mous for their invaluable contribution in the construction and testing of the experimental system and the production of the software.

8. REFERENCES.

1. J. Terauds, Biomechanics cinematography and high speed photography, Proceedings of the International Society for Optical Engineers, 1981.
2. R. Shapiro, C. Blow and G. Rash, Video Digitizing analysis system, Int. J. Sport Biomechanics, 3 80-86 (1987).

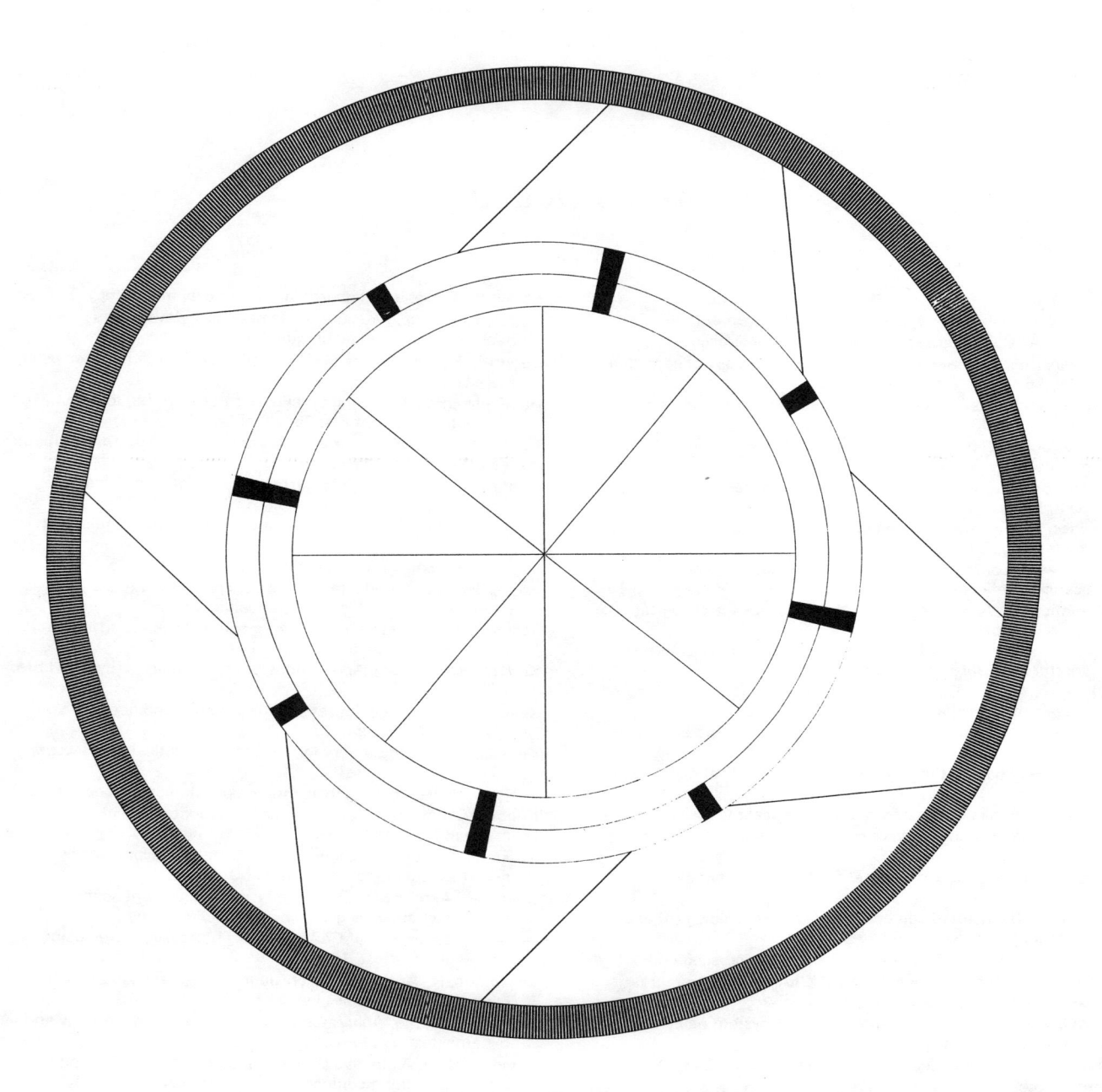

Figure 5. The rotating disk.

AUTHOR INDEX

Appel, Roland K., Scanning differential optical system for simultaneous phase and amplitude measurement, 54
Ash, Eric A., Computerized surface plasmon microscopy, 231
Bacallao, Robert, Confocal fluorescence microscopy of epithelial cells, 167
Bertero, M., Inverse problems in fluorescence confocal scanning microscopy, 33
Bescós, Julián, Reflectance and optical contrast of old manuscripts: wavelength dependence, 258
Bianco, Bruno, Computerized analysis of high resolution images by scanning acoustic microscopy, 237
Boccacci, P., Inverse problems in fluorescence confocal scanning microscopy, 33
Boyde, Alan, Confocal and conventional modes in tandem scanning reflected light microscopy, 98
Brakenhoff, G. J., Inverted confocal microscopy for biological and material applications, 169
———, Modeling of 3-D confocal imaging at high numerical aperture in fluorescence, 39
Cambiaso, Andrea, Computerized analysis of high resolution images by scanning acoustic microscopy, 237
Cavanagh, H. Dwight, In vivo confocal imaging of the eye using tandem scanning confocal microscopy, 122
Chou, C.-H., Imaging theory for the scanning optical microscope, 104
Corle, T. R., Phase imaging in scanning optical microscopes, 114
Damm, Tobias, Optimization of recording conditions in laser scanning microscopy, 69
Ebbeni, Jean, Overview of coherent optics applications in metrology, 2
Engelhardt, J., Inverted confocal microscopy for biological and material applications, 169
Feuerbaum, Hans Peter, Infrared laser scan microscope, 226
Fontein, P. F., Topography of GaAs/AlGaAs heterostructures using the lateral photo effect, 197
Fritz, J., Automated latch-up measurement system using a laser scanning microscope, 217
Fujimoto, A., Semiconductor laser digital scanner, 245
Gergely, Stephen, Novel opto-electronic method of position measurement, 263
Giusto, Daniele D., Computerized analysis of high resolution images by scanning acoustic microscopy, 237
Glindemann, Andreas, Measurement of the degree of coherence in conventional microscopes, 84
Grattarola, Massimo, Computerized analysis of high resolution images by scanning acoustic microscopy, 237
Hamed, A. M., Eccentricity errors combined with wavefront aberration in a coherent scanning microscope, 63
Hamilton, Douglas K., Confocal interference microscopy, 92
Hänninen, Pekka, Confocal fluorescence microscopes for biological research, 146
Hegedus, Zoltan S., Pupil filters in confocal imaging, 14

Hellmuth, T., Spherical aberration in confocal microscopy, 28
Helsztyński, Jerzy, Measurements of optical waveguides by a near-field scanning technique, 250
Hendriks, P., Topography of GaAs/AlGaAs heterostructures using the lateral photo effect, 197
Hu, Evelyn, Phase-shifting and Fourier transforming for submicron linewidth measurement, 76
Hunter, Ian, Apparatus for laser scanning microscopy and dynamic testing of muscle cells, 152
Hunter, Peter, Apparatus for laser scanning microscopy and dynamic testing of muscle cells, 152
Ihrig, C., Confocal laser scanning microscopy for ophthalmology, 147
Imanaka, K., Semiconductor laser digital scanner, 245
Jaque, Francisco, Reflectance and optical contrast of old manuscripts: wavelength dependence, 258
Jester, James V., In vivo confocal imaging of the eye using tandem scanning confocal microscopy, 122
Kapitza, Hans-Georg, Applications of the microscope system LSM, 173
Kaschke, Michael, Optimization of recording conditions in laser scanning microscopy, 69
Kino, G. S., Imaging theory for the scanning optical microscope, 104
———, Phase imaging in scanning optical microscopes, 114
Knebel, W., Inverted confocal microscopy for biological and material applications, 169
Kozek, Tadeusz W., Measurements of optical waveguides by a near-field scanning technique, 250
Kucernak, Anthony R., Scanning laser photocurrent spectroscopy of electrochemically grown bismuth sulphide films, 202
———, Topography of GaAs/AlGaAs heterostructures using the lateral photo effect, 197
Lackmann, R., Automated latch-up measurement system using a laser scanning microscope, 217
Lafontaine, Serge, Apparatus for laser scanning microscopy and dynamic testing of muscle cells, 152
Lemp, Michael A., In vivo confocal imaging of the eye using tandem scanning confocal microscopy, 122
Louis, Thomas A., Minority carrier lifetime mapping in gallium arsenide by time-resolved photoluminescence scanning microscopy, 188
Ma, Chang-ming, Theoretical and experimental research on super-resolution of microscopes: I. partially coherent illumination and resolving power, 45
Masters, Barry R., Scanning microscope for optically sectioning the living cornea, 133
Matthews, Hubert J., Confocal interference microscopy, 92
Monoto, Luis, Reflectance and optical contrast of old manuscripts: wavelength dependence, 258
Nielsen, Poul, Apparatus for laser scanning microscopy and dynamic testing of muscle cells, 152

Offside, Marlo J., Axial resolution of a confocal scanning optical microscope, 18
Otto, J., Circuit analysis in ICs using the scanning laser microscope, 211
Peat, Robert, Scanning laser photocurrent spectroscopy of electrochemically grown bismuth sulphide films, 202
———, Topography of GaAs/AlGaAs heterostructures using the lateral photo effect, 197
Pester, P. D., Photoluminescence and optical beam-induced current imaging of defects, 182
Peters, Reiner, Laser scanning microscopy to study molecular transport in single cells, 160
Pick, Reinhard, Confocal fluorescence microscopes for biological research, 146
Pike, E. R., Inverse problems in fluorescence confocal scanning microscopy, 33
Plies, E., Circuit analysis in ICs using the scanning laser microscope, 211
Quincke, J., Circuit analysis in ICs using the scanning laser microscope, 211
Rihs, Hans-Peter, Laser scanning microscopy to study molecular transport in single cells, 160
Rix, B., Automated latch-up measurement system using a laser scanning microscope, 217
Sauer, Heinrich, Laser scanning microscopy to study molecular transport in single cells, 160
Scholz, Manfred, Laser scanning microscopy to study molecular transport in single cells, 160
See, Chung Wah, Axial resolution of a confocal scanning optical microscope, 18
———, Scanning differential optical system for simultaneous phase and amplitude measurement, 54
Seidel, P., Spherical aberration in confocal microscopy, 28
Sekii, H., Semiconductor laser digital scanner, 245
Sheppard, Colin J. R., Confocal interference microscopy, 92
Shimura, M., Semiconductor laser digital scanner, 245
Siegel, A., Spherical aberration in confocal microscopy, 28
Smith, Robin W., Theoretical and experimental research on super-resolution of microscopes: I. partially coherent illumination and resolving power, 45
Somekh, Michael G., Axial resolution of a confocal scanning optical microscope, 18
———, Scanning differential optical system for simultaneous phase and amplitude measurement, 54

Stamm, Uwe, Optimization of recording conditions in laser scanning microscopy, 69
Stelzer, Ernst H. K., Confocal fluorescence microscopes for biological research, 146
———, Confocal fluorescence microscopy of epithelial cells, 167
Storz, Clemens, Confocal fluorescence microscopes for biological research, 146
Stricker, Reiner, Confocal fluorescence microscopes for biological research, 146
Syson, Andrew J., Novel opto-electronic method of position measurement, 263
Takagi, T., Semiconductor laser digital scanner, 245
Tedesco, Mariateresa, Computerized analysis of high resolution images by scanning acoustic microscopy, 237
van der Voort, H. T. M., Modeling of 3-D confocal imaging at high numerical aperture in fluorescence, 39
———, Inverted confocal microscopy for biological and material applications, 169
Wade, Glen, Phase-shifting and Fourier transforming for submicron linewidth measurement, 76
Wijnaendts-van-Resandt, R. W., Confocal laser scanning microscopy for ophthalmology, 147
———, Inverted confocal microscopy for biological and material applications, 169
Wilke, Volker, Applications of the microscope system LSM, 173
Williams, David E., Scanning laser photocurrent spectroscopy of electrochemically grown bismuth sulphide films, 202
———, Topography of GaAs/AlGaAs heterostructures using the lateral photo effect, 197
Wilson, T., Photoluminescence and optical beam-induced current imaging of defects, 182
Wolter, J., Topography of GaAs/AlGaAs heterostructures using the lateral photo effect, 197
Xiao, G. Q., Imaging theory for the scanning optical microscope, 104
Xu, Yiping, Phase-shifting and Fourier transforming for submicron linewidth measurement, 76
Yeatman, Eric M., Computerized surface plasmon microscopy, 231
Ziegler, Eberhard, Infrared laser scan microscope, 226
Zinser, G., Confocal laser scanning microscopy for ophthalmology, 147